STRUCTURAL
MECHANICS

STRUCTURAL MECHANICS

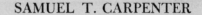

SAMUEL T. CARPENTER

*Professor of Civil Engineering
and Chairman of Department*

*Division of Engineering
Swarthmore College
Swarthmore, Pennsylvania*

ROBERT E. KRIEGER PUBLISHING COMPANY
HUNTINGTON, NEW YORK
1977

Original Edition 1960
Fourth Printing, September, 1966
Reprint 1977 with corrections

Printed and Published by
ROBERT E. KRIEGER PUBLISHING CO., INC.
645 NEW YORK AVENUE
HUNTINGTON, NEW YORK 11743

Printed in the United States of America

Library of Congress Cataloging in Publication Data

Carpenter, Samuel T.
    Structural mechanics.

    Reprint of the edition published by Wiley, New York.
    Includes index.
    1. Structures, Theory of. 2. Mechanics.
I. Title.
[TA645.C27   1976]   624'.171   75-31671
ISBN 0-88275-363-0

Preface

This textbook has been compiled after a quarter century of teaching structural analysis and structural mechanics to undergraduate engineering students. Although the text is primarily intended for use in undergraduate instruction, a number of the modern topics may be of value to graduate students as well as to practicing engineers.

The title *Structural Mechanics* was selected as being descriptive, owing to the emphasis placed on the mathematical treatment of structural theory bearing on indeterminate structures, as well as the introduction of topics from Advanced Mechanics of Materials. Topics in the latter category include Finite Difference Approximations, Fourier Series, Structural Dynamics, and Stability. The book also includes a brief presentation of Newmark's numerical method for deflections and buckling. The subject of beam columns is considered, and the modifications of moment distribution for axially loaded members are developed.

The intent of this book is to present the fundamental principles so as to provide a sound background for further extended study in the field of structures. In view of the need for a more mathematical treatment, the use of calculus and, in many sections, differential equations is stressed. This is all within the capacity and background of the undergraduate student and serves to keep these valuable mathematical tools in continuous use.

I teach the major topics covered in this book in a one-semester course, utilizing three lecture classes and one three-hour problem laboratory per week. The prerequisite for the course is an elementary structural theory course in determinate structures, with a brief introduction to indeterminate structures and moment distribution. In the prerequisite course, all of the material of Chapter 2 has been taught except finite difference approximations, Newmark's method, Castigliano, and series methods. The subject matter has also been taught in two semesters without the problem laboratory. In that event, I have often augmented the text material with selected topics from slab and shell theory and ultimate design theory. Slab theory can be easily introduced and dealt with by finite differences as well as by series since these mathematical tools have been developed in the text. The instructor will undoubtedly be able to exercise many other options in fitting the book to his needs.

I am indebted to all in my profession who have contributed to the literature in this important field. The many critical questions that former students have asked, while seeking to clarify the basic principles in their own minds, have been of immeasurable help in setting the format of this book. I also wish to acknowledge the sound teaching of my former teacher, Professor Clyde T. Morris, The Ohio State University, as well as the inspiration provided by my former colleague, the late Professor Scott B. Lilly of Swarthmore College.

I owe special acknowledgment to Mrs. Frances Wills who typed all of the manuscript several times and who was always sympathetic. Charles G. Thatcher, formerly Professor of Mechanical Engineering at Swarthmore College, developed all of the working drawings from which the finished figures were produced. I will never be able to repay the debt which I owe for the educational opportunities and moral support provided by my Mother and Father. Finally, I am grateful for the encouragement and endless patience of my wife, Mary.

SAMUEL T. CARPENTER

Swarthmore College
February 1960

Contents

chapter 4 Moment Distribution

chapter 5 Slope Deflection Method

chapter 6 Elastic Center and Column Analogy Methods

chapter 13 Axially Loaded Members and Beam Columns

Introduction

1

1-1. INDETERMINATE STRUCTURES

Structures may be thought of as all works of man that transmit and resist the forces of nature and in so doing serve a functional or useful purpose. The applications of structures, in the air, on the sea, and on the land, are too numerous to mention, hence a comprehensive study of structural mechanics is an essential for every engineer.

A statically indeterminate structure is a structure that cannot be analyzed by the principles of static equilibrium alone. This problem is presented in many structures, on account of functional requirements, limitations on types of framing, need for stiffness, and often by the continuity introduced by the material used, such as reinforced concrete. If an indeterminate structure is to be analyzed, ways and means of determining additional equations beyond the equations of statics must be established. Immediately it is evident that statically indeterminate structures will be more difficult to solve than determinate ones.

A statically indeterminate structure is classified in general by determining how indeterminate it is, by establishing the "degree of indeterminacy." The degree of indeterminacy simply states the number of equations to be established beyond the static equations. A structure may be indeterminate with regard to outer forces only, internal forces only, or a combination of external and internal forces. To illustrate indeterminacy with regard to external forces, refer to Fig. 1, where a beam is supported at three points. If we bear in mind that points of application of reactive forces are known, a routine counting of the unknown characteristics of these forces would indicate four unknown quantities, namely, magnitude and direction of R_A and magnitude for both V_B and V_C. By accepting the three well-known equations of static equilibrium as $\sum F_x = 0$, $\sum F_y = 0$, and $\sum M = 0$ for a

1

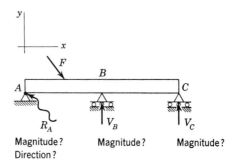

FIG. 1–1

co-planar force system, it is found that the four unknowns exceed the three available equations of static equilibrium by one. Hence, the degree of indeterminacy is one, or the beam is said to be indeterminate to the first degree with respect to reactive forces.

An alternative approach to ascertain the degree of indeterminacy would be to remove selected outer reactive forces until the structure is reduced to a statically determinate and stable *base structure*. Figure 2, a repetition

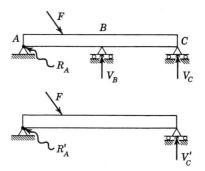

FIG. 1–2

of Fig. 1, illustrates this approach. The reaction at B can be removed and beam is still stably supported, but no other reaction may be removed. In this instance V_B may be termed the *redundant* force, which, in this interpretation, physically means that V_B is superfluous for establishing static equilibrium. The original structure may be classed as having one redundant or alternatively termed indeterminate to the first degree as described previously. This method is of no greater value than the first procedure since both lead to the same number of necessary extra equations beyond those of static equilibrium, but it does represent a method of checking.

In general, for a frame composed of continuous members and rigidly framed joints, there is a chance that every cross-section of every member may be subjected to a general internal force and bending moment system, defined as shear, axial thrust, and bending moment. Figure 3a is a two-legged framed bent and it is desired to determine the degree of indeterminacy. Again the two approaches can be used, but first use the method of counting the unknowns and comparing with the number of static equations available for solution. The moments and forces shown at A and D are the reaction component possibilities at these restrained ends and they are six in number. Six unknowns would require six equations to solve completely, but since only three independent equations can be derived from statics the structure is indeterminate to the third degree. The alternate approach is indicated in Fig. 3b, where, to make structure

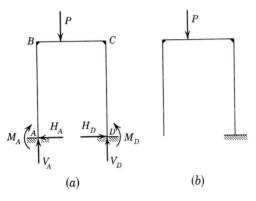

(a)   (b)

FIG. 1-3

statically determinate, it is cut at A. This physically removes the shear, thrust, and moment, at A. These quantities are termed the *redundants*, and since they are three in number the original structure is said to be indeterminate to the third degree.

Figure 4a is a complex rigid-jointed structure that is indeterminate with respect to both reactive and inner components. The second approach, namely the reduction of the structure to a statically determinate *base structure* by removal of redundants, is of great value in this instance. An analysis proceeds by cutting the members of the frame as necessary until a static calculation of forces in all members is possible. Figure 4b shows the structure cut three times; and by remembering that it is feasible for a shear and a normal axial force plus a bending moment to have existed in original structure at each section, it may be found that structure

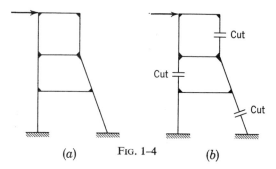

(a)   Fig. 1-4   (b)

has 3 × 3 or 9 redundant quantities; thus this structure would be classed as an indeterminate structure to the ninth degree.

Pin-connected trusses with loads applied at panel points are stable assemblages of two-force members. From earlier fundamental courses it will be recalled that the minimum number of members, b, for rigidity of a coplanar truss system was shown to be equal to $2n-3$, where n equals the number of pinned joints. Figure 5a illustrates an application of this

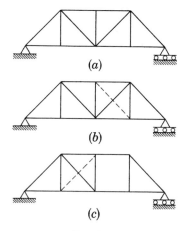

(a)

(b)

(c)

Fig. 1-5

equation, where $n = 8$, $b = 2 \times 8 - 3 = 13$ members. Referring to Fig. 5b, where the dotted member has been added, there are 14 members, or one more member than is needed for a rigidly framed truss system. As will be shown later, this is also equivalent to stating that the dotted member is redundant or the stresses in truss members are indeterminate to the first degree. Figure 5c represents the omission of one diagonal member and the addition of an alternate dotted member, keeping the total number of members equal to 13. This emphasizes that caution must be exercised in applying the equation ($2n - 3$) as the satisfying of it is not sufficient within itself to establish that the structure is rigidly framed, since the pin-jointed framing of Fig. 5c is unstable with one panel free to collapse, causing the entire truss to collapse.

The determination of the total indeterminacy of a trussed structure must include an examination of outer forces as well as inner forces. In Fig. 6a, the structure has been loaded with a general force system and supported so that the reactions are determinate. It should be particularly noted that for a generalized loading and external stability the reaction components can never number less than three. A computation of all external and internal forces could be undertaken by isolating each joint of the structure

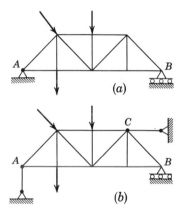

(a)

(b)

FIG. 1–6

and setting up two equations of static equilibrium for the concurrent forces at each joint. The total number of such equations would be $2n$ and the total number of unknowns would be the number of two-force members plus the number of reaction components, or $b + 3$. For static determination of all stresses and reaction forces ($b + 3$) can not be greater than the $2n$ static equations available, or $b = 2n - 3$. It should be noted that $b = 2n - 3$ was also a requirement for a rigidly framed truss system, hence a trussed structure that is framed stably and a trussed structure that is determinate with regard to all forces have the same requirement for the number of reaction components and internal members. This is important since internal members may be replaced by reaction components, or reaction components removed from one position and replaced at another point, provided that this is done with full recognition of stability. Figure

$6b$ indicates a change in reaction forces, with a two-force member being introduced at A, and since this substitution destroys the horizontal reaction potential at A, another two-force member has been added at C. Thus, the number of reaction components remains at three and the structure is still externally stable. These alternate arrangements indicate the broad choice of framing and supporting restraints that are possible, while still maintaining the trussed structure in a stable and force resistant form. The fact that $(2n - 3)$ indicates the minimum number of members and reaction components should now be clear; it should also be understood that members and the supporting components must be selected intelligently.

A broad view of analysis for degree of indeterminacy of truss structures may now be developed. This broader view stems from the same approach expressed previously for frames, namely, the procedure of removing redundant quantities until a determinate stable *base structure* remains. Figure $7a$ will serve to illustrate. If V_B and the member marked X are

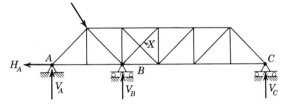

FIG. 1-7

removed, the internal forces and the remaining reactions may be found from the equations of statics. The *base structure* so obtained also satisfies all conditions for stability. V_B and S_X, the stress in member X, would be termed the redundant quantities and the original structure would be classed as indeterminate to the second degree, once with regard to outer forces and once with regard to inner forces. The same result would also be obtained by counting the number of joints and the number of members and reaction components; thus, $n = 12, b = 22$, reaction components $= 4$. The unknowns number 26 and the available equations by the joint to joint equilibrium processes would be $2 \times 12 = 24$. $26 - 24 = 2$, or there are two more unknowns than equations of static equilibrium, or an indeterminacy of the second degree exists as previously determined.

1-2. EQUATIONS FROM GEOMETRY

Although equations of statics are always essential, it has been established that for indeterminate structures more unknowns exist than can be determined from the equations of static equilibrium alone.   It is then imperative that additional conditions be established for writing additional equations.   These additional conditions must come from a consideration of the geometry of the structure and the force or moment conditions which restrain an elastic structure so it conforms to these geometric requirements. For each redundant quantity a geometric condition must be employed, such as a condition of deflection, rotation, or an elastic change in length. The establishment of equations to satisfy these conditions is certainly the prime objective of a study of indeterminate structures, and structural mechanics, and it is this area of the subject that calls for analytical work of high order.

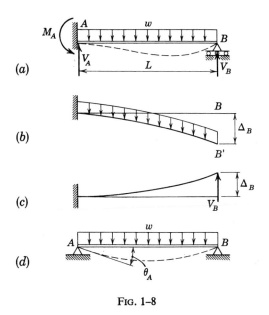

FIG. 1-8

As an example of geometric conditions, beam AB of Fig. 8a, fixed against rotation at A and simply supported at B, may be studied geometrically, and it can be stated that no rotation or deflection may occur at A and no vertical deflection can occur at B.   These geometrical facts can be explicitly termed the boundary conditions and become the basis for formulating the necessary extra equations.   The number and type of geometric conditions depend on the degree of indeterminacy and the selection of redundants.   Beam AB is indeterminate to the first degree. (The student should verify this.)   First by selecting V_B as a redundant and removing the support from beam, the beam will deflect downward at B, a distance of $\varDelta_B$, as shown in Fig. 8b, and this is calculable. If the true magnitude of V_B were known and applied to B', point B' would deflect up $\varDelta_B$ and be returned to point B. This physical return of B' to B leads to the geometric condition that requires the resultant deflection at B due to all forces to be equal to zero, which would be the condition used to establish the extra equation. If M_A had been selected as the redundant and removed from beam in Fig. 8a, the deflected geometry of Fig. 8d would result and the rotation of the tangent at A, θ_A, is calculable. M_A is the moment required to rotate the tangent of Fig. 8d back to the horizontal. The geometric condition thus requires the resultant rotation at A to be equal to zero.

Although there are many methods for calculating the final reactions, shears, and moments of indeterminate structures, all of them depend at some point on a consideration of the geometry of the deflected or distorted structure. The foregoing discussion has emphasized geometry to reveal the fundamentals of ascertaining indeterminacy or redundancy, and in so doing has approached the problem from the standpoint of basic principles. These principles serve to remove much of the camouflage around the problems associated with indeterminate structures. These principles are summarized as:

(a) Identification of indeterminate structure by removing selected redundant restraints which reduce structure to a *base structure* that can be dealt with by static equilibrium equations alone.

(b) Identification of boundary conditions of original structure in terms of deflection, rotation, or member distortion.

(c) The calculation of the relevant displacements or rotations of *base structure* at points where redundants were removed.

(d) The calculation of the values of the necessary redundant restraints to restore the geometry of the *base structure* to comply with required boundary conditions.

1–3. ADVANTAGES AND LIMITATIONS OF INDETERMINATE STRUCTURES

The selection of the basic form of structure to satisfy a given purpose is usually made after a study of many factors. The basic form of the structure challenges the creative and analytical powers of the engineer or architect, and rarely can costs, construction methods, and materials be eliminated from consideration. It can be easily seen that the selection of a structure leads to a synthesis of many related factors.

The indeterminate structure has many advantages and disadvantages compared to a statically determinate form; however, one of the largest disadvantages, the difficulty of analytical solution, has largely disappeared with the discovery of simplifying theoretical procedures and electronic computation. Nevertheless, the fact that a trial design of members must exist before a full analysis can take place slows the design process. An indeterminate structure is generally a stiffer structure for a given weight of material than a statically determinate structure, and it is generally recognized that in the case of overloads the indeterminate structure can often furnish a compensation by redistributions within the structure. An entirely new field of structural design, termed *ultimate design*, is coming into use, which adapts the properties of materials and the capacity for readjustment through yielding of the materials so that a sounder view can be reached concerning true structural safety.

Many instances exist in which an indeterminate structure should not be used or, if used, adjustments in the analysis should be made, such as when foundation conditions are so unsatisfactory that excessive settlement may be expected. Foundation movement modifies the basic structural geometry accepted in establishing the extra equations beyond statics. If this modification is severe or unpredictable the designer may either change the form of structure to one less affected by foundation movement or accept the fact that a statically determinate structure should be used. This problem may be visualized by considering a two-span continuous bridge with the center support or pier resting on compressible soil and the end abutments supported on solid rock. If the soil under the pier consolidates or compresses under load in an excessive or unforeseen manner, the center pier may settle away from the bridge girders. This, in effect, converts the two-span continuous bridge into a single span bridge accompanied by severe or catastrophic stress modifications. If a review of foundation conditions prior to design should indicate the possibility of this latter condition occurring, the designer would certainly change the form of the structure to either two simple beam spans, one longer span eliminating the center pier, or to an arch span if rock at abutments is sound. The final selection of the type of structure also involves a critical examination of all economic factors.

REFERENCES

1. Sutherland, H., and Bowman, H. L., *Structural Theory*, fourth edition, New York: John Wiley, 1950.
2. Michaels, Leonard, *Contemporary Structure in Architecture*, New York: Reinhold, 1950.

3. Timoshenko, S. P., *History of Strength of Materials*, New York: McGraw-Hill, 1953.
4. Westergaard, H. M., "One Hundred Fifty Years Advance in Structural Analysis," *Trans. Am. Soc. Civil Engineers*, **94**, 226–240 (1930).
5. Wilbur, J. B., and Norris, C. H., *Elementary Structural Analysis*, New York: McGraw-Hill, 1948.

Deflection Theory

2

2-1. INTRODUCTION

The importance of the deflected geometry of a structure has been established. For statically determinate structures the fundamental disciplines of static equilibrium and free body analysis were the basic tools. It is now essential to add to these disciplines of statics the concepts of deformation, rotation, and deflection of elastic structures in order to solve indeterminate problems.

There are many methods for calculating slopes and deflections of members undergoing flexure. Several methods have already been encountered by the student in mechanics of materials and first courses in structural theory; however, since they will be frequently introduced into this text, a review of them is important. In addition to methods already known by the student, new concepts dealing with strain energy will be introduced as well as additional numerical techniques of dealing with elastic curves. This chapter will also discuss several methods for determining the deflections of trusses.

2-2. BASIC FLEXURE AND GEOMETRY

The major assumption concerning the deformation of a beam cross-section is that an initially plane cross-section remains plane during bending. This assumption associated with basic principles of elastic strain and certain loading limitations leads to the well-known general flexure formula

$$\sigma = \frac{Mc}{I}$$

where σ = unit stress in pounds per square inch on an outer fiber of beam
in bending

M = bending moment in inch pounds

I = moment of inertia of the cross-sectional area about the
neutral axis in inches to the fourth power units

c = distance to an outer fiber from the neutral axis in inches

In Fig. 1a *mm* and *nn* represent two vertical planes dx apart passing
through the unloaded beam *AB*.   When load *P* is applied the elastic beam
deflects, as shown by the elastic curve.   The two vertical planes rotate
since the lower fibers of beam are stretched and the top fibers compressed.
A free body of a differential length of the loaded beam is shown in Fig. 1b.

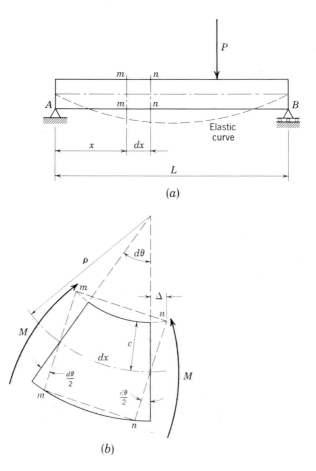

(a)

(b)

FIG. 2–1

The bending moments M are shown, and although shearing forces are present they are not indicated. The rotated planar cross-sections are drawn in their proper relation to the original vertical cross-sections mm and nn.

The relative rotation of the two cross-sections in Fig. 1b is indicated by the small angle $d\theta$ in radian units. The radius of curvature of the elastic curve for the free body element is denoted as ρ. If dx along the elastic curve is treated as a short arc, we may state that $\rho \, d\theta = dx$.

The geometry of the elastic curve as represented by $d\theta$ and ρ may now be related to the elastic deformations produced by the bending action. In Fig. 1b let Δ represent one-half of the shortening of the top fiber, then $c \cdot \dfrac{d\theta}{2} = \Delta$. Since the outer fiber unit stress is σ the unit strain on the outer fiber is σ/E where E is the modulus of elasticity in pounds per square inch. The unit strain times one-half the undeformed lengths of the top fiber must also equal Δ. Consequently, we may write

$$\Delta = \frac{\sigma}{E} \cdot \frac{dx}{2}$$

or

$$c \frac{d\theta}{2} = \frac{\sigma}{E} \cdot \frac{dx}{2}$$

Substituting Mc/I for σ, we obtain

$$c \frac{d\theta}{2} = \frac{Mc}{EI} \cdot \frac{dx}{2}$$

and by canceling common terms,

$$d\theta = \frac{M \, dx}{EI}$$

Furthermore, since $\rho \, d\theta = dx$, or $d\theta = dx/\rho$, we find by substitution that

$$\frac{1}{\rho} = \frac{M}{EI} \tag{2-1}$$

The latter equation, expressing curvature, $1/\rho$, is fundamental to all deflection methods no matter how many other principles are involved.

The elastic curve is defined as the deflected configuration of the neutral plane containing the neutral axes of the individual cross-sections of the loaded beam. It is also usual to speak of this curve as the elastic line for which there is a determinable equation. To determine an equation of the elastic curve or line, additional geometric relations are established in terms of a coordinate system.

Figure 2 represents the coordinate geometry of an elastic curve. The origin is clearly marked and the direction of the axes shown. From the geometry, where ds is a differential length along the curve,

$$\rho \, d\theta = ds$$

but, for the usual elastic curve as found for structures, the slopes are small and ds is approximately equal to dx, hence

$$\rho \, d\theta = dx$$

and by employing equation 2-1 we obtain

$$d\theta = \frac{dx}{\rho} = \frac{M}{EI} \, dx \qquad (2\text{-}2)$$

where $d\theta$ is the small angle between the tangents to the elastic curve at m and n in radian units and represents the differential change in slope in the differential length dx.

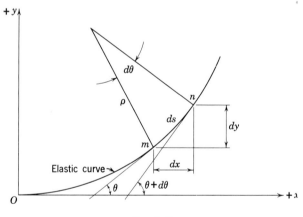

FIG. 2–2

From calculus as dx approaches zero length, the slope $= \tan \theta = \dfrac{dy}{dx}$ or in radian units for small angles

$$\theta = \frac{dy}{dx}$$

differentiating both sides with respect to x

$$\frac{d\theta}{dx} = \frac{d^2y}{dx^2}$$

Substituting $d\theta = (M/EI)\,dx$ from equation 2-2

$$\frac{(M/EI)dx}{dx} = \frac{d^2y}{dx^2}$$

which simplifies to

$$\frac{d^2y}{dx^2} = \frac{M}{EI}$$

which may be rewritten as

$$EI\frac{d^2y}{dx^2} = M \qquad\qquad (2\text{-}3)$$

Equation 2-3 is termed the differential equation of the elastic curve and forms the basis of the double integration method for determining slope and deflection.   Equation 2-3 assumes that the direction of positive y is

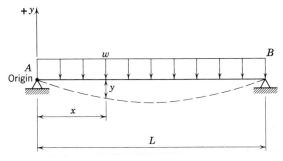

FIG. 2–3

upward, where the sign of the moment is based on the usual beam convention for positive moment* for flexural members, and if y is assumed to be positive downward equation 2-3 should be written as $EI\dfrac{d^2y}{dx^2} = -M$, with the sign of M still taken according to the usual convention.   This latter modification is necessary in order to establish the proper algebraic sign for $\dfrac{d^2y}{dx^2}$ when y is positive downward.

Examples

A beam AB of uniform cross-section supporting a uniform load of w lb per unit of length is shown in Fig. 3.   An analysis for slope and deflection should generally be preceded by drawing a sketch of the elastic curve. This is helpful in obtaining the feel of the problem in terms of deformation

---

* Positive moment when lower beam fibers are in tension; negative moment when upper beam fibers are in tension.

and a wrong concept of the elastic curve can be corrected later. The *origin of coordinates and direction of axes* should be explicitly shown on the sketch, and in this instance the origin is taken as point A. Once this decision is made, the boundary conditions of slope and deflection of the elastic curve may be noted and written down as

$$y = 0, \text{ when } x = 0 \text{ and } x = L$$

$$\frac{dy}{dx} = 0, \text{ when } x = \frac{L}{2}, \text{ due to symmetry}$$

It then remains to establish the bending moment M in terms of x and the loading and to ascertain that such equations for M are continuous functions. For the present problem, $M = (wL/2) x - wx^2/2$, and this expression is continuous for all values of x from $x = 0$ to $x = L$. Then

$$EI \frac{d^2y}{dx^2} = M = \frac{wLx}{2} - \frac{wx^2}{2}$$

The first integration leads to

$$EI \frac{dy}{dx} = \frac{wLx^2}{4} - \frac{wx^3}{6} + C_1$$

where C_1 is a constant of integration which, if divided by EI, is equal to the slope of the elastic curve at A or at the origin. The constant can be evaluated from the condition $\frac{dy}{dx} = 0$, when $x = \frac{L}{2}$

$$C_1 = -\tfrac{1}{24} wL^3$$

A second integration results in

$$EIy = \frac{wLx^3}{12} - \frac{wx^4}{24} - \frac{1}{24} wL^3 x + C_2,$$

where C_2 is a constant of integration which, if divided by EI, is equal to deflection at the origin and is equal to 0 since $y = 0$ when $x = 0$. Thus, the final equation of the elastic curve is

$$y = \frac{wLx^3}{12EI} - \frac{wx^4}{24EI} - \frac{1}{24} \frac{wL^3 x}{EI}$$

$$y = -\frac{w}{24EI} [x^4 - 2x^3 L + xL^3]$$

The maximum deflection occurs when $x = L/2$ or

$$y \text{ max} = -\frac{5wL^4}{384EI}$$

Figure 4 represents a beam in which M is not expressible as a continuous function for all values of x. Two expressions for moment will suffice and the principal difficulty is in the evaluation of the constants of integration.

For $x = 0$ to a, $M = Px$. For $x = a$ to $x = 3a$, $M = Pa$. Owing to symmetry, these two expressions for moment will be sufficient for a solution. The total evaluation of the constants must depend upon the mutual geometric conditions of the elastic curve at point C as determinable from the separate equations.

For $x = 0$ to $x = a$

$$EI\frac{d^2y}{dx^2} = Px$$

$$EI\frac{dy}{dx} = \frac{Px^2}{2} + C_1$$

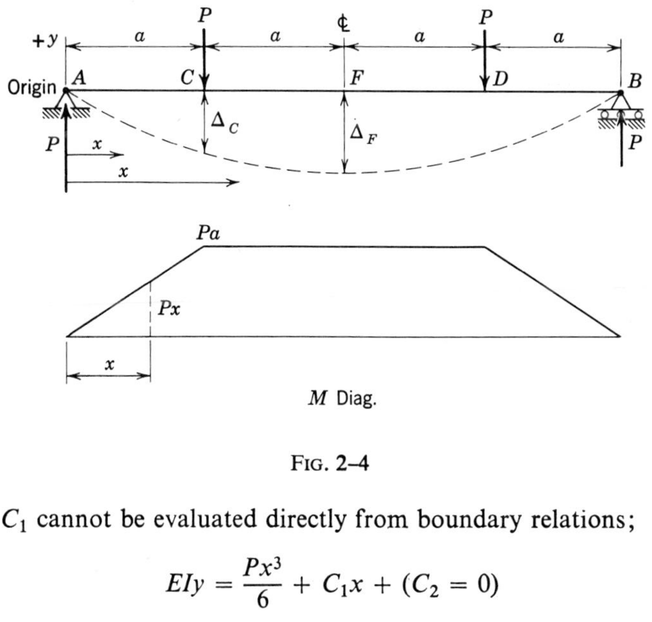

Fig. 2–4

where C_1 cannot be evaluated directly from boundary relations;

then
$$EIy = \frac{Px^3}{6} + C_1x + (C_2 = 0)$$

where $C_2 = 0$, since $y = 0$ when $x = 0$.

The second expression for moment can now be utilized and

$$EI\frac{d^2y}{dx^2} = Pa$$

$$EI\frac{dy}{dx} = Pax + C_3$$

when $x = 2a$, $\dfrac{dy}{dx} = 0$ from symmetry, requiring $C_3 = -2Pa^2$,

then
$$EIy = \frac{Pax^2}{2} - 2Pa^2x + C_4$$

where C_4 can not be evaluated directly.

With C_1 and C_4 unknown, two additional independent geometric conditions are needed for solution, and these exist at point C on the continuous elastic curve, where the slope and deflection at point C must be equal for the left- and right-hand segments.

The equality of slopes, with $x = a$, provides

$$\frac{Pa^2}{2} + C_1 = Pa^2 - 2Pa^2$$

$$C_1 = -\frac{3}{2}Pa^2$$

and the equality of deflections, with C_1 substituted, leads to

$$\frac{Pa^3}{6} - \frac{3}{2}Pa^3 = \frac{Pa^3}{2} - 2Pa^3 + C_4$$

$$C_4 = +\frac{1}{6}Pa^3$$

With the constants evaluated, the equations of the elastic curve are fully determined.

Upon proper substitution, the slopes and deflections may be evaluated and we obtain

$$\theta_A = -\frac{3}{2}\frac{Pa^2}{EI}$$

$$\theta_C = -\frac{Pa^2}{EI}$$

$$\Delta_C = -\frac{8}{6}\frac{Pa^3}{EI}$$

$$\Delta_F = -\frac{11}{6}\frac{Pa^3}{EI}$$

A beam loaded unsymmetrically is shown in Fig. 5. The discontinuity in the bending moment curve necessitates two separate applications of $EI\dfrac{d^2y}{dx^2} = M$. In the first equation, x will vary from O to a, and in the second equation, x will vary from a to L. Two of the constants of integration must be evaluated by recognizing that the elastic curve is continuous through C, and that slope as well as deflection at C are equal for either set of equations.

The differential equation for bending moment, $EI\dfrac{d^2y}{dx^2}$, is a transitional equation, interconnecting and relating bending moment and its derivatives, shear and loading intensity, with the geometry of the elastic curve. The following table presents the sequence of derivatives from the deflection curve, whose ordinates are y, to the loading intensity curve, whose ordinates are w. A sequence of derivatives, from the slope curve to the loading intensity curve, is also indicated. The table also shows all other derivative equivalents derived from static equilibrium principles relating bending moment, shear, and load intensity. The load intensity is positive when loading acts upward, and positive deflection, y, is considered to be positive upward, the x coordinate considered to be increasing in positive value to the right. This sign convention may be changed at will as long as consistency exists.

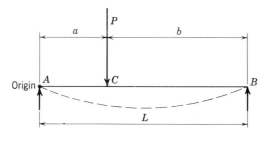

Fig. 2–5

The relationships of the table indicate, if $y = f(x)$ is known, that all other relations could be obtained by differentiation. They further reveal that if any of the beam functions is given as a function of x, then all other functions may be obtained by either differentiation or by integration. Differentiation is the easier of the two operations; however, integration is generally involved along with the determination of the constants of integration. This is best demonstrated by an example.

Let a simply supported beam of span L be loaded with a uniform load of w acting downwards. The origin is at the left-hand end of span, and V_O, M_O, θ_O, and y_O indicate boundary values at the origin, or when $x = 0$. V_O equals the end reaction $wL/2$; M_O and y_O equal zero.

Interrelationships of Loading Intensity, Shear, Moment, Slope, and Deflection

Quantity	Symbol	Derivative Equivalents				Common Integral Relations
Load Intensity	w	$\dfrac{dV}{dx}$ or	$\dfrac{d^2M}{dx^2}$ or	$EI\dfrac{d^3\theta}{dx^3}$ or	$EI\dfrac{d^4y}{dx^4}$	$V = \int_0^x w\,dx + V_O$
Shear	V	$\dfrac{dM}{dx}$ or	$EI\dfrac{d^2\theta}{dx^2}$ or	$EI\dfrac{d^3y}{dx^3}$		$M = \int_0^x V\,dx + M_O$
Moment	M	$EI\dfrac{d\theta}{dx}$	$EI\dfrac{d^2y}{dx^2}$			$\theta = \int_0^x \dfrac{M\,dx}{EI} + \theta_O$
$EI \times$ Slope	$EI\theta$	$EI\dfrac{dy}{dx}$				$y = \int_0^x \theta\,dx + y_O$
$EI \times$ Deflection	EIy					

Consequently, by successive integrations starting with the load curve we obtain

$$EI\frac{d^4y}{dx^4} = -w$$

$$V = EI\frac{d^3y}{dx^3} = -\int_0^x w\,dx + V_O$$

$$V = -wx + \frac{wL}{2}$$

$$M = EI\frac{d^2y}{dx^2} = -\int_0^x wx\,dx + \int_0^x \frac{wL}{2}\,dx + (M_O = 0)$$

$$M = -\frac{wx^2}{2} + \frac{wLx}{2}$$

$$\theta = \frac{dy}{dx} = -\frac{1}{EI}\int_0^x \frac{wx^2}{2}\,dx + \frac{1}{EI}\int_0^x \frac{wLx}{2}\,dx + \theta_o$$

$$\theta = -\frac{wx^3}{6EI} + \frac{wLx^2}{4EI} + \theta_o$$

when $x = \frac{L}{2}$, $\theta = 0$ $\therefore$ $\theta_O = -\frac{wL^3}{24EI}$

$$y = -\frac{1}{EI}\int_0^x \frac{wx^3}{6}\,dx + \frac{1}{EI}\int_0^x \frac{wLx^2}{4}\,dx - \frac{1}{EI}\int_0^x \frac{wL^3\,dx}{24} + (y_O = 0)$$

which simplifies to

$$y = -\frac{wx^4}{24EI} + \frac{wLx^3}{12EI} - \frac{wL^3x}{24EI}$$

$$y = -\frac{w}{24EI}[x^4 - 2x^3L + xL^3]$$

2-3. MOMENT-AREA METHOD

The moment-area method, presented in 1873 by Charles E. Greene, provides a basis for substituting numerical integration for mathematical integration when dealing with functions of the elastic curve. It is particularly useful in problems where the bending moment relation is discontinuous and, although equation of the elastic curve can be established by moment-area principles, the chief merit of the method is bypassing of such equations.

The relations employed by the moment-area method are stated as:

Relation No. 1   The angle between the tangents drawn to any two points on the elastic curve, provided beam was initially straight, is equal to the area under the M/EI curve between these two points.

Relation No. 2   The deflection of a point A on an elastic curve from a tangent drawn to point B of the same elastic curve, provided beam was initially straight, is equal to the statical moment taken around A of the area under the M/EI curve between A and B treated as a load.

These relations are established in *Mechanics of Materials* (see ref. 5)*, but the intent of the above statements are additionally shown in Fig. 6.

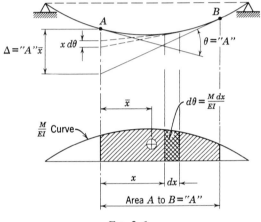

FIG. 2-6

The fundamental on which these relations are primarily based is expressed by equation 2-2, where for a differential length of beam dx the differential change in slope is

$$d\theta = \frac{M}{EI} dx$$

or the total angle between the tangents drawn to A and B is

$$\theta = \int_A^B \frac{M}{EI} dx = A = \text{area under } \frac{M}{EI} \text{ curve} \quad \text{(Relation No. 1)}$$

Letting $d\Delta = x\,d\theta$, where $d\Delta$ is a differential deflection at point A, owing to the flexure of one dx element, then

$$d\Delta = \frac{M}{EI} dx \cdot x$$

---

* References are located before Problems at the end of each chapter.

The summation of the differential deflections may be stated as

$$\varDelta = \int_{A}^{B} x\left(\frac{M}{EI} \cdot dx\right) = \bar{x}A = \begin{cases} \text{moment of area under } M/EI \\ \text{curve between } A \text{ and } B \\ \text{about } A. \end{cases} \quad \text{(Relation No. 2)}$$

The question of signs is important and can be understood by referring to Fig. 7. Figure 7a represents a segment of an elastic curve under the influence of positive bending moment. Two reference tangents have been drawn to curve at points A and B. The angle between the tangents, or the angular change in slope, is a counterclockwise angle when going from left to right and a clockwise angle when going from right to left. For the positive bending moment assumed, the area under the M/EI diagram

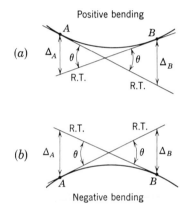

FIG. 2-7

between A and B is positive. A rule can be formulated, but, in the author's opinion, a sketch similar to Fig. 7a is more dependable when in doubt. The sign of deflection is more definite as in Fig. 7a; $\varDelta_A$ or $\varDelta_B$ implies that the point on elastic curve lies above the reference tangent for positive bending. The opposite of these observations is true for negative bending, as shown in Fig. 7b.

Before applying the moment-area relations, a geometrical analysis of the problem must be made in terms of the slope and deflection quantities needed and the slopes and deflections necessary to achieve this result. The key to geometrical analysis lies in the wise selection of the reference tangents (indicated on sketches as R.T.) drawn tangent to the elastic curve. Although a reference tangent may be drawn to an elastic curve at any point, a complete analysis generally requires that such tangents be drawn to the elastic curve at the supports where the deflection is zero or at points on elastic curve where the slope is zero. The general approach is best illustrated by the following examples.

Examples

Figure 8 is the same beam as shown in Fig. 4. A sketch of the elastic curve is shown with a horizontal reference tangent drawn to elastic curve at the center line of the span. To find Δ_F, geometric analysis indicates it is equal to the deflection of point A from the reference tangent. To find Δ_C find Δ_1, a deflection from R.T., and subtract from Δ_F.

The moment diagram should always be drawn in the simplest possible manner, and the separate areas represented along with the appropriate moment arms to the centroids of the areas involved. The analysis, selection of reference tangents, drawing of simplified moment diagrams, and the identification of appropriate areas and centroids are all analytical steps to a successful solution. Unless all these steps can be clearly and properly made, it is useless to perform the algebra and arithmetic, which are often tedious and certainly unprofitable if the initial analysis is wrong.

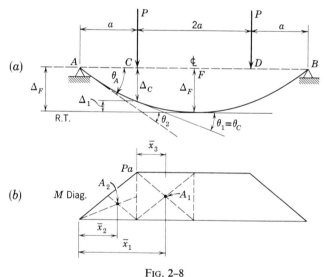

FIG. 2–8

Referring to Fig. 8 and taking moments about A of areas A_1 and A_2 of the moment curve and recognizing that EI is constant, we obtain

$$EI\Delta_F = A_1\bar{x}_1 + A_2\bar{x}_2$$

$$EI\Delta_F = Pa \cdot a \cdot \left(a + \frac{a}{2}\right) + \frac{Pa \cdot a}{2}\left(\frac{2}{3}a\right)$$

$$= \frac{11}{6}Pa^3$$

From Fig. 8

$$\Delta_C = \Delta_F - \Delta_1 = \Delta_F - \frac{A_1 \bar{x}_3}{EI}$$

$$EI\Delta_C = \frac{11}{6} Pa^3 - Pa \cdot a \cdot \frac{a}{2} = \frac{8}{6} Pa^3$$

Using Relation No. 1, and the geometric analysis shown in Fig. 8,

$$EI\theta_C = EI\theta_1 = A_1 = Pa \cdot a = Pa^2$$

$$EI\theta_A = EI\theta_2 = A_1 + A_2 = Pa^2 + \frac{Pa^2}{2} = \frac{3}{2} Pa^2$$

The deflection and slope values check those previously found by the double integration method.

A second example is shown in Fig. 9. Owing to lack of symmetry the reference tangent No. 1 is first drawn to elastic curve at the support B. An analysis of the geometry, using this reference tangent, can then be made to find θ_B and Δ_A. Point C, at a support and on the elastic curve, does not deflect, hence if Δ_1 is known, θ_B can be determined. It also follows from proportion that $\Delta_2 = \frac{1}{3}\Delta_1$. The deflection of point A from the R.T. is designated as Δ_3. Then, by geometry $\Delta_A = \Delta_2 - \Delta_3$.

After the above process of geometric analysis has shown the procedure to be feasible, then the relations of the moment-area method are applied as follows, treating the beam as having a uniform value of EI.

Employing reference tangent No. 1

$$EI\Delta_1 = A_1 \bar{x}_1 + A_2 \bar{x}_2$$

$$EI\Delta_1 = \left(\frac{12a \cdot 3a}{2}\right)\left(\frac{2}{3} \times 3a\right) - \left(\frac{15a \cdot a}{2}\right)\left(2a + \frac{2}{3}a\right)$$

$$= 16a^3$$

$$EI\Delta_2 = \left(\frac{a}{3a}\right)16a^3 = \frac{16}{3}a^3$$

$$EI\Delta_3 = -\frac{3a \cdot a}{2}\left(\frac{2}{3}a\right) = -a^3 \qquad \text{(Minus sign means point on elastic curve is below R.T.)}$$

since $\Delta_A = \Delta_2 - \Delta_3$

$$EI\Delta_A = \frac{16}{3}a^3 - a^3 = \frac{13}{3}a^3$$

$$\theta_B = \frac{\Delta_1}{3a} = \frac{16a^3}{EI} \div 3a = \frac{16}{3EI}a^2$$

$$\theta_A = \theta_B - \theta_1 = \frac{16a^2}{3EI} - \frac{1}{EI}\frac{3a \cdot a}{2}$$

$$\theta_A = \frac{23a^2}{6EI}$$

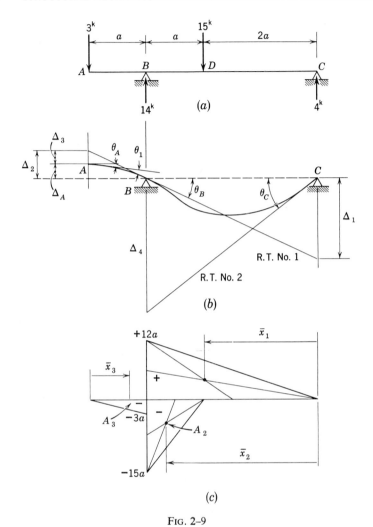

FIG. 2–9

If another reference tangent is drawn to the elastic curve at point C, Δ_4 can be computed and θ_C determined. The detailed calculations are

$$EI\Delta_4 = \frac{12a \cdot 3a}{2}\,(a) - \frac{15a \cdot a}{2}\left(\frac{a}{3}\right) = \frac{31a^3}{2}$$

and

$$EI\theta_C = \frac{EI\Delta_4}{3a} = \frac{31}{6}a^2$$

Slope and deflection at other points on the elastic curve may be found by continuing the application of the two basic moment-area relations.

2–4. CONJUGATE-BEAM METHOD

The moment-area method may be used to derive the procedure known as the conjugate-beam method. To apply the moment-area method, attention was given to the geometrical analysis of the elastic curve, whereas in the conjugate-beam method the geometrical quantities of slope and deflection of the elastic curve are found without a prior geometric analysis. To many engineers this is an advantage, but to others its automatic procedures (once the sign convention is understood) cloaks the feel for structural deformation, although in reality the moment-area method and the conjugate-beam method are mutually exchangeable. In any event, the method should definitely be in the student's vocabulary and it will be referred to later in this book.

The basis of the method is easily explained by making use of geometric relationships and analogies. Figure 10a represents a real beam AB wherein the angular deformation of only one differential element dx at C is to be considered. If the moment acting on the differential element is M, then the angular change in slope in radian units in the dx length is $d\theta = \dfrac{M\,dx}{EI}$. In Fig. 10b the angle is depicted as a sharp kink of $d\theta$, while the remainder of the beam remains as two intersecting straight lines. The beam is then in its deflected configuration as a result of this local kinking, and the geometrical relations of Fig. 10b will lead to values of the end slopes, θ_A and θ_B, as well as to a value for deflection at any section. It can then be noted from Fig. 10c that if $d\theta$ is treated as a downward load on the beam at C, such a load will produce end reactions at A and B equal to the end slopes θ_A and θ_B. Figure 10d is the shear diagram where the values of shear are equal to the slope of deflection curve of Fig. 10b. Figure 10e is the bending moment diagram produced by the angular load $d\theta$ where the value of bending moment is equal to the ordinates of deflection curve of Fig. 10b. It is to be noted that positive bending moment due to the angular load is plotted downward in keeping with actual physical conditions and in agreement with the convention of treating positive deflections as downward deflections.

The foregoing may now be restated. The differential angle change $d\theta$ at C may be interpreted as an elastic loading over a short length dx, and M/EI may then be interpreted as the intensity of loading per unit of length or the rate of angle or slope change along the elastic curve. The beam AB of Fig. 10c is spoken of as the conjugate beam and the shear and moment diagrams which follow refer to the conjugate beam. The analogy between shear and slope, and moment and deflection, of the real and conjugate beams has been demonstrated.

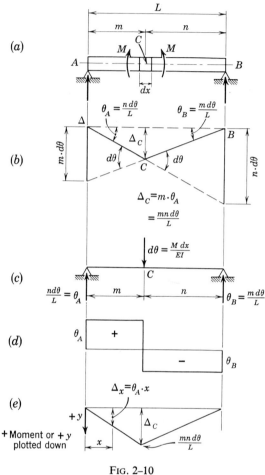

FIG. 2–10

It now remains to generalize the conjugate-beam analogy for a beam subjected to a continuous series of differential angle changes. Since superposition is valid, the conjugate beam is loaded with the M/EI diagram for the real beam where the ordinates represent the rate of angle change at every section. *The student will note that the M/EI loading for the conjugate beam is the same as the M/EI diagram in the moment-area method.*

In summary, the two important conjugate-beam relations for a given section, where the conjugate beam is loaded with the M/EI diagram, are:

1. The shear in the conjugate beam is equal to the slope of elastic curve of the real beam.

2. The bending moment in the conjugate beam is equal to the deflection of the elastic curve of the real beam.   Maximum deflection in the real beam occurs at sections of zero shear in the conjugate beam.

The usual sign convention is to treat down loads as negative and upward loads as positive.   When the moment curve of the real beam is positive, the moment curve should be treated as producing a downward load on the conjugate beam.   A negative moment curve should be treated as an upward load on the conjugate beam.   The sign of the bending moment for the conjugate beam is then established as positive or negative by the usual sign convention for bending moments.   The acceptance of the foregoing sign usage will mean that positive bending moment in the conjugate beam agrees with downward deflection, hence consider the positive y axis as directed downward.   Positive shear in the conjugate beam means positive slope for the real beam if the y axis is taken positive downward and the positive x axis extends from left to right.

Examples

The beam of Fig. 11 has been used before to illustrate double integration and moment-area methods.   The problem will now be solved by the conjugate-beam method.

After the bending moment diagram for the real beam has been determined, it is placed on the conjugate beam of Fig. 11c as a loading diagram.   A downward load is indicated because of the positive bending moment.   The reaction at A or θ_A, due to symmetry, is equal to one-half of the load on this conjugate beam, hence $\theta_A = 3Pa^2/2EI$.   To determine the deflection and slope at C, use the free body diagram of Fig. 11d to determine shear and moment in the conjugate beam at C.   The detailed calculations may now be made as follows.

$$Q_1 = \frac{Pa}{EI} \times \frac{a}{2} = \frac{Pa^2}{2EI}$$

$$\theta_C = \frac{3Pa^2}{2EI} - \frac{Pa^2}{2EI} = \frac{Pa^2}{EI}$$

$$m_C = \Delta_C = a \times \frac{3}{2}\frac{Pa^2}{EI} - \frac{a}{3} \times \frac{Pa^2}{2EI}$$

$$\Delta_C = +\frac{8}{6}\frac{Pa^3}{EI} \quad (+\text{ sign indicates downward deflection})$$

A free body similarly taken from A to F, will permit Δ_F to be determined.

In the example of Fig. 11, the conjugate beam is similar to the real beam and has the same simple supporting conditions.   The deflection at A and

B is zero and, according to our analogy, the moment of the conjugate beam, representing deflection, must also be zero at A and B. In general, the supporting and internal conditions of the conjugate beam must be so selected and arranged to provide for zero moment at points where zero deflection exists in real beam and for a moment at points when deflection

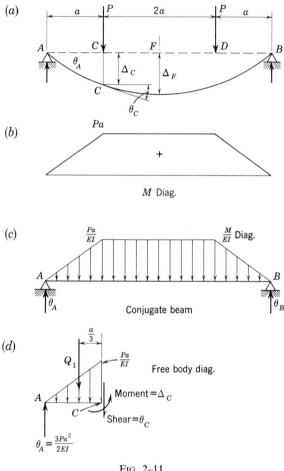

FIG. 2–11

is possible. Furthermore, since shear in the conjugate beam represents slope, according to our analogy, then the supporting and internal conditions must also provide for a shear at points where rotation is possible. To provide for zero deflection, a hinge must be introduced, which of course also possesses the capability of transmitting a shearing force. In the

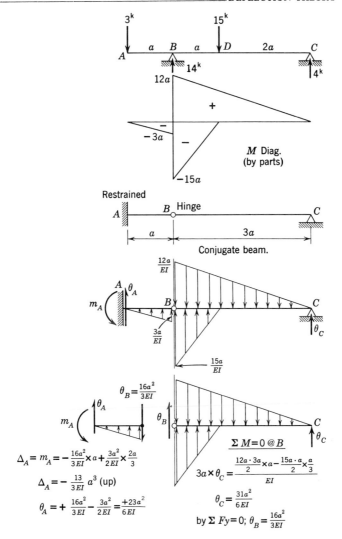

FIG. 2–12

event of a free end, where both deflection and slope are possible, then the end condition of conjugate beam must be a fixed or restrained support in order to develop both moment and shear in the conjugate beam under the conjugate-beam loading.

The beam of Fig. 12 represents the general problem. According to the principles discussed, A of the conjugate beam is restrained since A of the real beam is a free end. A hinge is inserted at B, a point of zero deflection, and C remains simply supported as it was in real beam.   These necessary conditions are independent of the loading on the real beam.   Standard procedures are then followed in solving the problem of Fig. 12 with all computations shown.   Free body and equilibrium principles suffice to determine not only the deflection and slope at A, but the slope at B and C as well.

2–5. CONJUGATE-BEAM METHOD—NUMERICAL PROCEDURE

A numerical procedure for applying the conjugate-beam method was developed by N. M. Newmark (see ref. 3) and has many advantages in making advanced studies of deflection and buckling.   The fundamental nature of the method should be understood so that it becomes a part of the student's vocabulary, and only the basic aspects are discussed here.

The approach of the conjugate-beam method was to load the conjugate beam with the M/EI diagram.   The numerical method also takes this as the starting point, followed by a step-by-step procedure for determination of shear and moment, which have been demonstrated to be the equivalent of slopes and deflections.   The step-by-step procedure first involves the reduction of the variable M/EI loading on the beam to a series of equivalent concentrated loads.   Figure 13a represents a loaded real beam and Fig. 13b the standard moment diagram.   Figure 13c indicates the conjugate beam with the M/EI curve placed on top of a series of simple span stringers of equal span length "a".   The stringers are supported at each panel point by a floor beam.

The stringer reactions are transmitted directly to the conjugate beam as concentrated loads through the floor beams, and since these loads are derived from the M/EI loading, they physically represent concentrated angle changes, or abrupt changes in slope of the elastic curve.   Geometrically this procedure will be found to be the equivalent of defining the elastic curve as a series of straight lines, or a deflection polygon, with the deflection at the intersection or panel points equal to the deflection of the real beam.

To simplify the numerical work, only the coefficients of common factors are shown in Fig. 13.   The common factor to the right of the diagrams contains the quantities of force, length, and EI, as required, and the relevant arithmetical coefficients.   The intervals along the beam are equal and are all valued at "a" in this example.   Although four intervals are used in this problem, the number of intervals is entirely arbitrary and will generally depend on how many points on the deflection curve are to be evaluated.

As has been explained, the ordinates of M/EI curve represent the intensity of loading on the stringers. The stringer and floor beam reactions are easily obtained in this example by considering the equilibrium of each stringer as a simple beam and are shown in Fig. 13d as the concentrated angle changes. (The student should check the values shown.)

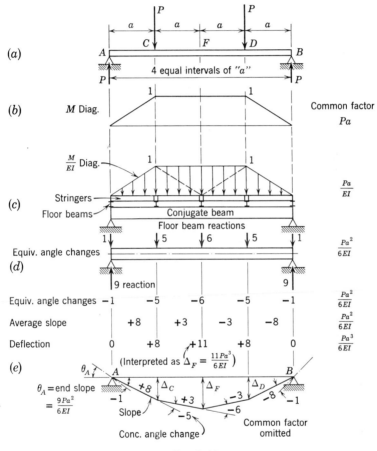

FIG. 2–13

The systematic calculations follow immediately below Fig. 13d, and start by listing the concentrated angle changes that must bear a negative sign since they are the equivalent of downward forces. The next line of the table represents the average slope of each segment of the elastic curve, and from the conjugate-beam analogy these are the equivalent of conjugate-beam shear. Figure 13e indicates the physical representation of these slopes.   Since deflection is the equivalent of moment and the area under a shear curve between two sections equals change in moment, the co-efficients of deflection are obtained quickly by starting at the left and accumulating successive areas between intervals as coefficients of $Pa^3/6EI$. It should be noted that as a result of the conventions followed, the deflection is positive and means downward deflection.   Note the polygonal deflection diagram of Fig. 13e.   The final results at C and F check the previous solutions by other deflection methods.

Example

The beam of Fig. 14 may be taken to illustrate additional considerations in applying the numerical method to a general problem.   Since the problem has been solved before by other methods, the solution can be easily followed.   The interval selected is "a" and the separate stringers (not shown) are of span "a", and floor beams (not shown) support the stringers at each panel point.   The M/EI loading curve constructed by parts produces the concentrated loads of Figs. 14c and 14d, where the values on the latter figure represent the net loads, or concentrated angle changes delivered to the conjugate beam at each panel point.   The routine tabular solution follows immediately below the loaded beam of Fig. 14d, and starts with an assumed average slope of convenient value for the first interval.   In this example a $+ 3$ slope was chosen.   The average slopes for the successive intervals follow quickly by the method of panel shears. Trial deflections (by method of moments) have been computed by arbitrarily assuming the deflection at A to be zero.   From the shears, or average slopes, the trial deflections at the panel points along the beam are tabulated. The calculations indicate that points B and C each have a trial deflection that is not equal to zero as is required to satisfy actual supporting conditions, C having moved up and B having moved down.   This situation may be corrected to the true geometrical conditions by physically rotating and also translating the beam vertically in its deflected form so that the deflections at B and C are zero.   This rotation and translation may be mathematically accomplished by introducing linear corrections as shown in the tabulation, the requirement being that the final values of Δ_B and Δ_C be zero.   The linear correction is -3 at B and $+66$ at C.   These corrections establish the slope of linear correction line as $\frac{69}{3}$ or 23 units per panel.   Corrections to the deflections and the trial deflections are algebraically added to obtain the true deflections shown.   True average slopes may be determined by starting at C and working towards the left since the deflection at C is zero and the deflection of the first panel point to the left is known.   The diagram at the bottom of Fig. 14 depicts the final results. All of the computed quantities may be easily identified in this diagram.

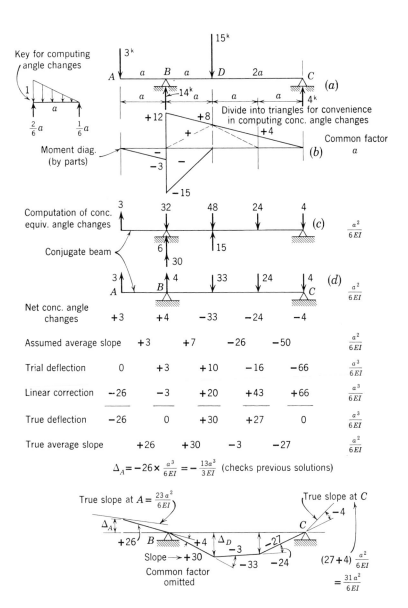

FIG. 2–14

Although the above examples have demonstrated the numerical method for beams where the bending moment curve was a linear function, it is just as valid for curved bending moment diagrams. The only added difficulty is in the calculation of the concentrated angle changes. Special formulas to cover such cases were developed by Newmark in his original paper. For normal deflection problems no time is usually saved by this method, but for buckling and the combination of transverse and axial loading the method has great merit. The latter type of problem is discussed in Chapter 13.

2–6. BASIC STRAIN ENERGY METHOD

Work and energy relations are found in many fields of science, and the field of structural mechanics has adapted many of these relations and concepts to determine elastic slopes and deflections. The energy methods represent a unique and mathematical application of the basic law of conservation of energy, and it is accepted that the form or manner in which energy appears or is stored is dependent on the physical problem. For elastic structures, external work is done on the structure by physically applied external forces or moments, and an equal amount of potential energy is stored through the action of internal forces and the elastic strain mechanisms of the structure. In the discussion to follow the loads are assumed to be slowly applied so that dynamic influences may be omitted.

The work done by a constantly applied force is equal to the magnitude of the force multiplied by the distance through which the point of application of the force moves in the direction of the force. If the force is variable, an integration is necessary to determine the work done, or it is said that the work done is equal to the area under the curve representing force and displacement. The sign of work is positive if the displacement is in the direction of the force, and if displacement and direction of the force are oppositely directed the work done is said to be negative.

A prismatic elastic bar is shown in Fig. 15 having a length L in inches, cross-sectional area A in square inches, and the material of elastic modulus E in pounds per square inch. It is desired to transfer a load P from a crane to the end of the bar at B. The crane lowers the load P slowly until the load just contacts the bearing plate at B. At this instant no force has been applied to the bar, but with further lowering of the load the resisting force in the bar is slowly built up to the value of P. At this time the load is fully applied to the bar and the crane has accomplished its purpose. The bar has deformed an amount $\Delta = PL/AE$, and the external work done on the bar by P equals $P \cdot \Delta/2$. For a full understanding it should be realized that the loss in potential energy of the load P must equal $P \cdot \Delta$, or twice the external work done on the bar, and hence it appears that a discrepancy exists until it is recognized that one-half of the loss in potential energy is utilized in doing work on the crane.   The strain energy stored in bar equals the net change in potential energy.

The internal strain energy stored in the bar is equal to the external work done on the bar.   The triangle abc of Fig. 15 may be considered to be a force-deformation diagram for the bar, where force in the elastic bar is directly proportional to deformation.   The area of triangle abc is $P \cdot \Delta/2$, or $P^2L/2AE$, and represents the strain energy stored internally in the bar.   Later this expression will be used in developing the theory relating to deflection of trusses.

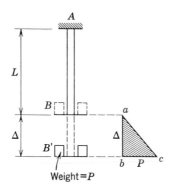

Fig. 2–15

Work and strain energy concepts may be associated to determine beam deflections.   The strain energy relations for a beam may be developed by referring to Fig. 16.   The beam of this figure is slowly loaded with an external load P and the beam deflects under the load by an amount Δ with load doing external work on the beam equal to $P \cdot \Delta/2$.   The internal strain energy stored within the beam must equal the external work.   To account for the physical action of the elements of the beam, assume that a differential element of beam dx long is removed from the beam.   This element is shown enlarged in Fig. 16c, the solid lines indicating the plane cross-sections before bending and the dotted lines indicating the deformed shape of the element after bending.   The beam element is assumed to be subjected to a uniform moment of M and, owing to this moment, $d\theta = M\,dx/EI$, where $d\theta$ is the rotation in radian units of one cross-section relative to the other.   All of the factors for computation of the external work on this particular elemental section are now available. Recalling that the work performed by a couple is equal to the value of the couple multiplied by the angle through which the couple rotates, and also recognizing that the value of bending moment at the section varies linearly from zero to M during the application of the load P, we see then that the external work on the particular element is $\dfrac{M \cdot d\theta}{2}$. Since the element is unique, the external work on the element must be equal to the internal

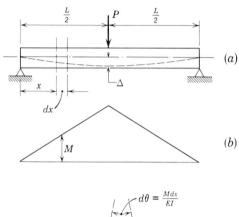

(a)

(b)

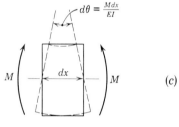

(c)

FIG. 2–16

strain energy stored within the element. Thus, the internal strain energy for the differential element is

$$dU = \frac{M}{2} \cdot \frac{M \cdot dx}{EI} = \frac{M^2 \, dx}{2EI}$$

and, summing the differential work for all differential elements of the beam, we obtain

$$U = \int \frac{M^2 \, dx}{2EI} \tag{2-4}$$

where U represents the total stored internal strain energy due to bending in the entire beam. Since the internal strain energy must equal the external work, the deflection of beam under the load P may be computed.

Returning to Fig. 16a, the strain energy is

$$U = \int \frac{M^2\,dx}{2EI} = 2 \int_0^{L/2} \frac{(Px/2)^2\,dx}{2EI}$$

$$U = \frac{P^2}{4EI} \int_0^{L/2} x^2\,dx = \frac{P^2L^3}{96EI}$$

The external work may now be equated to the internal strain energy as follows.

$$\frac{P \cdot \varDelta}{2} = \frac{P^2L^3}{96EI}$$

or

$$\varDelta = \frac{PL^3}{48EI}$$

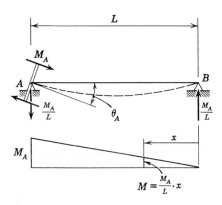

FIG. 2–17

The method of basic work and strain energy may also be applied to determine the slopes of the tangents to elastic curves. Figure 17 depicts a simple beam loaded slowly with an end couple M_A. The couple rotates through the slope angle θ_A, as the couple increases in value from zero to M_A, and external work is done equal to $M_A \cdot \theta_A/2$. As before, the internal bending strain energy must equal the external work. By equation 2-4 we obtain

$$U = \int \frac{M^2 dx}{2EI} = \int_0^L \frac{(M_A \cdot x/L)^2\,dx}{2EI}$$

$$U = \frac{M_A{}^2}{2EI \cdot L^2} \int_0^L x^2\,dx = \frac{M_A{}^2 L}{6EI}$$

we may now equate the external work to the internal strain energy and
obtain

$$\frac{M_A \cdot \theta_A}{2} = \frac{M_A^2 L}{6EI}$$

$$\theta_A = \frac{M_A L}{3EI}$$

a result which is easily proved by other deflection methods.

Although the basic work and energy method has merit and is of great
importance, it also has a particular limitation.   Deflection may be
computed only at the loaded point in the direction of the applied load and
the calculation of rotation limited to the point on the elastic curve where
an external moment is applied.   Symmetrical cases may also be analyzed.
For these reasons the basic work and energy method as presented has no
value in defining or calculating intermediate values of deflection or slope
along an elastic curve.   A general total energy approach may be developed
to permit a determination of slope and deflection at any section but it will
not be discussed in this book since it is similar to the unit load-unit couple
method which follows.

2-7. UNIT LOAD–UNIT COUPLE METHOD

To circumvent the limitations of the total work or energy method, an
additional relation of work and strain energy may be obtained by considering two systems of loads, one system an auxiliary loading and the
other system the actual loads on structure.

Figure 18a is a beam undergoing deflection owing to load P at B.   The
deflection of beam at A or Δ_A, is to be determined.   First consider the
beam loaded with an auxiliary vertical load of 1 lb at A, in Fig. 18c.
This load may be termed the *unit load* and although this load produces a
small deflection, hence doing external work on the beam, this deflection
and work are unrelated to the problem of finding Δ_A due to P.   The
important part played by the unit load is to establish a bending moment
"m" at each section of the beam.   These moments define the "m"
diagram in contrast to the "M" diagram for the real loading on beam.
It should be re-emphasized that the unit loading is placed on the beam
before the real loading and hence the bending moment "m" exists prior
to the superposition of "M" due to the real loading.

The fundamentals of work and strain energy as they apply to the unit
load approach may now be established.   With the unit load already on
beam at A, apply the load P at B.   The constant unit load will move down
Δ_A, the deflection at A caused by P, and the unit load therefore does

additional external work on the beam equal to $1 \cdot \Delta_A$. This work is a particularly defined fraction of the total external work done by the two load systems and is often termed *virtual work* where Δ_A would represent a virtual displacement. To formulate a deflection equation, this specific fraction of the total external work must be equated to an appropriate expression for internal strain energy. This can be made explicit by

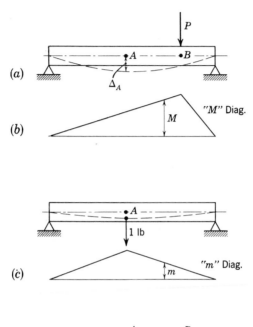

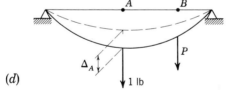

FIG. 2–18

considering an unstrained differential element of the beam as shown in Fig. 19a. When the unit load is placed at A the differential element is subjected to a moment "m" and the plane cross-sections rotate relative to one another by a differential angle $d\phi$. See Fig. 19b. The internal potential strain energy stored in the element is equal to the external work done on it by "m" and this strain energy is shown by the area of triangle *abc* of Fig. 19e. Figure 19c indicates the deformation of the differential element under the presence of " M " due to P. Recognizing that this latter angular deformation is superimposable upon that caused by the unit load, we see that Fig. 19d indicates the superimposed conditions. The angular deformation curve plotted against total moment extends from point c to d of Fig. 19e. The area of triangle cde represents the potential strain energy due to " M " alone, or $\dfrac{M^2\ dx}{2EI}$ as previously determined, but the important area of Fig. 19e is the rectangular area $bcef$. It is this particular area that

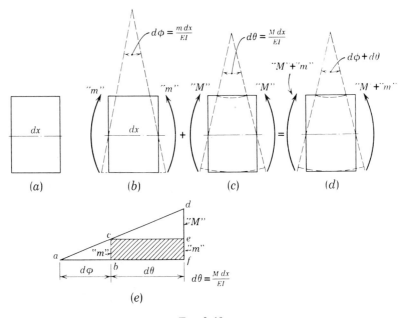

FIG. 2–19

is relevant to this method, and its appearance and its identification are solely the result of having the unit load on the beam before the real loading was applied. This area is often termed the *virtual strain energy* as it appears during a virtual displacement produced by P and since its very existence and identification depend on the prior existence of the unit load. This rectangular area in terms of the nomenclature is equal to $\dfrac{Mm\ dx}{EI}$. The equation expressing the equality of the total external virtual work to the total internal virtual strain energy stored in the beam follows as

$$1 \cdot \Delta_A = \int \frac{Mm\ dx}{EI} \tag{2-5}$$

with the integration to be carried out over the entire length of the beam. Figure 18d indicates the deflected elastic curve of beam under the action of the unit load and the load P, where Δ_A is due to P. The student should note that multiplication by 1 lb on the left-hand side of equation 2-5 is necessary to preserve the units of work.

As an example of this method, compute the vertical deflection of the free end A of the cantilever beam of Fig. 20a, which has a uniform cross-section. The first step in analysis is to place a 1-lb vertical load at A as in Fig. 20c and then draw the "m" diagram of Fig. 20d. The real loading

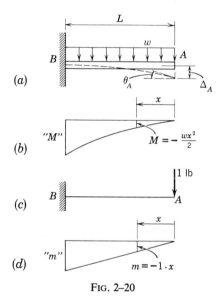

FIG. 2–20

of Fig. 20a is then superimposed on the unit loading of Fig. 20c, thus causing A to move down Δ_A, or, more importantly to this method, the one-pound load does external virtual work on the beam equal to $1 \cdot \Delta_A$. The moment "M" due to the real loading is shown graphically in Fig. 20b.

The remaining step is to evaluate the internal virtual strain energy by integration methods. An excellent and advisable procedure is to show how the variable x is measured directly on the moment sketches, and on the same figures indicate the basic formulation of "m" and "M" as a function of this variable and the loading. In the example

$$m = -1 \cdot x = -x$$

$$M = -wx \cdot \frac{x}{2} = -\frac{wx^2}{2}$$

Then by equation 2-5

$$1 \cdot \Delta_A = \int \frac{Mm \, dx}{EI} = \frac{1}{EI} \int_0^L \left(\frac{-wx^2}{2} \right)(-x) \, dx$$

$$1 \cdot \Delta_A = + \frac{wL^4}{8EI}$$

It should be particularly noted that to preserve the concept of work the left-hand side of the above equation was written as $1 \cdot \Delta_A$, meaning a force of 1 lb times a distance Δ_A, or the work done by the 1-lb force. The sign resulting from the integration in this case was positive, which, when interpreted as positive work, means that point A displaced under the influence

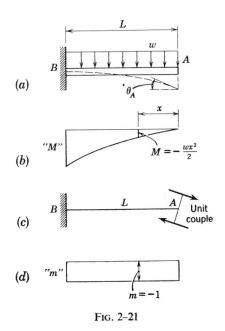

FIG. 2–21

of the real loading in the direction of the 1-lb load. If the result had been negative the interpretation of sign would have meant that point A moved in a direction opposite to the direction of the 1-lb load, or that the 1-lb load had done negative work.

The principles of this method may be extended to permit the determination of slopes of the elastic curve. Figure 21a is a repeat of the cantilever beam of Fig. 20a. In this instance we seek to determine θ_A, the slope of elastic curve at A. Since rotation is wanted, the external virtual work term must involve a couple instead of a force. Instead of a unit load to produce "m", a unit couple, say 1 ft-lb, is first applied at A as shown in

Fig. 21c.   The "m" diagram is quite simple in this case and is a rectangle as given in Fig. 21d.   After the unit couple is applied to the beam at the point where the slope is to be determined, the real loading is then applied, and the unit couple forced to rotate through angle θ_A in radians.   In mathematical terms and equating virtual external work to virtual strain energy,

$$1 \cdot \theta_A = \int \frac{Mm\,dx}{EI} = \frac{1}{EI} \int_0^L \left(\frac{-wx^2}{2}\right)(-1)\,dx$$

$$\theta_A = +\frac{wL^3}{6EI}$$

where the positive sign of result indicates that the tangent to the curve at A rotates in the same direction as the unit couple under the action of the real loads.

An interesting equivalence may be noted concerning this method and the moment-area method.   In writing the general integral $\int \frac{Mm\,dx}{EI}$, place in this integral the value of "m" in terms of x and let "M" stand as a symbol only.   Then, for the beam of Fig. 20a

$$1 \cdot \Delta_A = \int \frac{Mm\,dx}{EI} = \int_0^L \frac{M(-x)\,dx}{EI}$$

$$\Delta_A = \int_0^L -\frac{Mx}{EI}\,dx$$

and with rearranged terms

$$\Delta_A = -\int_0^L (x)\left(\frac{M\,dx}{EI}\right)$$

If the term $\frac{M\,dx}{EI}$ is treated as a differential area under the M/EI diagram, then the integral may be interpreted as the moment of the area of the M/EI diagram about A.   This is precisely the same as the principle used for determining deflections by the moment-area method and can be restated as $\Delta_A = A\bar{x}$; where A is the area under M/EI curve and $\bar{x}$ is the distance to its centroid measured from end of beam.

The equivalence of the two methods may be also shown by referring to Fig. 21 where θ_A was to be determined

$$1 \cdot \theta_A = \int \frac{Mm\,dx}{EI} = \int_0^L \left(\frac{M}{EI}\right)(-1)\,dx$$

or

$$\theta_A = -\int_0^L \frac{M}{EI}\,dx$$

which is interpreted as explained above as the area under the M/EI

diagram between A and B.   Although this concurrence or equivalence of method is important, and is not unexpected, since the fundamental relationship $1/\rho = M/EI$ has been accepted and indirectly incorporated in both the energy and moment-area procedures, the fundamentals of the analysis for each method are different.   In applying the moment-area method, a careful analysis must be made of the full geometry of the elastic curve, whereas in the subject method the geometric analysis of the entire curve is bypassed.   If mathematical integration can be eliminated in favor of arithmetical integration the advantages of the unit load or unit couple approach are greatly enhanced.   A few examples should clarify the subject method and the variations in its method of application.

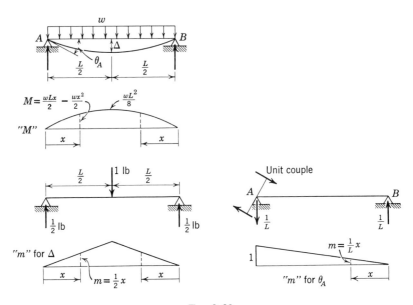

FIG. 2–22

Additional Examples—Unit Load-Unit Couple Method

A uniformly loaded beam is shown in Fig. 22.   The vertical deflection at the center of span and the slope at A are required.   The M diagram is sketched and, as the next step in determining Δ, a 1-lb vertical load is placed at the center line of span.   Although the unit load must theoretically be considered as applied first, it is generally more convenient to develop the M diagram first.   The "m" diagram is shown, and since it is discontinuous but symmetrical the algebraic expression for moment "m" may be written with x varying from an origin at either the left or the right, and extending to the center of the span.   The ordinates of the "M"

diagram may be represented by a continuous function with origin at either support.

$$1 \cdot \varDelta = \int_0^L \frac{Mm\,dx}{EI} = \frac{2}{EI} \int_0^{L/2} \left(\frac{wLx}{2} - \frac{wx^2}{2} \right) \left(\frac{x}{2} \right) dx$$

and after integration and simplification

$$\varDelta = \frac{5wL^4}{384EI}$$

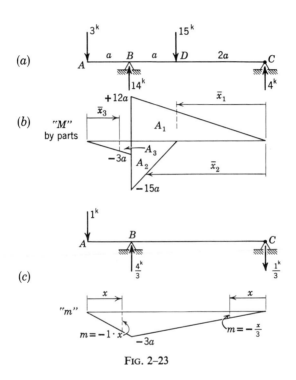

FIG. 2–23

To determine θ_A, a unit couple is applied at A, providing the triangular "m" diagram of Fig. 22. The expression for "m" is written as a continuous function for all values of x measured from the right-hand support. Consequently

$$1 \cdot \theta_A = \int \frac{Mm\,dx}{EI} = \frac{1}{EI} \int_0^L \left(\frac{wLx}{2} - \frac{wx^2}{2} \right) \left(\frac{x}{L} \right) dx$$

$$\theta_A = \frac{wL^3}{24EI}$$

Figure 23a represents a beam used as an example of other methods and it will be used again to illustrate the general aspects of arithmetic integration. The vertical deflection at A is to be determined. The "M" diagram is drawn by parts. The "m" diagram as developed by a 1-kip load at A is also shown in Fig. 23c. It is desired to solve by arithmetic integration, hence "M" will be kept in general terms while "m" is written in the easiest manner possible, changing origins for x as required to obtain simple and continuous functions for "m". Two origins are shown in Fig. 23c. It is to be noted that the algebraic expression for "M" is avoided by the following procedure:

$$1 \cdot \varDelta_A = \int \frac{Mm\ dx}{EI} = \int_C^B \left(-\frac{x}{3}\right)\left(\frac{M}{EI}\ dx\right) + \int_A^B (-x)\left(\frac{M}{EI}\ dx\right)$$

with the integrals and their limits interpreted arithmetically by referring to Fig. 23b as

$$EI\varDelta_A = -\tfrac{1}{3}\bar{x}_1 A_1 - \tfrac{1}{3}\bar{x}_2 A_2 - \bar{x}_3 A_3$$

$$= -\frac{1}{3}(2a)\left(+\frac{12a \cdot 3a}{2}\right) - \frac{1}{3}\left(2a + \frac{2a}{3}\right)\left(-\frac{15a \cdot a}{2}\right) - \left(\frac{2a}{3}\right)\left(-\frac{3a \cdot a}{2}\right)$$

which simplifies to

$$EI\varDelta_A = -12a^3 + \frac{20}{3}a^3 + a^3 = -\frac{13}{3}a^3$$

$$\varDelta_A = -\frac{13}{3EI}a^3$$

The negative sign indicates that point A deflects upward since the result represents negative work done by the 1-kip force when the real loading is applied.

The student should once more recognize that the final arithmetical work is identical to that of the moment-area method, and that the significant difference of the two approaches is in the steps taken in analysis.

2–8. CASTIGLIANO'S THEOREM

Another fundamental energy approach for determining deflections is the mathematical method based upon Castigliano's theorem. The development of the theorem follows.

In Fig. 24a, let P_1, P_2, and P_3 represent a general system of concentrated loads placed on a beam with $\varDelta_1$, $\varDelta_2$, and $\varDelta_3$ equal to the vertical deflections as shown. The beam is in a stable configuration and the external work done on the beam or the strain energy stored in the beam is

$$U = \frac{P_1 \cdot \varDelta_1}{2} + \frac{P_2 \cdot \varDelta_2}{2} + \frac{P_3 \cdot \varDelta_3}{2} \qquad (2\text{-}6)$$

For the next step of the development consider that the load P_1 is increased by a differential amount dP_1 while loads P_2 and P_3 remain constant. Owing to the increase in P_1 the deflection under each of the loads increases differentially, as shown in Fig. 24b.

Since Δ_1 was due to all loads on beam, the partial derivative of Δ_1 with respect to P_1 mathematically states the rate of change of Δ_1 with respect to P_1. Hence we may write

$$\frac{\partial \Delta_1}{\partial P_1} \cdot dP_1 = d\Delta_1 \tag{2-7}$$

and similarly

$$\frac{\partial \Delta_2}{\partial P_1} \cdot dP_1 = d\Delta_2; \quad \frac{\partial \Delta_3}{\partial P_1} \cdot dP_1 = d\Delta_3$$

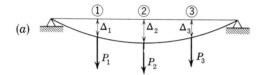

(a)

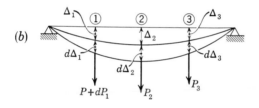

(b)

FIG. 2–24

The differential change or variation in external work due to increasing P_1 by dP_1 is

$$dU = \frac{\partial U}{\partial P_1} \cdot dP_1 = \frac{d\Delta_1 \cdot dP_1}{2} + P_1 \cdot d\Delta_1 + P_2 \cdot d\Delta_2 + P_3 \cdot d\Delta_3$$

If the first term on the right-hand side of the above equation is dropped, since it is a term of the second order, and if the partial derivative relations of equation 2-7 are substituted for the differential deflections, then

$$\frac{\partial U}{\partial P_1} \cdot dP_1 = P_1 \cdot \frac{\partial \Delta_1}{\partial P_1} \cdot dP_1 \cdot + P_2 \frac{\partial \Delta_2}{\partial P_1} \cdot dP_1 + P_3 \frac{\partial \Delta_3}{\partial P_1} \cdot dP_1$$

dividing both sides by dP_1

$$\frac{\partial U}{\partial P_1} = P_1 \frac{\partial \Delta_1}{\partial P_1} + P_2 \frac{\partial \Delta_2}{\partial P_1} + P_3 \frac{\partial \Delta_3}{\partial P_1} \tag{2-8}$$

but as previously stated by equation 2-6,

$$U = \frac{P_1 \cdot \Delta_1}{2} + \frac{P_2 \cdot \Delta_2}{2} + \frac{P_3 \cdot \Delta_3}{2}$$

and when differentiated with respect to P_1

$$\frac{2\partial U}{\partial P_1} = \Delta_1 + P_1 \frac{\partial \Delta_1}{\partial P_1} + P_2 \frac{\partial \Delta_2}{\partial P_1} + P_3 \frac{\partial \Delta_3}{\partial P_1} \tag{2-9}$$

Subtracting from equation 2-9 the previously determined relation of equation 2-8, the following fundamental relation is obtained:

$$\frac{\partial U}{\partial P_1} = \Delta_1 \tag{2-10}$$

This relation states in mathematical terms that if the partial derivative of U with respect to a given force is taken, the value of this derivative is

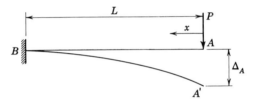

FIG. 2–25

equal to the deflection in the direction of the force at the point of application of the force.   This is Castigliano's theorem and it is generally termed his first theorem.   The theorem may also be extended to the determination of rotation or slope in the direction of a couple if the expression for U incorporates the couple.   In applying the theorem the student is reminded that partial derivatives are involved.

An example will clarify the approach of this method.   Let Fig. 25a represent a cantilever beam loaded with a vertical force P at A.   The vertical deflection at A may be found by the above theorem since the deflection is in the direction of P the applied force at A.   First establish the expression for U as

$$U = \int_0^L \frac{M^2\, dx}{2EI} = \frac{1}{2EI} \int_0^L (Px)^2\, dx$$

or

$$U = \frac{P^2 L^3}{6EI}$$

then by equation 2-10 we obtain

$$\Delta_A = \frac{\partial U}{\partial P} = \frac{PL^3}{3EI}$$

It is to be noted that the integration was performed before the differentiation, but these processes could be reversed if correct mathematical procedures are always followed.

An interesting relation results if the partial differentiation is performed in a general manner as follows, if M is a function of the external forces on the structure. The partial derivative with respect to a particular force P is indicated.

$$U = \int \frac{M^2 \, dx}{2EI}$$

$$\Delta = \frac{\partial U}{\partial P} = \int \frac{M \frac{\partial M}{\partial P} \, dx}{EI} \tag{2-11}$$

where $\frac{\partial M}{\partial P}$, stating the rate of change of M with respect to the force P, is analogous to m produced by the unit load in the unit load method. Hence, after differentiation, the integrations are identical by either this method or the approach identifiable as $1 \cdot \Delta = \int \frac{Mm \, dx}{EI}$, or the unit load method. For this reason, and for the reason that mathematical difficulties are often encountered in working with total derivatives, the $\int \frac{Mm \, dx}{EI}$ method is often preferred. *The student, in recognizing the similarity of the final integration steps, should also bear in mind the difference in approach.* The Castigliano approach is entirely mathematical, with the equivalent of "m" showing up automatically through the mathematical processes, whereas the unit load approach necessitates the formulation of "m" from an appropriate unit loading. Frequently, the two methods are used to establish independently the basic integrals as a check. This point of view will be developed later in the book when dealing with the calculation of redundants.

Finally, it must be realized that the method has its limitations from the standpoint that deflection and slopes can not be computed at all points along an elastic curve, with ease, as can be done by the moment-area method. Modifications such as use of auxiliary loads at points where deflections are required have been established, but such modifications accomplish no more than the unit load method.

To use this theorem, deflections must be directly proportional to load. In the event that this is not true, complementary energy must be used, which is often designated as U^*. U as defined by equation 2-6 is the summation of total area under the separate load deflection curves plotted for each individual load and its relevant deflection, where load is plotted vertically and deflection plotted as the abscissa. The complimentary energy U^* is the summation of the areas above the load deflection curve. Complimentary energy will not be considered in this book, but it is recommended that it be briefly referred to in the available source material.

2-9. TRUSS DEFLECTIONS—GRAPHICAL METHOD

Truss deflections are due to the changes in length of the individual elastic members of the truss under load. Since a truss normally consists of an assemblage of two-force members forming a series of triangles, a simple graphical procedure has been devised known as the Williot diagram for determining truss deflections.

The elements of the procedure may be first illustrated by referring to Fig. 26. AB and AC are two-force members and are pin-connected at A, and pin-connected at B and C. The vertical load P will produce a tensile stress in AC, thus elongating this member beyond its unstressed length AC. Member AB will be in compression and hence will shorten. ΔL, the lengthening or shortening of a two-force member is SL/AE, where S is the total force in member in pounds, L its unstressed length in feet or inches, A the cross-sectional area in square inches, and E, the modulus of elasticity in pounds per square inch. The units given are the usual ones, but they may be varied to suit a given problem. The graphical solution always requires the prior calculation of the changes in length of the members which are small and frequently no larger than $\frac{1}{2000}L$.

Assuming that ΔL for AB and BC are known in magnitude, the deflected position A' of point A, Fig. 26a, can be determined by first disconnecting the members at point A in their original unstrained length and then laying off the ΔL of each individual member to the scale of the drawing, as shown in Fig. 26b. (ΔL is greatly exaggerated in drawing.) This establishes the deformed geometric length for AB and AC, and, with these lengths as radii and B and C as centers, two arcs can be drawn to reconnect the truss at joint A. The intersection of the two arcs is at point A', the deflected position of A, and $A'B$ and $A'C$ are the true positions in space of members AB and AC. Point A' is to the right and down, relative to A.

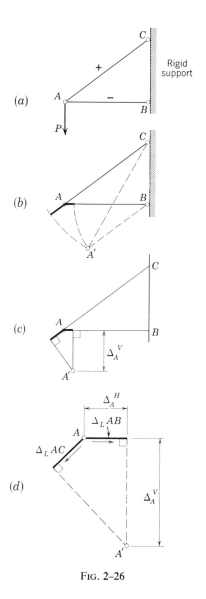

FIG. 2–26

The deflections of Fig. 26b are greatly exaggerated since ΔL was indicated out of proportion. Such exaggeration points out the impracticability of actually drawing the deformations to the same scale as the structure. Several factors permit the drawing of the deformations to a larger practic-able scale separately from the structure. In the first place, the deflections will be small, hence the actual movement of A to A' is small and *the tangent to the arcs* may be substituted for the arcs of Fig. 26*b*.   Figure 26*c* indicates these tangents and they are drawn perpendicular to the original orientation of the members.   Point A' will be in the same position as heretofore except for minor differences of the second order.   To make this latter procedure entirely feasible as a drawing board method, any appropriate scale for deformations may be used, as shown in Fig. 26*d*.   It is to be noted that the deformations are laid off parallel to the relevant members and that perpendiculars are drawn from the ends of the lines representing deformation.   The directions in which the deformations are laid out are identified by arrows.   The intersection of these perpendiculars represents the displaced position of A relative to its original position.   The relative motion of A may be measured in terms of its total movement from A to A' or by means of the horizontal and vertical components of this movement. Figure 26*d* is known as a Williot diagram.

Figure 27 is a modification of Fig. 26, in that point B is now free to move and a member BC and a load Q has been added.   Owing to P and Q, the three elastic members AB, AC, and BC change length.   The sign of stress which also represents the sign of ΔL is shown on the figure.   The deflection of A is known to be influenced by the deformation of BC as well as by deformation of AC and AB.   Although this situation is more complicated than before, it is subject to a similar geometric solution.   In general, it is essential to select a reference point and a reference bar.   In this instance, point C will be selected as the reference point since it does not move and the member BC selected as the reference member since it does not rotate.   Disconnect the undeformed structure at point A and then let point B move down the ΔL of member BC.   On the drawing board this operation is the equivalent of drawing BC in its new geometric length with the C end held in its original position.   This graphical construction is shown in Fig. 27*b*, and where the member AB is permitted to move down parallel to its original position.   The deformations of AC and AB may now be laid off and the reconnected position of A at A' found by drawing the perpendiculars as previously explained.

The geometric construction may now be simplified to a Williot diagram procedure by noting that the downward movement of point A of member AB in Fig. 27*b* is equal to the ΔL of member BC.   Making use of this fact, Fig. 27*c* may be drawn to a convenient deformation scale.   The step-by-step solution is as follows: select a reference member BC and a reference point; either end of the reference member will do, in general, but here point C is selected.   The ΔL of the reference bar is then laid off to a deformation scale as follows: Establish the reference point C conveniently and, since B moves down relative to C, draw ΔL for BC downward from C.   This ΔL

is shown as *CB* in Fig. 27c.   It is important to visualize that the members meeting at *A* have been disconnected.   The next step is to draw the *ΔL* of the member *AB* in its proper direction.   Relative to the located point *B*, the *A* end of the member *AB* moves to the right since *AB* shortens.   The

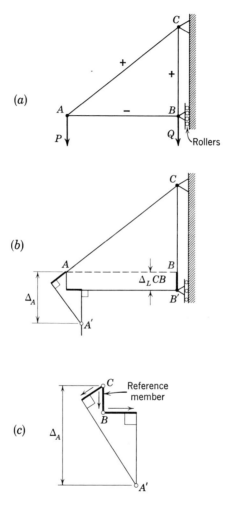

Fɪɢ. 2–27

direction of the relative movement is shown by the arrow.   The *A* end of member *AC* moves downward and to the left, parallel to *AC* since *AC* stretches.   Hence, the *ΔL* of *AC* is laid off parallel to *AC* from *C*.   Point *A'*, the reconnected location of point *A*, is found by drawing the perpendiculars.   The resulting diagram of Fig. 27c is the Williot diagram. It is necessary to emphasize that this diagram is a relative motion diagram from which the relative motion in space of any point may be measured from any other point.

A four-panel truss, symmetrically framed and symmetrically loaded, is shown in Fig. 28a.   The signs on the members represent the sign of the stress as well as the sign of deformation, positive for tension and member elongation, negative for compression and member shortening.   The deflections of all joints of the truss are to be determined.   As before, select a reference member and a reference point.   For the first solution to be described the vertical member U_2L_2 has been selected as the reference member since, owing to the complete symmetry of framing and loading, it

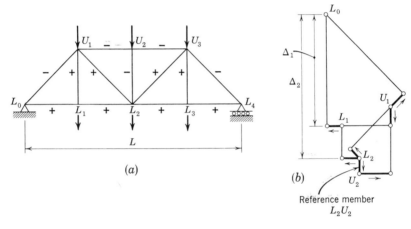

(a)

(b)

Reference member L_2U_2

FIG. 2-28

does not rotate as the truss deflects.   Either end of this member may be selected as the reference point and L_2 has been selected.   With a convenient scale for deformation the Williot diagram of Fig. 28b may now be constructed.   The procedure remains the same but it may be helpful if it is outlined once more.   The construction commences by laying off the ΔL of U_2L_2 to scale with U_2 placed below L_2 since U_2 moves downward relative to L_2.   Actually this initial step implies that the triangle $U_1L_2U_2$ was disconnected at U_1 while member U_2L_2 was permitted to change length. The deformations for members U_1U_2 and U_1L_2 are then laid off to scale in the appropriate relative directions from U_2 and L_2 respectively. Perpendiculars are erected to the direction of these latter deformations and their intersection locates the reconnected relative displaced position of joint U_1.

With U_1 now located and with L_2 as previously located, the next triangle to be dealt with is $L_1U_1L_2$. With this triangle disconnected at joint L_1 the $\varDelta L$ of the appropriate members are laid off and the above procedure repeated to establish the relative position of L_1. A repetition of the procedure through the last remaining triangle $L_0U_1L_1$ will establish L_0 on the Williot diagram. On account of symmetry, the truss to the right of the center vertical need not be considered. The results of the Williot diagram of Fig. 28b must be interpreted as relative movements, and the main purpose of the diagram may be to determine only the vertical deflections of the lower chord panel points. Attention must now be directed toward the geometric or deflection conditions at the supports, which may be explicitly defined as the boundary conditions. L_0 is held in position by a fixed shoe, while L_4 is supported on rollers or a rocker permitting only a horizontal movement at L_4. These conditions are the boundary conditions, and since vertical deflections are relative to a support fixed in vertical position, these deflections can be scaled directly from Fig. 28b, using L_0 as the reference point. Thus the downward deflection of L_1 and L_2 may be established. The vertical deflection of U_1 and U_2 may be similarly established. The horizontal movement of the same joints may be also measured horizontally from L_0. This relative motion aspect of the Williot diagram should not escape the student's understanding as it could be established from the diagram that L_1 has moved vertically upward with respect to L_2 and to the left relative to L_2. Other relative movements may be similarly measured.

The conditions of symmetry for the truss of Fig. 28a, although a common situation, do not represent the general case. Unsymmetrical framing, variable loading, and complex conditions at the supports require a more general procedure. Consider the truss of Fig. 29a, noting the supports at L_0 and L_3 with the structure cantilevered one panel beyond L_3. The primary difficulty in determining the deflections graphically is that no single member can now be selected as a reference member with a certainty that it does not rotate. This difficulty may be resolved by selecting any member as the reference member and eventually correcting for its rotational effects. To avoid complexity in the description to follow, the lower chord member L_0L_1 will be selected as the reference member, and L_0 will be selected as the reference point. Figure 29b may now be started if the $\varDelta L$ of all the members are known. The start is made by drawing the $\varDelta L$ of the reference member with the ends of this line properly identified. Then the detailed procedure of working through the structure, triangle by triangle, is again followed. It may be noted that the accuracy of solution depends on drafting techniques, and for that reason circles are generally drawn around the terminals of all scaled lengths and intersections to

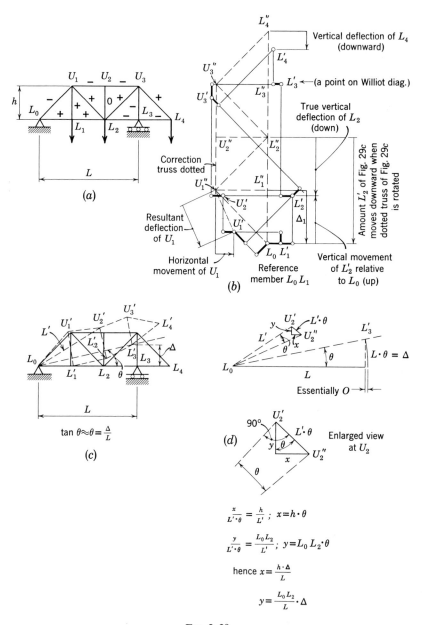

$$\tan \theta \approx \theta = \frac{\Delta}{L}$$

(c)

$$\frac{x}{L' \cdot \theta} = \frac{h}{L'} \; ; \; x = h \cdot \theta$$

$$\frac{y}{L' \cdot \theta} = \frac{L_0 L_2}{L'} \; ; \; y = L_0 L_2 \cdot \theta$$

hence $x = \dfrac{h \cdot \Delta}{L}$

$$y = \frac{L_0 L_2}{L} \cdot \Delta$$

FIG. 2–29

preserve these points.   The heavy lines represent scaled values of ΔL, and the light solid lines are the perpendiculars.

The diagram resulting from using L_0L_1 as the reference bar is again termed the Williot diagram.   However, the true vertical and horizontal deflections can not yet be measured since it was assumed that L_0L_1 did not rotate.   The construction so far will permit only the determination of the distorted outline of the truss drawn from L_0L_1 as the base.   Figure 29c shows the distorted outline based on the relative movements of Fig. 29b (a greatly exaggerated and distorted view; all movements are actually small when compared with structural dimensions).   The important feature to note is that L_3 has moved to L'_3, or that a line between L_0 and L_3 has rotated upward through a small angle θ.   This angle is calculable by observing that Δ, the vertical movement of L_3 to L'_3, may be obtained from Fig. 29b.   To obtain true corrected deflections, point L'_3 must be rotated downward to its true support level.   This amounts to rotating the entire distorted outline of Fig. 29c through the angle θ about L_0 with all joints translating through an arc.   Figure 29d indicates the geometry of this translation and the horizontal and vertical components of the translation for the upper chord joint U_2.   Since the angle of rotation is small, the x component of the movement is proportional to the height of U_2 above L_0 and the vertical movement downward or the y component is directly proportional to the horizontal distance from L_0 to U_2.   A similar set of geometric relationships will be found for all other joints.   These components must be considered in the nature of a correction to the relative deflections.

The corrections are automatically made and combined with the relative deflections in the proper algebraic sense if a correction truss, shown dotted in Fig. 29b, is drawn.   The correction truss is similar geometrically to the actual truss and is drawn so that all of its members make an angle of 90° with the actual truss.   This is true since any line from L_0 of the actual truss to any other point on the truss is perpendicular to the movement of that point when rotated, and the movement of the point is directly proportional to its radial distance from L_0.   Furthermore, the overall dimensions of the correction truss constructed to the deformation scale are controlled by the supporting conditions of the actual truss.   For example, L_0, has no vertical or horizontal movement, hence L''_0 of the correction truss is placed in coincidence with L_0.   Joint L_3 has no vertical deflection, hence L''_3 of the correction truss will be on a horizontal line through L_3.   With these conditions the dotted correction truss may be drawn.   The concept of the correction truss was credited to Mohr, and the combined diagram of Fig. 29b is usually termed the Williot-Mohr diagram.   The combination of movements for L_2 is illustrated.   In general, a measurement *from* a point on the correction truss *to* the similarly identified point on the original Williot diagram provides the magnitude and direction of the resultant displacement.   This is indicated on Fig. 29*b* for U_1.

2–10. TRUSS DEFLECTIONS—UNIT LOAD METHOD

Although the graphical method for determining truss deflections is extremely valuable as a procedure for obtaining all deflections of a truss, an analytical method is preferred in many solutions pertaining to indeterminate structures.   Virtual work and strain energy principles, or the

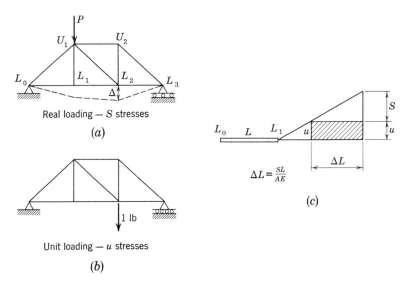

Real loading — *S* stresses

(*a*)

$$\Delta L = \frac{SL}{AE}$$

(*c*)

Unit loading — *u* stresses

(*b*)

Fɪɢ. 2–30

unit load method, are generally applied.   Basic principles remain the same as presented for the unit load method applied to beams; however, the truss stores its internal strain energy by virtue of the axial deformation of the members.

Assume that the truss of Fig. 30 is loaded with a load P at U_1 and that the vertical deflection of L_2 is desired.   Let it be assumed that the truss members have been designed and that the stresses produced by P, termed S, are known for all of the members.   The ΔL or change in length of each member due to the load system can now be determined.   These ΔL values, once computed, are retained for use in the further steps of the solution.

The theoretical development starts with the placement of a 1-lb vertical unit load at L_2, the panel point at which the deflection is required on the otherwise unloaded truss of Fig. 30b.   The 1-lb load causes a determinable axial stress in each truss member.   The stresses produced by this unit load are generally called the u stresses, so called since u is the first letter of the word unity.   Although the u stresses cause each stressed member to change length, these deformations are not needed.   As the second step of the theoretical development add the real loading P to the truss of Fig. 30b, noting that the 1-lb load which is already on the truss at L_2 will move down, thus doing external virtual work on the truss equal to $1 \cdot \Delta$.   This external virtual work must be equated to the internal virtual strain energy. Figure 30c shows one member L_0L_1 with its force deformation-diagram. A study of this diagram will indicate that the rectangular area equal to $u \cdot \Delta L$ represents the relevant strain energy since u is produced by the unit loading and ΔL by the real loading.   It is to be particularly noted once again that this area is revealed and identifiable for the sole reason that the unit load was placed on structure before the real loading was introduced.   Since all members stressed by P will deform, then a finite summation for all members of the product $u \cdot \Delta L$ may be written as $\sum u \cdot \Delta L$. It thus follows that

$$1 \cdot \Delta = \sum u \cdot \Delta L$$

which is generally restated as

$$1 \cdot \Delta = \sum \frac{SuL}{AE}, \quad \text{where } \frac{SL}{AE} = \Delta L \qquad (2\text{-}12)$$

If the horizontal deflection is required at a particular joint, the unit load is placed horizontally at that joint to determine the u stresses.   The student should once more note that the multiplication by 1 lb on the left-hand side of equation 2-12 preserves the units of work.

Deflections may be computed, no matter what the source of ΔL. Normally its source is the external load on truss, but the change in member length may be due to temperature or errors in fabrication.   For truss of Fig. 31 compute the vertical deflection of L_2 if the diagonal U_1L_2 has been fabricated $\frac{1}{2}$ in. short of its true geometric length.   Place a 1-lb load at L_2 as shown and compute the u stress in member U_1L_2 only since no other member is assumed to change length.   The value of u is $+\frac{1}{3} \times \frac{25}{15} = +\frac{5}{9}$ lb.   Then

$$1 \cdot \Delta = u \cdot \Delta L$$

$$1 \cdot \Delta = (+\tfrac{5}{9})(-\tfrac{1}{2})$$

$$\Delta = -\tfrac{5}{18} \text{ in. (up)}$$

The value of Δ would also be the same if this same member U_1L_2 should contract $\frac{1}{2}$ in., owing to temperature effect.

As a further example of this method, consider the truss of Fig. 31 with a 90-kip load at L_1 as in Fig. 31b, where the vertical deflection at L_2 is to be calculated. In a problem involving many members, a tabular solution is an expeditious way of recording all quantities. The table below is the tabulation for this problem; the values of S are for the 90-kip loading and the values of u for a unit load of one kip at L_2. The areas of the steel members are the actual cross-sectional areas. Notice the importance of sign of stress.

Truss Deflection
(at L_2)

Member	L, ft	A, sq in.	L/A	S, kips	u, kips	SuL/A
L_0L_1	20	4	5	$+80$	$+\frac{4}{9}$	$+\frac{1600}{9}$
L_1L_2	20	4	5	$+80$	$+\frac{4}{9}$	$+\frac{1600}{9}$
L_2L_3	20	4	5	$+40$	$+\frac{8}{9}$	$+\frac{1600}{9}$
L_0U_1	25	10	2.5	-100	$-\frac{5}{9}$	$+\frac{1250}{9}$
U_1U_2	20	8	2.5	-40	$-\frac{8}{9}$	$+\frac{800}{9}$
U_2L_3	25	10	2.5	-50	$-\frac{10}{9}$	$+\frac{1250}{9}$
U_1L_1	15	5	3.0	$+90$	0	0
U_1L_2	25	10	2.5	-50	$+\frac{5}{9}$	$-\frac{625}{9}$
U_2L_2	15	3	5.0	$+30$	$+\frac{2}{3}$	$+\frac{900}{9}$

$$\sum \frac{SuL}{A} = +\frac{8375}{9}$$

with $E = 30{,}000$ ksi $\Delta = \dfrac{8375}{9 \times 30{,}000} = 0.031$ ft

A contrast of this method with the Williot-Mohr procedure will clearly show that, whereas the Williot-Mohr diagram enables the deflections of all joints to be determined, the $\sum \dfrac{SuL}{AE}$ method requires a separate calculation for each deflection needed.

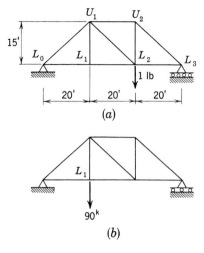

FIG. 2–31

2–11. DEFLECTIONS BY FOURIER SERIES

In many deflection problems the discontinuity of the bending moment curve causes difficulties in integration and determination of constants. In some situations only approximate solutions are desired. The Fourier series or trigonometric series permits a continuous type of function to substitute approximately for a discontinuous function or, in many instances, to approximate a continuous function. Deflection problems are of the latter type since all general deflection curves are continuous.

There are many forms of trigonometric series, the simplest being either the sine or cosine series. The beam of Fig. 32a has a load P at the center line, producing a symmetrical elastic curve. To fit this curve by a sine series, let

$$y = a_1 \sin \frac{\pi x}{L} + a_2 \sin \frac{2\pi x}{L}$$

$$+ a_3 \sin \frac{3\pi x}{L} + \cdots + a_n \sin \frac{n\pi x}{L} \qquad (2\text{-}13)$$

which would involve the determination of the coefficients $a_1, a_2, a_3, \cdots a_n$.

The above expression, taken term by term, is shown in Fig. 32b. The half-sine waves always have their maximum ordinate equal to their coefficient, with the sign of the ordinate determined by the sign of the trigonometric function.

In general,

$$y = \sum_{n=1}^{n=\infty} a_n \sin \frac{n\pi x}{L} \qquad (2\text{-}14)$$

where n, an integer, equals the number of half-sine waves in the length L. In the example problem, owing to symmetry, only the odd values of n are valid, whereas for unsymmetrical situations both odd and even values of n are involved.   For the example problem, with odd values of n

$$y = \sum_{n=1,3,5}^{n=\infty} a_n \sin \frac{n\pi x}{L}$$

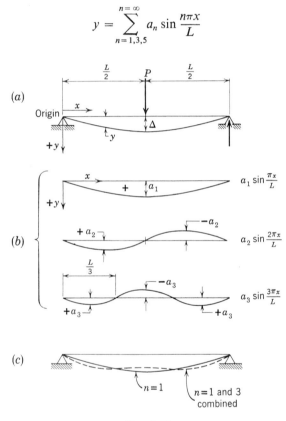

FIG. 2–32

The more terms of the series used the closer the approximation is to the true elastic curve.   Figure 32c indicates combination of curves for $n = 1$ and $n = 3$.

The use of the series expressions 2-13 and 2-14 requires the determination of the value of the coefficients to suit the problem.   Resort is usually

made to energy methods for determining the coefficients. From previous
articles the total strain energy in flexure is stated as

$$U = \int \frac{M^2\,dx}{2EI} \tag{2-15}$$

letting $\qquad\qquad EI\frac{d^2y}{dx^2} = M$

the strain energy is restated as

$$U = \frac{EI}{2}\int_0^L \left(\frac{d^2y}{dx^2}\right)^2 dx \tag{2-16}$$

To determine $\left(\dfrac{d^2y}{dx^2}\right)^2$ assume

$$y = \sum_{n=1}^{n=\infty} a_n \sin\frac{n\pi x}{L}$$

The successive differentiations follow as

$$\frac{dy}{dx} = \sum_{n=1}^{n=\infty} \frac{\pi n a_n}{L}\cos\frac{n\pi x}{L}$$

$$\frac{d^2y}{dx^2} = -\sum_{n=1}^{n=\infty} \frac{\pi^2 n^2 a_n}{L^2}\sin\frac{n\pi x}{L}$$

then

$$\left(\frac{d^2y}{dx^2}\right)^2 = \frac{\pi^4}{L^4}\left[\sum_{n=1}^{n=\infty} n^2 a_n \sin\frac{n\pi x}{L}\right]^2 \tag{2-17}$$

If the term in brackets of equation 2-17 is squared, two separate series
with different types of general terms may be identified. The first series
represents the sum of the products of like terms with a typical and general
term

$$n^4 a^2_n \sin^2\frac{n\pi x}{L} \tag{2-18}$$

The second series represents the sum of the products of unlike terms,
identified here by using subscripts n and m (representing integers) to
differentiate general unlike coefficients where n and m also represent the
unlike integer number of half-sine waves. For example, n could be 1 and
$m = 2$, or $n = 2$ and $m = 1$, hence there would be twice as many terms of

unlike values of n and m as there would be terms of like values of n and m. Each term of the second series has the general form

$$2n^2m^2a_na_m \sin \frac{n\pi x}{L} \sin \frac{m\pi x}{L} \qquad (2\text{-}19)$$

Although the calculus is too long for detailed development, the student may easily verify that integration of terms of the square of the trigonometric function of equation 2-18 from O to L will yield

$$\int_0^L \sin^2 \frac{n\pi x}{L} \, dx = \frac{L}{2} \qquad (2\text{-}20)$$

and the integration of the trigonometric terms in the form of equation 2-19 between O and L will always yield a value of zero. This leads to great simplification since substitution of these results in equations 2-16 and 2-17 yields

$$U = \frac{\pi^4}{4L^3} \cdot EI \sum_{n=1}^{n=\infty} n^4 a^2_{\,n} \qquad (2\text{-}21)$$

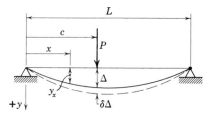

FIG. 2–33

We are now in a position to establish a procedure for the determination of the coefficients to fit a given deflection problem. For this purpose, the general problem of Fig. 33 will be used where the load P is at a distance c from the left reaction. Assume that the deflection curve is a trigonometric series, with any particular term equal to $a_n \sin \frac{n\pi x}{L}$, and by partial differentiation of equation 2-21 determine the rate of change in strain energy with respect to the general coefficient a_n

$$\frac{\partial U}{\partial a_n} = \frac{\pi^4 EI}{4L^3} (2n^4 a_n)$$

The change or variation in internal strain energy due to a small change in a_n of δa_n is

$$\delta U = \frac{\partial U}{\partial a_n} \cdot \delta a_n = \frac{\pi^4 EI}{2L^3} n^4 a_n \cdot \delta a_n$$

The small change in a_n will displace the load by $\delta\varDelta$ causing load P to do additional external work of $P \cdot \delta\varDelta$, where $\delta\varDelta$ equals $\delta a_n \sin \dfrac{n\pi c}{L}$. With change in external work equal to change in internal strain energy,

$$P \cdot \delta a_n \sin \frac{n\pi c}{L} = \frac{\pi^4 EI}{2L^3} n^4 a_n \, \delta a_n$$

$$a_n = \frac{2PL^3}{\pi^4 n^4 EI} \sin \frac{n\pi c}{L} \tag{2-22}$$

Equation 2-22 is the general equation for determining the general coefficient, hence the deflection curve in a form of a Fourier series may be written by combining equations 2-14 and 2-22 as

$$y = \frac{2PL^3}{\pi^4 EI} \sum_{n=1}^{n=\infty} \frac{1}{n^4} \sin \frac{n\pi c}{L} \sin \frac{n\pi x}{L} \tag{2-23}$$

To illustrate the use of equation 2-23, assume that c equals $L/2$ and $x = L/2$, then the deflection at the center line as calculated from the first three terms of the series would be

$$y = \frac{2PL^3}{\pi^4 EI} \left[\frac{1}{1^4} \sin^2 \frac{\pi}{2} + \frac{1}{2^4} \sin^2 \pi + \frac{1}{3^4} \sin^2 \frac{3\pi}{2} \right]$$

$$y = \frac{2PL^3}{\pi^4 EI} \left[1 + 0 + \frac{1}{81} \right]$$

Sufficient accuracy may be obtained using one term of the series for this case. With the first term only

$$y = \frac{2PL^3}{\pi^4 EI} = \frac{PL^3}{48.7EI}$$

which may be compared with the exact value of $PL^3/48EI$.

2–12. DEFLECTIONS BY FINITE DIFFERENCE APPROXIMATIONS

Frequently deflection problems may arise wherein integration of the differential equation by ordinary methods may be either difficult or impossible. Recourse may then be made to the method of finite differences where the analytical differential equations are replaced by appropriate finite difference equations.

The governing differential equation $\dfrac{d^2y}{dx^2} = \dfrac{M}{EI}$ is of the second order where $\dfrac{d^2y}{dx^2} = \dfrac{d}{dx}\left(\dfrac{dy}{dx}\right)$. It will be shown that a finite difference approximation exists for $\dfrac{dy}{dx}$ and $\dfrac{d^2y}{dx^2}$.

Referring to Fig 34, let the curve shown describe y as an $f(x)$. Points along the curve are identified in relation to a central point o and are spaced at equal horizontal intervals of Δx where $\Delta x = h$.

Let $\dot{y} = \text{slope} = \left(\dfrac{dy}{dx}\right) = \lim_{\Delta x \to 0} \dfrac{\Delta y}{\Delta x}$

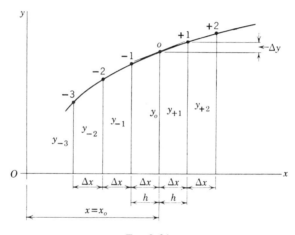

FIG. 2–34

For point o the finite change in y between points 1 and o is

$$\Delta y = y_1 - y_o$$

and $$\dot{y}_o \approx \frac{\Delta y}{\Delta x} \approx \frac{y_1 - y_o}{h}$$

As an alternate and improved approximation we may assume that the tangent to the curve at o is parallel to the secant joining points $+1$ and -1. Then the slope at o is approximately

$$\dot{y}_o \approx \frac{y_1 - y_{-1}}{2\Delta x} \approx \frac{y_1 - y_{-1}}{2h} \tag{2-24}$$

This will be termed the first finite difference approximation at o. The smaller the interval h, the better the approximation is.

Letting
$$\ddot{y} = \frac{d^2y}{dx^2} = \frac{d}{dx}\left(\frac{dy}{dx}\right)$$

$$= \lim_{\varDelta x \to 0} \frac{\varDelta^2 y}{\varDelta x^2}$$

In finite terms $\varDelta^2 y$ symbolizes second differences of $\varDelta y$ whereas $\varDelta y$ by itself symbolizes a first difference.   For point o

$$\varDelta^2 y = (y_1 - y_o) - (y_o - y_{-1})$$
$$\varDelta^2 y = y_1 - 2y_o + y_{-1}$$

$$\ddot{y}_o \approx \frac{\varDelta^2 y}{\varDelta x^2} \approx \frac{y_{+1} - 2y_o + y_{-1}}{h^2} \qquad (2\text{-}25)$$

$\ddot{y}_o$ will be termed the second finite difference approximation.

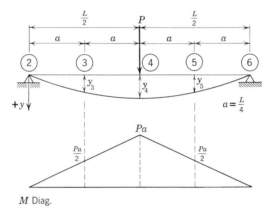

M Diag.

FIG. 2–35

A sufficient background of the finite difference approximations has now been determined to enable a deflection problem to be solved.   An example will clarify.   In Fig. 35 let it be desired to find the deflection at the one-quarter points and center line of the span.   Also assume that the positive direction of y is downward.   The solution is started by marking off points on curve at equal finite intervals a.   The bending moment curve is drawn with moments computed at these finite intervals.

The governing differential equation of the elastic curve is

$$\frac{d^2y}{dx^2} = -\frac{M}{EI}$$

since y is positive downward.   This equation is now rewritten in terms of finite differences in the general form of equation 2-25 as

$$\frac{y_{+1} - 2y_o + y_{-1}}{a^2} = -\frac{M}{EI}$$

or

$$y_{+1} - 2y_o + y_{-1} = -\frac{a^2 M}{EI}$$

or reorienting terms

$$y_{-1} - 2y_o + y_{+1} = -\frac{a^2 M}{EI} \qquad (a)$$

Equations may now be written interrelating the deflections of points along the elastic curve of Fig. 35 with bending moment.
   At point 3

$$y_{-1} = y_2 = 0$$
$$y_o = y_3$$
$$y_{+1} = y_4$$

hence from general equation (a)

$$0 - 2y_3 + y_4 = -\frac{Pa^3}{2EI} \qquad (b)$$

At point 4

$$y_3 - 2y_4 + y_5 = -\frac{Pa^3}{EI}$$

but since $y_5 = y_3$

$$2y_3 - 2y_4 = -\frac{Pa^3}{EI} \qquad (c)$$

Equations (b) and (c) may now be solved simultaneously giving

$$y_3 = +\frac{Pa^3}{EI} = \frac{PL^3}{64EI}$$

and

$$y_4 = +\frac{3Pa^3}{2EI} = \frac{3PL^3}{128EI} = \frac{PL^3}{42.7EI}$$

The result for y_4 may be compared with the exact deflection of $PL^3/48EI$. The large error is due to the use of only four finite intervals.   A larger number of intervals would reduce the error but, at the same time, would add considerable labor in solving additional equations.

A modification of the procedure to eliminate simultaneous equations is known as the relaxation method developed by Southwell.*   The procedure is quite general and is easily extended to the solution of many physical problems.

The procedure and intent of the relaxation method may best be explained by a deflection example.   First rewrite

$$\frac{d^2y}{dx^2} = -\frac{M}{EI}$$

as
$$\frac{d^2y}{dx^2} + \frac{M}{EI} = 0$$

If this fundamental differential equation is exactly satisfied there would be a zero residual.   If not satisfied, the value of the left-hand side is termed the residual or R.

To incorporate finite differences, rewrite the general equation 2-25 as

$$y_{-1} - 2y_o + y_{+1} + a^2M/EI = 0$$

If all the y's are equal to zero, the residual R would be a^2M/EI. For a deflection problem, we could start with an assumption that all y's are zero and thus by relaxation adjust this flat deflection curve upward or downward to reduce R to zero at each selected point on the elastic curve.

Figure 36a is the same problem as Fig. 35.   The objective is to determine deflection at points 3, 4, and 5 by the relaxation approach.   Figure 36b is the moment diagram and Fig. 36c is a table for recording and dealing with residuals and deflections.   The columns at each point are headed R for residual and y for deflection.   The first line of Fig. 36c records the initial values for R equal to a^2M/EI based on the assumption of zero deflection at all points.   Values are stated as coefficients of Pa^3/EI.   The second line is achieved by assuming that point 4 moves down one unit while all other points are restrained with zero deflections.   This establishes a basis for calculating a corrective residual at points 3, 4, and 5.

At point 4 the corrective residual is

$$R = y_3 - 2y_4 + y_5$$
$$R = 0 - 2 \times 1 + 0 = -2$$

At points 3 and 5 the corrective residual is

$$R = y_2 - 2y_3 + y_4$$
$$R = 0 - 2 \times 0 + 1 = +1$$

_____

* R. V. Southwell, *Relaxation Methods in Theoretical Physics*, Oxford University Press, 1946.

These corrective residuals are listed in line 2 and combined with the initial residual to provide the results of the first cycle in line 3. Note that the boundary points 2 and 6 receive no attention as both the residual and the deflection are zero for these points. It may be noted that at the end of the first cycle the sign of the residual has changed at point 4 and that the residuals at points 3 and 5 have been increased. The relaxing process

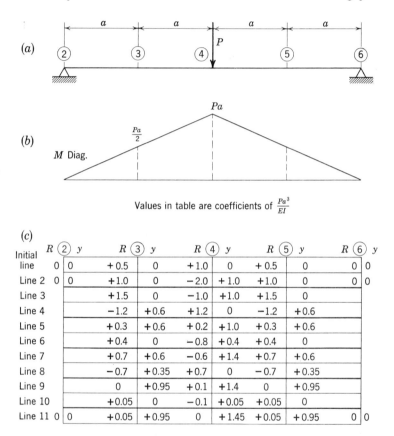

Values in table are coefficients of $\frac{Pa^3}{EI}$

(c)

	R ② y		R ③ y		R ④ y		R ⑤ y		R ⑥ y	
Initial line	0	0	+0.5	0	+1.0	0	+0.5	0	0	0
Line 2	0	0	+1.0	0	−2.0	+1.0	+1.0	0	0	0
Line 3			+1.5	0	−1.0	+1.0	+1.5	0		
Line 4			−1.2	+0.6	+1.2	0	−1.2	+0.6		
Line 5			+0.3	+0.6	+0.2	+1.0	+0.3	+0.6		
Line 6			+0.4	0	−0.8	+0.4	+0.4	0		
Line 7			+0.7	+0.6	−0.6	+1.4	+0.7	+0.6		
Line 8			−0.7	+0.35	+0.7	0	−0.7	+0.35		
Line 9			0	+0.95	+0.1	+1.4	0	+0.95		
Line 10			+0.05	0	−0.1	+0.05	+0.05	0		
Line 11	0	0	+0.05	+0.95	0	+1.45	+0.05	+0.95	0	0

FIG. 2–36

calls for judgment in selecting the appropriate changes in values of y for the successive cycles in order to have the residuals approach zero. Taking advantage of symmetry the residuals will be reduced by taking a +0.60 for y at points 3 and 5. The changes in the values of the residuals are recorded in line 4. They are computed by the general formula $y_{-1} - 2y_o + y_{+1}$. Line 5 summarizes the combinations of line 3 and line 4. In line 6 an additional deflection of +0.4 is assumed at point 4 and changes or corrections in all residuals are computed. The remainder of the table is repetitive, ending in this example with line 11. A small residual still remains at points 3 and 5, with the final deflection at point 4 equal to

$$\frac{1.45Pa^3}{EI} = \frac{1}{44.1}\frac{PL^3}{EI}$$

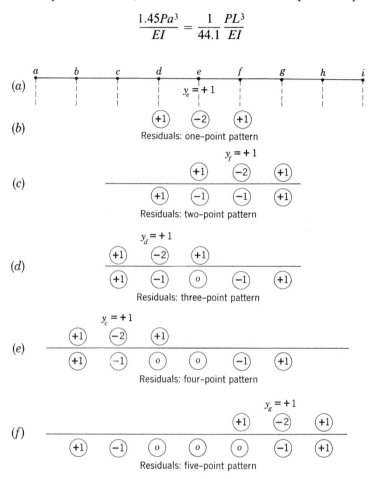

FIG. 2–37

The accuracy of the answer depends on the energy of the computor and the number of intervals.

To circumvent much of the work of point by point relaxation, a line relaxation process is employed in which the deflections are changed by a unity amount or a multiple thereof at one or more points and changes in the residuals computed at a group of points in one operation. This can be explained by referring to Fig. 37. In Fig. 37a a line of points is shown.

Point e is displaced downward one unit ($y_e = +1$), but all adjacent points are held without deflection. The general equation $y_{-1} - 2y_o + y_{+1}$ is used to obtain the residuals given in Fig. 37b as residuals for a one-point pattern of displacement. Point f is then displaced $y_f = +1$, providing a change in residuals following the one-point pattern. These residuals combined with those of the previous one-point pattern yield the combined residuals known as the two-point pattern for a simultaneous displacement of points e and f by unity. The procedure is duplicated by successively displacing d, c, and g, and obtaining in order the residuals for three-point,

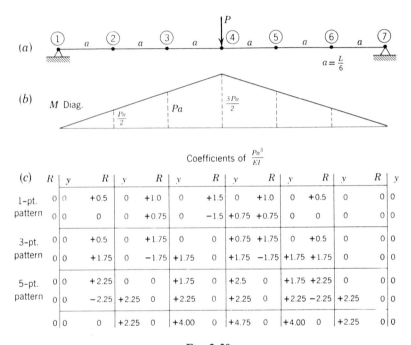

Coefficients of $\frac{Pa^3}{EI}$

(c)	R	y	R	y	R	y	R	y	R	y	R	y	R	y
1-pt pattern	0	0	+0.5	0	+1.0	0	+1.5	0	+1.0	0	+0.5	0	0	0
	0	0	0	0	+0.75	0	-1.5	+0.75	+0.75	0	0	0	0	0
3-pt pattern	0	0	+0.5	0	+1.75	0	0	+0.75	+1.75	0	+0.5	0	0	0
	0	0	+1.75	0	-1.75	+1.75	0	+1.75	-1.75	+1.75	+1.75	0	0	0
5-pt pattern	0	0	+2.25	0	0	+1.75	0	+2.5	0	+1.75	+2.25	0	0	0
	0	0	-2.25	+2.25	0	+2.25	0	+2.25	0	+2.25	-2.25	+2.25	0	0
	0	0	0	+2.25	0	+4.00	0	+4.75	0	+4.00	0	+2.25	0	0

FIG. 2–38

four-point, and five-point patterns. With this simple chart, relaxations may be made by expeditious grouping of points, taking advantage of the fact that displacements may be made without changing every residual.

Figure 38a is the same problem as solved heretofore, with the finite interval equal to $L/6$. The tabulations in Fig. 38c are made in terms of coefficients of Pa^3/EI for convenience. The first line of the tabulation records the initial residual. A progressive system of eliminating all residuals follows with the first step being the elimination of the residual at point 4, by taking the first estimate of y_4 as $+0.75$ in line 2. Line 3 is a summary of residuals after this first step. To eliminate the residuals at points 3 and 5, assume increments of y_3, y_4, and y_5 to be each equal to $+1.75$. Line 5 is a summary after this three-point relaxation. The method should now be apparent, as it can be noted that the successive steps of the solution are forcing the residuals to vanish towards the end supports where the deflection is zero. It remains to perform the five-point relaxation of line 6 to eliminate all residuals. The final deflections at points 2, 3, and 4 by this method are successively 2.25, 4.00, and 4.75 times Pa^3/EI. The central deflection may be restated as $PL^3/45.5EI$ for comparison with the true theoretical answer of $PL^3/48EI$.

Problems involving uniform load would involve same procedure. Beams with overhanging ends require a modification in procedure which is not taken up in this text.

CONCLUDING REMARKS

This chapter has provided a review of several deflection methods and has also presented several additional arithmetical and mathematical methods. The principles on which they are based will be used in the remainder of this book. The student should review the principles as he finds necessary throughout his study of the following chapters.

REFERENCES

1. Anderson, P., *Statically Indeterminate Structures*, Chapters 2 and 3, New York: Ronald Press, 1953.
2. Hoff, N. J., *The Analysis of Structures*, Part 1, New York: John Wiley, 1956.
3. Newmark, N. M., "Numerical Procedures for Computing Deflections, Moments and Buckling Loads," *Trans. Am. Soc. Civil Engineers*, **108**, 1161 (1943).
4. Parcel, J. I., and R. B. B. Moorman, *Analysis of Statically Indeterminate Structures*, Chapter 2, New York: John Wiley, 1955.
5. Popov, E. P., *Mechanics of Materials*, Chapter 11, New York: Prentice-Hall, 1952.
6. Seely, F. B., and J. O. Smith, *Advanced Mechanics of Materials*, second edition, Chapters 2, 13, 14, 15, 16, New York: John Wiley, 1952.
7. Southwell, R. V., *Theory of Elasticity*, Chapters 1 and 2, Oxford: Clarendon Press, 1936.
8. Timoshenko, S., *Strength of Materials*, Part II, pp. 46–53, New York: Van Nostrand, 1956.
9. Wang, C. K., *Applied Elasticity*, Chapters 6 and 7, New York: McGraw-Hill, 1953.

PROBLEMS

2-1. A steel plate $\frac{1}{4}$ in. thick by 6 in. wide is to be bent to form an arc of a circular curve. If the $\frac{1}{4}$ in. dimension is in the radial direction, what is the maximum bending stress in the plate if bent to a 10-ft radius?

(For all of the following problems treat EI as uniform throughout the beam.)

2-2. A cantilever beam is loaded as shown. Using an origin at the right end, derive the equation of the elastic curve.

$$Ans. \ y = -\frac{w}{24EI}(x^4 - 4xL + 3L^4) - \frac{T}{2EI}(x - L)^2.$$

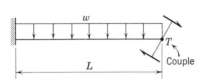

Prob. 2-2   Prob. 2-3

2-3. A cantilever beam is loaded as shown. Using an origin at the right end, derive the equation of the elastic curve.

$$Ans. \ y = -\frac{k}{120EI}(x^5 - 5xL^4 + 4L^5).$$

2-4. Derive the equation of the elastic curve between B and C.

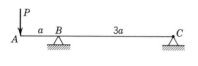
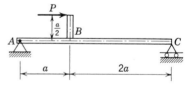

Prob. 2-4   Prob. 2-5

2-5. Derive the equations of the elastic curve and locate the position of the maximum deflection.

2-6. A beam is loaded with a sinusoidal loading curve as shown. Derive the equations for slope and deflection starting with the loading curve.

$$\left(EI\frac{d^4y}{dx^4} = -w\right)$$

Compare the maximum deflection with that of the beam loaded uniformly with a load of w_o lb per unit of length.

$$\textit{Ans. } y = - \frac{w_o L^4}{\pi^4 EI} \sin \frac{\pi x}{L}.$$

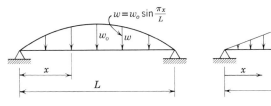

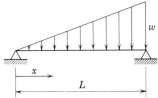

Prob. 2–6   Prob. 2–7

2–7. A beam is loaded as shown.   Derive the equation of the elastic curve. Locate the position of maximum deflection.   *Ans. x = 0.52L.*

2–8. Determine the deflection at A by either the double integration method or the moment-area method.   *Ans. $EI\Delta_A = 5616$ kip-ft^3.*

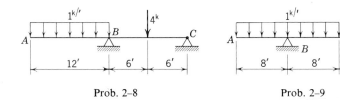

Prob. 2–8   Prob. 2–9

2–9. Determine the deflection of A and C by the moment-area method.

2–10. Do problem 2-2 by the moment-area method.

2–11. Do problem 2-3 by the moment-area method.

2–12. Do problem 2-4 by the moment-area method.

2–13. Determine the horizontal deflection of A and the vertical deflection of C by the moment-area method.   *Ans. $\Delta_A = \dfrac{5Pa^3}{4EI}$.*

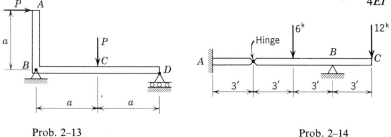

Prob. 2–13   Prob. 2–14

2–14. Determine the vertical deflection of C.

2–15. The beam *ABC* is quite flexible and thin. Determine an analytical expression for the deflection at *A* or *C* in terms of *EI, L, R,* and *P*.

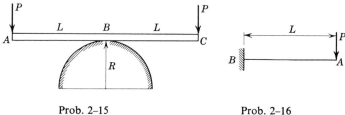

Prob. 2–15   Prob. 2–16

2–16. Determine the deflection at *A* by the conjugate-beam method and check by the moment-area method.

2–17. Do problem 2-4 by the conjugate-beam method.

2–18. For the beam shown, determine the necessary ratio of the forces P/Q to produce zero deflection at *C*. Use the conjugate-beam approach.

Ans. P/Q = $\frac{1}{3}$.

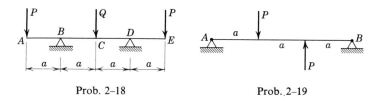

Prob. 2–18   Prob. 2–19

2–19. For the beam *AB* determine the slope at *A* by the following methods: (*a*) double integration (*b*) moment-area (*c*) conjugate-beam.

Ans. $\theta_A = Pa^2/9EI$.

2–20. Solve problem 2-8 by the Newmark numerical method.

2–21. Solve problem 2-19 by the Newmark numerical procedure.

2–22. Determine the deflection at *A* by the Newmark numerical procedure.

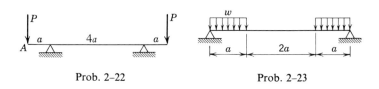

Prob. 2–22   Prob. 2–23

2–23. Determine the deflection at the center line by the Newmark numerical procedure and check by the moment-area method.

2–24. Solve problem 2-16 by the basic work and strain energy method.

2–25. Find the vertical deflections at the loading points for problem 2–19 by the basic work and strain energy method.

2–26. Solve problem 2–22 by the basic work and strain energy method.

2–27. A couple is applied at center of span in the vertical plane of a beam. Determine the rotation at the center line by basic work and energy methods. Check your answer by an alternate method of your choice.

Ans. $\theta = Ta/6EI.$

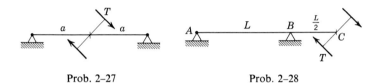

Prob. 2–27   Prob. 2–28

2–28. A couple is applied at end of beam. Determine the rotation at that point by basic work and energy methods. Check your answer by an alternate method of your choice. *Ans.* $\theta = 5\,TL/6EI.$

2–29. (*a*) Determine the deflection at center line by the unit load method. (*b*) Determine the slope at the reactions by the unit couple method.

Ans. (*a*) $PL^3/48EI$; (*b*) $PL^2/16EI.$

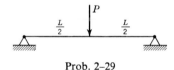

Prob. 2–29

2–30. For the beam of problem 2–22 determine the deflection at A, making full use of symmetry, by the unit load method.

2–31. By the unit load-unit couple method determine: (*a*) the horizontal deflection at B; (*b*) The horizontal deflection at D; (*c*) The rotation at D.

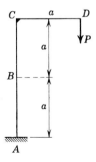

Ans. (*a*) $\Delta_B = Pa^3/2EI$; (*b*) $\Delta_D{}^H = 2Pa^3/EI$; (*c*) $\theta_D = 5Pa^2/2EI.$

Prob. 2–31

2–32. Determine the deflection at center line of span by the unit load method and check your answer by an alternate method of your choice.

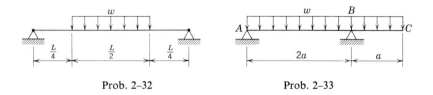

Prob. 2–32   Prob. 2–33

2–33. (a) Determine the rotation at B by the unit couple method; (b) Determine the deflection at C by the unit load method.

Ans. (a) $\theta_B = 0$; (b) $\Delta_C = wa^4/8EI$.

2–34. (a) Determine the horizontal displacement of point C; (b) Check your result for part (a) by an alternate method of your choice.

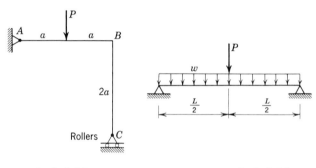

Prob. 2–34   Prob. 2–35

2–35. The beam is loaded with a uniform load and a load P. (a) Determine the deflection at the center line by the Castigliano theorem; (b) Let P approach zero in the expression found in part (a). What does the result represent?

2–36. For the beam of problem 2–19 determine the deflection at a load P by the Castigliano theorem.

2–37. Determine the vertical deflection at point A of the beam given in problem 2–4 by the Castigliano theorem.

2–38. Solve problem 2–27 by the Castigliano theorem.

2–39. Solve problem 2–28 by the Castigliano theorem.

2–40. Determine the vertical deflection of A by the Castigliano theorem. Check your answer by an alternate method of your choice.

Ans. $\Delta_A = 7Pa^3/3EI$.

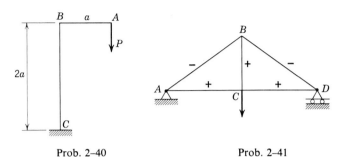

Prob. 2–40   Prob. 2–41

2–41. Sketch a free-hand Williot diagram for the symmetrical truss shown, assuming appropriate ΔL quantities consistent with signs of stress on members. Use BC as the reference member. Dimension the vertical deflection of C on your diagram.

2–42. Sketch a free-hand Williot diagram for the symmetrical truss shown, assuming the ΔL quantities consistent with signs of stress on members. Use DE as the reference member. Dimension the vertical deflection of C and the horizontal deflection of H on your diagram.

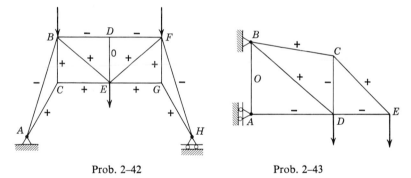

Prob. 2–42   Prob. 2–43

2–43. Sketch a free-hand Williot diagram using AB as the reference member. Assume the ΔL quantities consistent with signs of stress on members. Dimension the vertical and horizontal displacements of E.

2–44. Sketch a free-hand Williot diagram and then superimpose the Mohr correction truss. Indicate the vertical displacement of D and the horizontal movement of C on the resulting diagram. Assume ΔL of members as directed in previous problem. (See figure next page.)

2–45. Instructions the same as for problem 2–44. Indicate the vertical displacement of E on the final diagram. (See figure next page.)

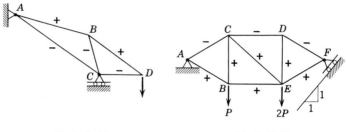

Prob. 2–44   Prob. 2–45

2–46. Construct the Williot and Mohr diagrams to scale, using ΔL of the individual members as equal to $\frac{1}{2000}$ of their length and of proper sign. The truss has been loaded symmetrically. Determine the vertical deflection of all lower chord panel points by (a) using FG as the reference member; (b) using CE as the reference member or an alternate member assigned by your instructor.

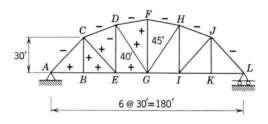

Prob. 2–46

2–47. Draw the Williot-Mohr diagrams and determine the deflection of panel point A. Assume that ΔL is $\frac{1}{2000}$ of length of members and of the sign indicated.

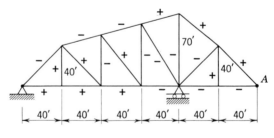

Prob. 2–47

2–48. For the trussed structure determine the vertical deflection at B by the unit load method. The area of each member is 10 sq in. $E = 30 \times 10^6$ psi.

Ans. 0:056 in.

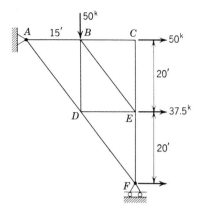

Prob. 2–48

2–49. (*a*) If point *A* is to be raised 1 in. by changing the length of an adjustable length member "*C*" only, what is the required change in length?; (*b*) How much will point *B* displace owing to the change in length calculated in part (*a*)?

Ans. (*a*) $-1/\sqrt{2}$ in.; (*b*) 1 in. down.

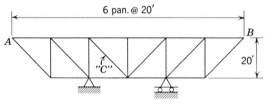

Prob. 2–49

2–50. Solve problem 2–46 by the unit load method.

2–51. Solve problem 2–47 by the unit load method.

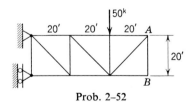

Prob. 2–52

2–52. Determine the rotation of member *AB* in radians. The area of each chord member is 10 sq in., and the area of each web member is 5 in.^{2}. $E = 30 \times 10^6$ psi.

Ans. $\frac{1}{1500}$ radian clockwise.

2–53. For the truss of problem 2–52, with load removed, determine the deflection of point *A* if the top chord has its temperature increased by 50° F. Coefficient of expansion = 6.5×10^{-6} per ° F.

2-54. Determine from basic principles the first three terms of a trigonometric series which will approximate the elastic curve for beam shown.

$$Ans.\ y = \frac{4PL^3}{\pi^4 EI} \sum_{n=1,3,5} \frac{1}{n^4} \sin \frac{n\pi b}{L} \sin \frac{n\pi x}{L}.$$

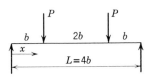

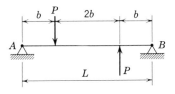

Prob. 2-54   Prob. 2-55

2-55. Approximate the elastic curve by one term of a sine series. *EI* is constant.

2-56. Determine three terms of a trigonometric series which will approximate the elastic curve. *EI* is constant.

$$Ans.\ y = \frac{2TL^2}{\pi^3 EI} \sum_{n=1,2,3} \frac{1}{n^3} \cos n\pi \sin \frac{n\pi x}{L}.$$

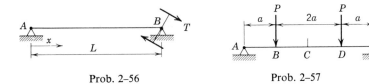

Prob. 2-56   Prob. 2-57

2-57. Determine the deflections at *B*, *C*, and *D* by the finite difference approximation approach. *EI* is constant.

2-58. Determine the deflections at *B*, *C*, and *D* by the finite difference approximation approach. *EI* is constant.

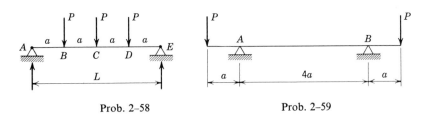

Prob. 2-58   Prob. 2-59

2-59. Determine the deflections at the one-quarter point and at midlength of span *AB* by the finite difference approximation method. *EI* is constant.

2-60. Do problem 2-58 by the relaxation method.

2–61. Do problem 2–59 by the relaxation method.

2–62. This problem provides an opportunity to review and interrelate the various theories and approaches. (*a*) Determine the equation for the elastic curve between *B* and *D* by the double integration method. (*b*) Determine the slope at *B* by the moment-area method. (*c*) Determine the vertical deflection at *C* by the unit load method. (*d*) Determine the vertical deflection at *C* by applying Castigliano's theorem.

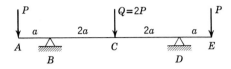

Prob. 2–62

Basic Principles
of Indeterminate Structures

3

3-1. INTRODUCTION

No single area of structural analysis has as many specialized methods as the field of continuous structures. Specialized methods have great value but should not be treated before a thorough study of basic general principles has been made. The extra equations needed to solve an indeterminate structure are derived from geometrical conditions; thus the key to analysis lies in a complete understanding of structural distortion and a visualization of the elastic curves.

3-2. BEAMS—INDETERMINATE TO FIRST DEGREE

Figure 1*a* represents a beam fully restrained (rotation of tangent prevented) at *A* and simply supported at *B*. The elastic curve of beam with a uniform *EI* is sketched as shown by dotted line. The author believes that the first step in the solution of essentially all indeterminate problems is to sketch the elastic curve, and the curve, if wrong, can be adjusted later. The next step in the solution should be the determination of the degree of indeterminacy or the number of redundants. This is easily ascertained by reducing the structure to a statically determinate structure, *or base structure*, by removing the forces or moments that are superfluous to static stability and equilibrium. In the current example, either V_B or M_A may be removed, but not both, and the structure remaining will be statically determinate; hence the beam is said to be statically indeterminate to the first degree or redundant to the first degree. The calculation of the redundants involves the deflections or rotations of the structure.

With V_B selected as the redundant and the reaction point removed, the determinate beam will deflect downward at B as shown in Fig. 1b. Δ_B may be computed by the moment-area procedure by employing a horizontal reference tangent to the elastic curve at A. The determinate moment diagram, or M_s, is shown. By the moment-area method we write

$$EI\Delta_B = -\frac{PL}{2} \cdot \frac{L}{2} \cdot \frac{1}{2} \left(\frac{L}{2} + \frac{2}{3} \cdot \frac{L}{2} \right)$$

$$EI\Delta_B = -\frac{5PL^3}{48}$$

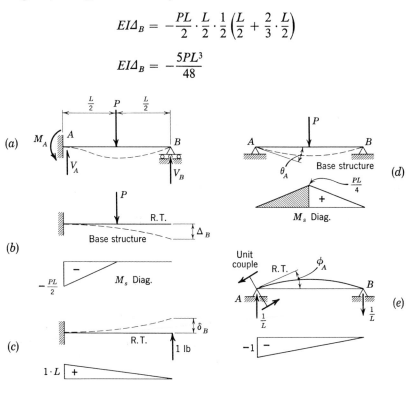

FIG. 3–1

The next phase of the solution is to note that the redundant V_B, if applied upward to the deflected beam of Fig. 1b, must restore the end B to its original position level with A. This can be accomplished physically by jacking B up an amount Δ_B, or may be accomplished mathematically by first finding the deflection δ_B produced by a 1-lb upward force. Then in algebraic terms, the resultant deflection at B, which must equal zero since the actual support at B does not deflect, may be expressed as $\Delta_s + \Delta_i = 0$, where Δ_s is the deflection of determinate *base structure* and Δ_i is the deflection produced by the redundant force. In terms of this problem we may write this geometric condition, often termed or defined as the *equation of consistent deformations*, as $\Delta_B + \delta_B \cdot V_B = 0$ from which we obtain

$$V_B = -\frac{\Delta_B}{\delta_B}$$

From Fig. 1c, and by the moment-area method and using a horizontal reference tangent,

$$EI\delta_B = +\frac{L \cdot L}{2} \times \frac{2}{3}L = +\frac{L^3}{3}$$

hence
$$V_B = -\frac{-\frac{5}{48}PL^3}{L^3/3} = +\frac{5}{16}P$$

The positive value for V_B indicates that the assumed upward direction of V_B is correct.

Since a choice existed between M_A and V_B as redundants, M_A could have been selected. With M_A removed, the determinate beam would deflect as shown in Fig. 1d and the tangent at A would rotate through an angle θ_A. The M_s diagram is drawn and by moment-area procedure, where θ_A is the angle between the end tangent and the tangent at center line of span, we obtain

$$EI\theta_A = \frac{PL}{4} \times \frac{L}{2} \times \frac{1}{2} = \frac{PL^2}{16}$$

M_A is the moment necessary to rotate the end tangent back to the horizontal position. The value of the moment can be determined by first computing the rotation ϕ_A produced by a unit couple at A. Calculations based on Fig. 1e and the moment-area method, with a reference tangent drawn to elastic curve at A, permits ϕ_A to be stated as

$$EI\phi_A = -\frac{L}{3}$$

The resultant rotation at A, which must equal zero, may be expressed as

$$\theta_s + \theta_i = 0$$

where θ_s is the rotation at A for the determinate *base structure* and θ_i is the rotation produced by the redundant moment. In terms of this problem we may write this geometric condition as $\theta_A + \phi_A \cdot M_A = 0$; hence

$$M_A = -\frac{\theta_A}{\phi_A} = -\frac{PL^2/16EI}{-L/3EI} = +\frac{3}{16}PL$$

The positive result for M_A indicates that the assumed direction of M_A is correct.

Figure 2 is a recapitulation of the results. Either M_A or V_B can be used in constructing the final shear and moment diagrams. Normally only M_A or V_B would be calculated, but in this case, since both were obtained, one furnishes a check on the other.

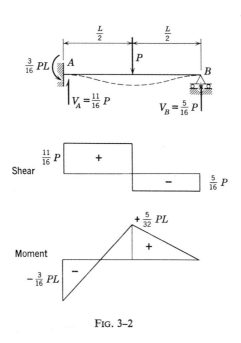

FIG. 3–2

Figure 3a represents a beam on three supports. The reactions at A, B, and C are unknown, and any one of the three reactions could be chosen as the redundant reaction, thereby reducing the beam to a statically determinate beam, or *base structure*, on two simple supports. The beam is then said to be indeterminate to the first degree with respect to the outer forces, and for the first solution V_B will be chosen as the redundant force. In Fig. 3b, the statically determinate structure deflects Δ_B at B. The moment-area procedure is followed in obtaining Δ_B after developing the M_s moment diagram in parts as shown and drawing a reference tangent to the elastic curve at C.

Referring to Fig. 3b:

$$EI\Delta_1 = 7.5 \times 24 \times \tfrac{24}{2} \times 8 - 6 \times 18 \times \tfrac{18}{2} \times 6$$

$$-\frac{72 \times 12}{3} \times \tfrac{12}{4} = 10{,}584 \text{ kip-ft}^3$$

By similar triangles

$$EI\Delta_2 = \tfrac{12}{24} \times EI\Delta_1 = 5292 \text{ kip-ft}^3$$

By a second application of moment-area procedures we obtain

$$EI\Delta_3 = 7.5 \times 12 \times \tfrac{12}{2} \times 4 - 6 \times 6 \times \tfrac{6}{2} \times 2 = 1944 \text{ kip-ft}^3$$

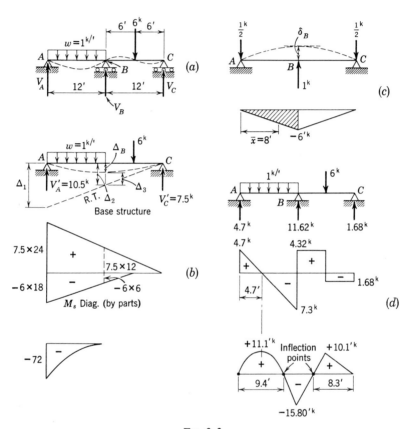

FIG. 3–3

Then

$$EI\Delta_B = EI\Delta_2 - EI\Delta_3 = 3348 \text{ kip-ft}^3$$

A unit load is now applied upward at B in Fig. 3c and $EI\delta_B$ computed.

$$EI\delta_B = -\frac{6 \times 12}{2} \times 8 = -288 \text{ kip-ft}^3 \text{ per kip}$$

Then it follows that B will be returned to its original undeflected level by V_B. Hence,

$$V_B = -\frac{\Delta_B}{\delta_B} = -\frac{3348}{-288} = +11.62 \text{ kips}$$

With V_B known, the values of V_A and V_C may be computed from equations of static equilibrium. Figure 3d denotes the final shear and moment diagrams. Observe that points of inflection, or sections of beam where there is zero moment, have been located.

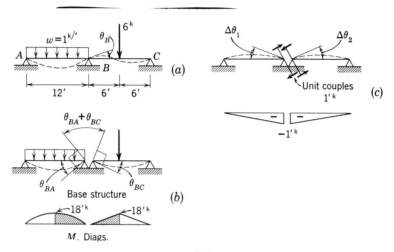

Fig. 3–4

An alternate approach to the solution of the preceding problem is presented in Fig. 4. The difference is in the choice of the redundant quantity. Although outer forces have been spoken of as redundants, the beam could be made determinate by severing the beam at B, thus destroying the continuity and the moment at B. The statically determinate base structure would now consist of the two simple spans AB and BC. The redundant required to maintain continuity is identified as M_B. Figure 4b indicates the two simple statically determinate spans with a sketch of their elastic curves and moment diagrams. The tangents to the respective elastic curves at B have rotated θ_{BA} and θ_{BC}. The following moment-area procedures will establish values for these rotations. Owing to symmetry the rotations are related to the cross-hatched areas indicated under the M_s moment curves.

$$EI\theta_{BA} = 18 \times 6 \times \tfrac{2}{3} = 72 \text{ kip-ft}^2$$
$$EI\theta_{BC} = 18 \times 6 \times \tfrac{1}{2} = 54 \text{ kip-ft}^2$$

The actual continuity of the original beam at B demands that the elastic curve of continuous structure for span BA and span BC have a common tangent line at B.   To satisfy this geometric condition the redundant moment M_B must be of such a value as to rotate the tangents at B of Fig. 4b to a common slope.   To achieve this condition, first determine the rotations of tangents due to unit couples applied to the base structure as shown in Fig. 4c.   The magnitude of the rotations from moment-area principles is

$$EI \, \Delta\theta_1 = EI \, \Delta\theta_2 = 4 \text{ kip-ft}^2 \text{ per kip-ft}$$

The geometrical equation can now be written based on angular rotations. Perpendiculars to the tangents in Fig. 4b are helpful in visualizing this equation, since the function of M_B is to make these perpendiculars coincide to satisfy the requirement of a common slope at B.   The equation expressing this condition is

$$\theta_{BA} + \theta_{BC} = M_B(\Delta\theta_1 + \Delta\theta_2)$$

from which we obtain

$$M_B = \frac{126/EI}{8/EI} = 15.75 \text{ ft-kips}$$

This value of M_B closely checks the previous value established for M_B in the bending moment diagram of Fig. 3d and would check exactly except for slide rule errors.   Reactions can now be obtained from the principles of static equilibrium.

No preference need be given at this time to either the use of moments or reactive forces as redundants.   Many special methods use moments to generalize the solutions; however, in general the use of both reactive forces and moments is encountered.   The principal objective at this stage is to emphasize the basic principles of solution, namely,

(a) reduce the given structure to a statically determinate condition, or base structure, by removing redundants;

(b) determine the slopes and/or deflections of the base structure;

(c) determine the values of the redundants to restore the base structure to the actual conditions of support and continuity of original structure.

3-3. BEAMS—INDETERMINATE TO THE SECOND DEGREE

The principles will be extended to cover the case of beams where two redundants must be removed to reduce the beam to statically determinate

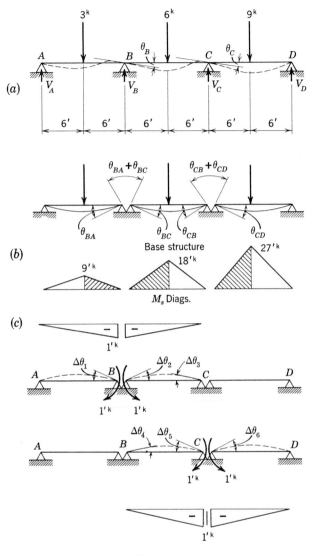

FIG. 3–5

conditions.   Figure 5a is a beam of uniform section continuous over four supports.   Exercising the option on selection of redundants, M_B and M_C have been selected.   The separate beam spans with continuity destroyed at B and C are redrawn in Fig. 5b.   The bending moment diagrams for the determinate conditions may then be constructed.   The slopes indicated are then obtained by moment-area or conjugate-beam procedures. As may easily be justified, these slopes are

$$EI\theta_{BA} = 27$$
$$EI\theta_{BC} = 54$$
$$EI\theta_{CB} = 54$$
$$EI\theta_{CD} = 81$$

Figure 5c illustrates the application of unit couples to the ends of simple spans at B and C and the moment diagrams pertaining to these moments. The conjugate-beam method was used to obtain the rotations of the end tangents and they are as follows:

$$EI\,\Delta\theta_1 = 4$$
$$EI\,\Delta\theta_2 = 4$$
$$EI\,\Delta\theta_3 = 2$$
$$EI\,\Delta\theta_4 = 2$$
$$EI\,\Delta\theta_5 = 4$$
$$EI\,\Delta\theta_6 = 4$$

The geometric equations based on re-establishing the continuity or common slope of beam at B and C by means of the redundant moments follow:

At B: $(\theta_{BA} + \theta_{BC}) = M_B(\Delta\theta_1 + \Delta\theta_2) + M_C \cdot \Delta\theta_4$

At C: $(\theta_{CB} + \theta_{CD}) = M_B \cdot \Delta\theta_3 + M_C (\Delta\theta_5 + \Delta\theta_6)$

With the substitution of the calculated geometric quantities and canceling of EI, the above two equations reduce to

$$8M_B + 2M_C = 81$$
$$2M_B + 8M_C = 135$$

A solution of these equations gives

$$M_B = 6.3 \text{ ft-kips}$$
and $$M_C = 15.3 \text{ ft-kips}$$

3–4. FIXED-ENDED BEAMS

A recurring assumption in structural theory stipulates that the end of a member is held securely against rotation. This restraint produces a fixed-ended condition and results in fixed-end moments being induced when the member is loaded.

Figure 6a illustrates a simple situation where both end A and end B are fixed, and where fixed-end moments M_A and M_B are due to the load P. If the end moments M_A and M_B were removed, the beam would be simply supported and statically determinate, thus M_A and M_B are redundant. Figure 6b is the moment diagram drawn by parts, the positive triangle representing the bending moment curve for the beam if simply supported. The negative portion of the diagram represents the moment produced by the redundant moments. Although alternate approaches exist, such as

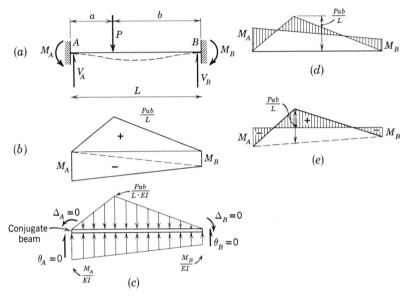

FIG. 3–6

illustrated previously, the present problem is particularly adapted for solution by the conjugate-beam method. Figure 6c shows the conjugate beam loaded with the M/EI diagrams. On account of the fixed-ended conditions, the rotation of the tangent at A and B or θ_A and θ_B must equal zero, and of course Δ_A and Δ_B equal zero. The end moments and end shears, or reactions, for the conjugate beam, representing respectively deflection and slope, must accordingly be equal to zero. The conjugate beam may then be said to be a beam in static equilibrium under the action of equal up and down loads, or physically it is similar to a floating structure in this case. The vertical equilibrium condition plus the moment equilibrium condition for the conjugate beam enable the two redundants to be determined.

Referring to Fig. 6c, the first equation dealing with vertical equilibrium of the conjugate beam is

$$\frac{1}{EI}\left(\frac{Pab}{L} \times \frac{L}{2} - \frac{M_A \cdot L}{2} - \frac{M_B \cdot L}{2}\right) = 0$$

and the moment equilibrium equation for the conjugate beam taking moments about B is

$$\frac{1}{EI}\left[\frac{Pab}{L} \cdot \frac{b}{2} \times \frac{2}{3}b + \frac{Pab}{L} \cdot \frac{a}{2}\left(b + \frac{a}{3}\right) - \frac{M_A L}{2} \times \frac{2}{3}L - \frac{M_B L}{2} \times \frac{1}{3}L\right] = 0$$

The solution and simplification of the above equations gives

$$M_A = \frac{Pab^2}{L^2} \quad \text{and} \quad M_B = \frac{Pa^2b}{L^2} \tag{3-1}$$

These two expressions are so universal for beams of uniform EI that the student would do well to memorize them.

Figure 6d illustrates the final moment diagram drawn by superimposing the negative portion of Fig. 6b over the positive curve. Such super-position amounts to a graphical addition of positive and negative values resulting in the uncanceled cross-hatched areas shown. Figure 6e is the conventional manner of indicating the moment diagram. An important observation to make is that the height of the diagram at the point of loading, measured from the dotted line joining the two end moments, is equal to the moment under load for the unrestrained simple end conditions. Thus, the end moments act to reduce the maximum positive moment in span.

Example

The equations $M_A = Pab^2/L^2$ and $M_B = Pa^2b/L^2$ may be used as a basis for determining the fixed-end moments for other loadings. Figure 7a represents a beam with a uniformly varying load which has a maximum intensity of loading of w lb per unit of length at A. In Fig. 7b, a differential length of beam dx long is loaded by a differential load P_1 equal to $wx\,dx/L$ which will make its individual and differential contribution to the total moments at A and B. A similar loading condition exists on all differential lengths, hence integral calculus may be used to summarize the total effect. The differential moment produced at A by P_1 is accordingly formulated by using the first of equations 3-1.

$$dM_A = \frac{\frac{wx}{L}dx(L-x)(x)^2}{L^2}$$

An integration over the entire span gives the total moment M_A, or

$$M_A = \frac{w}{L^3} \int_0^L (L - x)(x^3)dx = \frac{wL^2}{20}$$

similarly $$M_B = \frac{w}{L^3} \int_0^L (L - x)^2 \cdot (x^2)dx = \frac{wL^2}{30}$$

A similar analysis may be applied to determine fixed-ended moments for any other variable or partial span loadings.

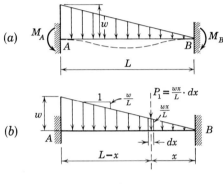

FIG. 3–7

3–5. THREE-MOMENT EQUATION

When a beam is continuous over more than three supports, the basic procedure of removing redundant forces or moments becomes laborious and repetitious. The three-moment equation developed in this section is a convenient way of interrelating the bending moments at the supports.

Figure 8a represents the general situation of any two adjacent spans of a multispan continuous beam. It is assumed that I_1 represents the moment of inertia of the span AB, and I_2 the similar value for span BC. The redundants are the bending moments in the beam at B, or M_B, along with M_A and M_C, the moment restraints furnished by the spans to left of A and to the right of C, respectively. Furthermore, these redundant moments are assumed to be positive in the usual beam convention sense. Figure 8b represents bending moment diagrams drawn in parts with the upper parts, identified with simple span conditions, shown separate from the bending moments due to the redundant moments. The development of the three-moment equation depends on the fact that the tangent to elastic curve at B (shown with a rotation of θ_B) is the common tangent to the continuous elastic curves of span AB and span BC at B. This geometrical fact enables the following equations to be written, noting of course that if Δ_A is below A, Δ_C must be above C. The moment-area procedure is used, and since the geometrical equations depend on an equality of θ_B, the common slope at B, the sign of either Δ_A or Δ_C must be adjusted by a negative sign.

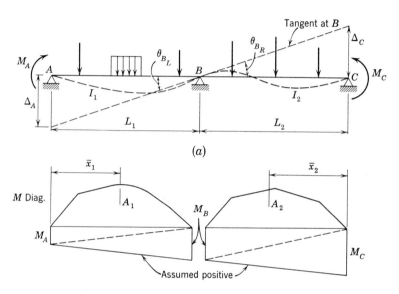

(a)

(b)

A_1 and A_2 denote areas, $\bar{x}_1$ and $\bar{x}_2$ denote distance to centroid

Fig. 3-8

From left-hand span AB From right-hand span BC

$$\theta_{B_L} = \theta_{B_R}$$

$$\frac{\Delta_A}{L_1} = -\frac{\Delta_C}{L_2}$$

Consequently, in terms of the moment diagrams of Fig. 3-8 we obtain

$$\frac{1}{L_1 \cdot EI_1}\left(A_1\bar{x}_1 + \frac{M_A \cdot L_1}{2} \times \frac{L_1}{3} + \frac{M_B L_1}{2} \times \frac{2L_1}{3}\right)$$
$$= -\frac{1}{L_2 EI_2}\left(A_2\bar{x}_2 + \frac{M_B L_2}{2} \times \frac{2}{3}L_2 + \frac{M_C L_2}{2}\cdot\frac{L_2}{3}\right)$$

and simplifying, the fundamental equation becomes

$$\frac{M_A L_1}{I_1} + 2M_B \left(\frac{L_1}{I_1} + \frac{L_2}{I_2}\right) + \frac{M_C L_2}{I_2} = -\frac{6A_1 \bar{x}_1}{L_1 I_1} - \frac{6A_2 \bar{x}_2}{L_2 I_2} \qquad (3\text{-}2)$$

which is known as the general form of the three-moment equation. The equation can be further simplified, if the moments of inertia of the two spans are equal, to the form

$$M_A L_1 + 2M_B(L_1 + L_2) + M_C L_2 = -\frac{6A_1 \bar{x}_1}{L_1} - \frac{6A_2 \bar{x}_2}{L_2} \qquad (3\text{-}3)$$

In application, the areas and centroids of the simple-span bending moment curves for the loading involved must be obtained. If the span

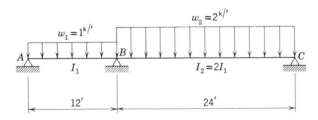

FIG. 3–9

AB is uniformly loaded with a uniform load of w_1, the general expression $6A_1\bar{x}_1/L_1 I_1$, since A_1 is the area under a parabolic moment curve, would become

$$\frac{6A_1 \bar{x}_1}{L_1 I_1} = \frac{6 \times (\tfrac{2}{3} w_1 L_1^3/8) \times L_1/2}{L_1 I_1} = \frac{w_1 L_1^3}{4I_1} \qquad (3\text{-}4)$$

Similarly, for a uniform load of w_2 on span L_2

$$\frac{6A_2 \bar{x}_2}{L_2 I_2} = \frac{w_2 L_2^3}{4I_2} \qquad (3\text{-}5)$$

Special cases involving concentrated loads can be dealt with by fundamental procedures for determining $A_1\bar{x}_1$ or $A_2\bar{x}_2$ as a separate calculation. For complex loadings the total $A_1\bar{x}_1$, or $A_2\bar{x}_2$ may be found by combining separately computed values for each load.

Examples
Figure 9 represents a two-span continuous beam with no moments applied at A and C, hence M_B remains as the only redundant.

The general equation reduces to

$$2M_B\left(\frac{12}{I_1} + \frac{24}{2I_1}\right) = -\frac{1 \times 12^3}{4I_1} - \frac{2 \times 24^3}{8I_1}$$

$$48M_B = -3888$$

$$M_B = -81 \text{ ft-kips}$$

The minus sign is the true sign of moment by the beam convention. It should now be clear why positive moment was assumed in the general derivation, as the negative sign indicates that the moment is opposite in direction to that assumed and that the sign is the correct one by the usual moment convention.

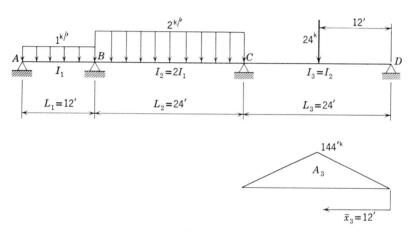

FIG. 3-10

Figure 10 represents a three-span beam where the span CD has been added to the spans of Fig. 9 and where the moment of inertia is not constant in all spans. The redundant moments are now M_B and M_C, and consequently two separate applications of the three-moment equation 3-2 are necessary. One equation is obtained by using the data of spans AB and BC, and the second equation by using the data of spans BC and CD.

For spans AB and BC

$$2M_B\left(\frac{12}{I_1} + \frac{24}{2I_1}\right) + M_C \cdot \frac{24}{2I_1} = -\frac{1 \times 12^3}{4I_1} - \frac{2 \times 24^3}{8I_1}$$

$$48M_B + 12M_C = -3888 \tag{3-6}$$

For spans BC and CD

with
$$\frac{6A_3\bar{x}_3}{L_3I_3} = \frac{6 \times 144 \times \frac{24}{2} \times 12}{24 \times 2I_1} = \frac{2592}{I_1}$$

$$\frac{M_B \times 24}{2I_1} + 2M_C\left(\frac{24}{2I_1} + \frac{24}{2I_1}\right) = -\frac{2 \times 24^3}{8I_1} - \frac{2592}{I_1}$$

$$12M_B + 48M_C = -3456 - 2592 = -6048 \qquad (3\text{-}7)$$

Equations 3-6 and 3-7 are repeated below and then solved simultaneously

$$48M_B + 12M_C = -3888$$
$$12M_B + 48M_C = -6048$$
$$M_C = -113 \text{ ft-kips}$$
$$M_B = -53.2 \text{ ft-kips}$$

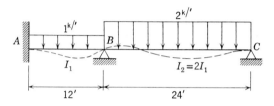

Phantom span

FIG. 3–11

With these moment values the reactions can be found and the shear diagram and moment diagram drawn.

Another example of the versatility of the three-moment equation may be illustrated by the problem of Fig. 11. The beam depicted in Fig. 11 differs from the beam of Fig. 9 in that end A is fully restrained with a fixed-end moment at A. The redundant moments are now M_A and M_B, necessitating two equations. To set up two equations by the subject method requires the addition of a phantom span of span length L_0 = zero. As this unloaded span length approaches zero, the two end tangents of the span's elastic curve coincide in a horizontal position which simulates the true restrained position of the horizontal tangent at the fixed end A.

The three-moment equation for the phantom span and span AB is

$$M_o\left(\frac{0}{I_o}\right) + 2M_A\left(\frac{0}{I_o} + \frac{12}{I_1}\right) + M_B \cdot \frac{12}{I_1} = -\frac{1 \times 12^3}{4I_1}$$

$$24M_A + 12M_B = -432$$

and for spans AB and BC, the equation is

$$\frac{M_A \cdot 12}{I_1} + 2M_B\left(\frac{12}{I_1} + \frac{24}{2I_1}\right) = -\frac{3888}{I_1}$$

$$12M_A + 48M_B = -3888$$

A solution of the two equations gives

$$M_A = +25.7 \text{ ft-kips} \qquad M_B = -87.4 \text{ ft-kips}$$

3–6. CONTINUOUS BEAMS—ENERGY METHODS

The Castigliano theorem relating strain energy, deflection, and slope offers another means of establishing the geometric equations for solution of an indeterminate structure.   This is best exemplified by an illustration.

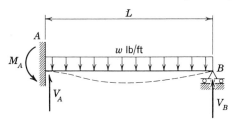

FIG. 3–12

Assume that V_B is chosen as the redundant reaction for the beam of Fig. 12.   With the bending moment equation expressed in terms of the reaction, uniform load, and x measured from B, the strain energy due to bending is

$$U = \int \frac{M^2 \, dx}{2EI} = \frac{1}{2EI} \int_0^L \left(V_B x - \frac{wx^2}{2}\right)^2 dx$$

Since the conditions of support require the deflection in the direction of V_B to be zero,

$$\frac{\partial U}{\partial V_B} = \Delta_B = 0$$

With the origin at B the following equation is determined by differentiating under the integral sign

$$\frac{\partial U}{\partial V_B} = \frac{1}{EI} \int_0^L \left(V_B x - \frac{wx^2}{2} \right) x \, dx = 0 \tag{3-8}$$

The indicated integration and solution results in

$$V_B = +\tfrac{3}{8} wL$$

It may also be observed that placing $\dfrac{\partial U}{\partial V_B}$ equal to zero is equivalent to establishing a condition based on a zero slope of an algebraic curve relating U and V_B where U is the ordinate and V_B the abcissa.   Points of zero slope will occur at maximum and minimum values of U or when U has a stationary value.   All redundant elastic structures, if in equilibrium, adjust their displacements in such a way as to minimize potential energy. This procedure of placing the partial derivative of U equal to zero is often referred to as the method of least work or Castigliano's second theorem.

The problem of Fig. 12 could also be solved by the Castigliano theorem treating M_A as the redundant.   First, writing the bending moment equation, with the origin for x at A,

$$M = -M_A + V_A \cdot x - \frac{wx^2}{2}$$

It is to be noted that this beam is redundant to the first degree and that V_A depends on M_A or is a function of M_A, the loading and span length. Thus, if M should be used in the above form, a total derivative (see a standard calculus text) would be needed in the step involving $\dfrac{\partial U}{\partial M_A}$.   The total derivative may of course be resorted to, but it is much more direct and less susceptible to error to use a static equation expressing V_A in terms of M_A, thus making M dependent solely on the redundant quantity.

By statics, with moments about B equal to zero, we obtain

$$V_A = \frac{M_A + wL^2/2}{L}$$

then
$$M = -M_A + \frac{M_A}{L} x + \frac{wL^2}{2L} \cdot x - \frac{wx^2}{2}$$

permitting U to be written as

$$U = \frac{1}{2EI} \int_0^L \left(-M_A + \frac{M_A x}{L} + \frac{wL^2}{2L} \cdot x - \frac{wx^2}{2} \right)^2 dx$$

Since the slope at $A = 0$, we obtain the following equation by partial differentiation of U with respect to M_A, or

$$\frac{\partial U}{\partial M_A} = \int_0^L \left(-M_A + \frac{M_A x}{L} + \frac{wL^2 x}{2L} - \frac{wx^2}{2} \right) \left(-1 + \frac{x}{L} \right) dx = 0$$

from which we obtain

$$M_A = +\frac{wL^2}{8}$$

Although there are not many situations where the same difficulty would occur, it is worth noting that the method will lead to erroneous results if there is not complete independence of the terms entering the moment equation. Such errors are generally due to a failure to use all of the available static equations to insure that the reaction or moment quantities in the equation represent only the redundant reactions or moments where the number of such redundants can never exceed the degree of indeterminacy. A full use of static equilibrium equations will prevent this source of error. It is for this reason that the unit load or unit couple method is often preferred to the subject method since in its application the taking of the derivative is avoided.

The method identified as the virtual work or unit load method is illustrated by using the beam of Fig. 13a, which is the same as the beam of Fig. 12. The sequential steps leading to a solution are modified to suit this method. First, remove the loads from the span and also the support at B to place the beam in a statically determinate condition, but unloaded. Then a unit vertical load acting up at B is applied to produce "m" moments as shown in Fig. 13b. Now—and this step is important for an understanding of the full philosophy of the method—the physical support at B is restored while the beam is under the effect of the 1-lb force. This action in a physical sense "locks" the "m" moments in the beam. The third and last physical step is now to apply or superimpose the actual loads to the span. These loads create the "M" moments as depicted graphically in Fig. 13c. While the actual loads are being applied during this final step the reaction at B increases to its value V_B, but the important physical consideration is that the 1-lb unit load is not permitted to move, since the reaction at B had been restored; hence it can do no external virtual work. This is equivalent to stating that $1 \cdot \Delta_B = 0$, since the deflection at B or $\Delta_B = 0$. In mathematical terms

$$1 \cdot \Delta_B = \int_0^L \frac{Mm \, dx}{EI} = \frac{1}{EI} \int_0^L \left(V_B x - \frac{wx^2}{2} \right)(x) \, dx = 0$$

Note that this equation checks equation 3-8 secured by the Castigliano approach after the differentiation step.   Here is one more demonstration that if

$$U = \int \frac{M^2}{2EI} \cdot dx$$

is differentiated in general form as

$$\frac{\partial U}{\partial P} = \int \frac{M \cdot \frac{\partial M}{\partial P}}{EI} \cdot dx$$

that $\frac{\partial M}{\partial P}$ is equivalent to "m" and V_B will again equal $\frac{3wL}{8}$.

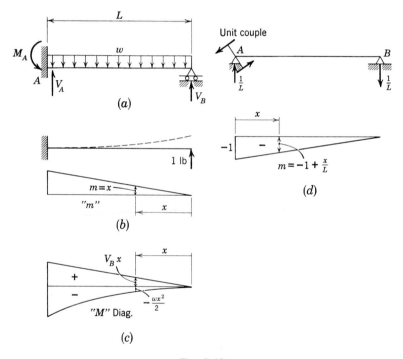

FIG. 3–13

If M_A is selected as the redundant a unit couple is first applied at A to the statically determinate unloaded beam of Fig. 13d.   The "m" diagram is drawn and the equation for "m" formulated.   Then A is clamped or fixed against rotation, thus locking in "m".   The final step is to superimpose the real loading, creating the "M" moments.   The moment "M" represents the total moment caused by the load and the reactions.

Physically, joint A does not rotate, hence the unit couple does no work while the loading is applied, therefore $1 \cdot \theta_A = 0$. In mathematical terms

$$1 \cdot \theta_A = \int_0^L \frac{Mm\,dx}{EI} = 0$$

which, with the origin for x at A, becomes

$$\int_0^L \left(-M_A + V_A x - \frac{wx^2}{2} \right)\left(-1 + \frac{x}{L} \right) dx = 0$$

Integration provides the following equation:

$$\frac{M_A}{2} - \frac{V_A L}{6} + \frac{1}{24} wL^2 = 0$$

Another equation is provided from statics by taking moments about an axis through B.

$$M_A - V_A L + \frac{wL^2}{2} = 0$$

The two equations solved simultaneously lead to values obtained before, which are

$$M_A = \tfrac{1}{8}wL^2$$
$$V_A = \tfrac{5}{8}wL$$

The $\int \dfrac{Mm\,dx}{EI}$ approach may be further illustrated by considering the beam of Fig. 14a. The beam is made statically determinate by removing the reaction at B and a unit load is installed upward on this unloaded beam at B, producing the "m" diagram of Fig. 14b. The support is then restored and the real loading applied. The "M" diagram of Fig. 14c has been drawn with V_A as the redundant instead of V_B to take full advantage of symmetry and the simplicity of formulating "m". This may appear inconsistent to the student, but let him be reminded that the final moments of this beam, no matter how they may be formulated mathematically, must lead to the same result, namely, that the elastic curve of the beam must have a zero deflection at B. Symmetry reduces the mathematical work, but the mathematical discontinuity in equation for M demands two separate integrals as shown in the following formulation.

$$1 \cdot \varDelta_B = \frac{2}{EI} \int_0^a (V_A x)\left(-\frac{1}{2}x \right) dx + \frac{2}{EI} \int_a^{2a} [V_A x - P(x-a)]\left[-\frac{x}{2} \right] dx = 0$$

Upon integration and simplification

$$V_A = \tfrac{5}{16}P$$

and by statics

$$V_B = \tfrac{11}{8}P$$

The final conventional moment diagram is shown in Fig. 14d.

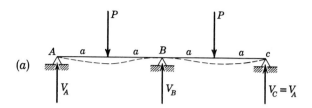

(a)

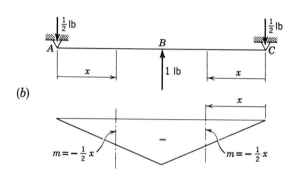

(b)

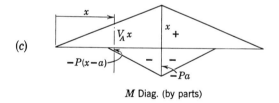

(c)

M Diag. (by parts)

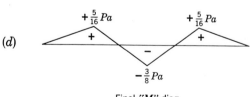

(d)

Final "M" diag.

FIG. 3–14

3–7. FRAMES

A structure composed of an interconnected assemblage of beams and columns is spoken of as a frame. If the connections of one individual member to another are assumed to be undeformable the frame is said to have rigid joints. This is a very common assumption which is rarely 100 per cent true; however, without it additional modifications in analysis are encountered. Figure 15 illustrates the geometry of a rigid joint before and after rotation of the joint. Without deformation of the joint material the tangents to elastic curve of the beam and the column must rotate through the same angle θ. The acceptance of equal rotation of the tangents is basic for all rigid joints. Member lengths are modified by the

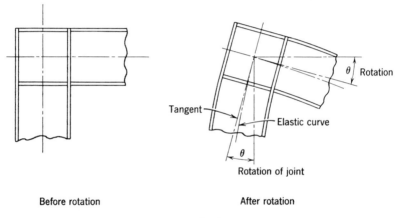

Before rotation   After rotation

FIG. 3–15

joint dimensions, but in this book the member length will be measured to the center of the joint.

The basic analysis of continuous frames is merely an extension of the same principles already encountered in dealing with continuous beams. The geometry is more complex but fundamental concepts for calculating deflection and slopes remain the same. Furthermore, the disciplines and procedures for selecting redundants and forming geometrical equations from boundary conditions are still essential. The fundamental principle of first reducing the structure to a statically determinate *base structure* still remains as the indispensible approach. Although special methods will later be presented, the elements of geometrical solution will be dealt with first.

3-8. FRAMES—INDETERMINATE TO FIRST DEGREE

Figure 16a represents a frame with all members assumed to have equal moment of inertia. The columns are hinged to their foundation and the joints of the frame at B and C are rigid. The rotation of these joints under the action of the load are θ_B and θ_C. The dotted lines represent the elastic curve.

Let it be required to determine the reactions at A and D. Symmetry establishes each of the vertical reactions as 5 kips. Owing to the fact that the distance between A and D remains fixed, horizontal reactions are induced at A and D as shown. Although equations of statics are informative to the point of recognizing that H_A is equal to H_D, the static equations must be augmented to determine their numerical value. As may be deduced from this fact, the structure is statically indeterminate. The solution is approached by making the frame statically determinate by removing the hinge at D and substituting rollers, which in effect eliminate the horizontal reactions. Since the horizontal forces have been removed to reduce the frame to its determinate condition, these automatically become the redundant forces.

Figure 16b is the statically determinate base structure and Δ_s represents the horizontal movement of the rollers produced by flexure, which may be calculated. The horizontal reaction is the force necessary to return D' to its true position at D. To compute Δ_s an analysis must be made of the elastic curve of the *base structure*, and as a preliminary step the bending moment diagram must be drawn as in Fig. 16c. As is the general practice in dealing with frames the outline of the structure is taken as the base line of the moment curve. In this book the ordinates of the bending moment diagram will be plotted on the side of member that is in tension. Furthermore, to be explicit about the diagram of Fig. 16c representing the moment of the statically determinate case, the moments will be termed M_s, the subscript s referring to the statically determinate condition.

Figures 16b and 16c must be related by a deflection method and the moment-area method will be illustrated first. Since the frame is symmetrical the rotation of rigid joints at B and C are equal numerically, but opposite in sign as shown. θ_1, the numerical value of this rotation from moment-areas and analysis of elastic curve of beam BC, becomes

$$EI\theta_1 = 12.5 \times 5 \times \tfrac{2}{3} = \tfrac{125}{3}$$

Joints B and C must translate to the right to enable the elastic curve of all members of frame to be consistent with the M_s moment. It should be noted that elastic curves of columns are still straight lines since M_s for columns was zero. In this instance the deflected frame of Fig. 16b was

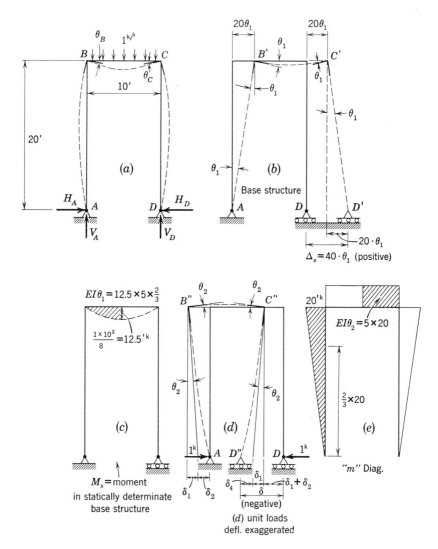

Fig. 3–16

visualized by starting at point A as a hinged point fixed in location and then traversing the elastic curve around the frame sketching each member from joint A to joint B, B to C, and finally from C to D. The outcome of this geometrical analysis shows that Δ_s equals $40 \cdot \theta_1$. The algebraic sign of Δ_s will be taken as positive since D moved to the right. This sign is optional as long as a consistent system is used.

The horizontal deflection of D in Fig. 16d produced by a 1-kip horizontal force has been termed δ. This deflection is computed by relating the geometry of Fig. 16d and the moments of Fig. 16e. Owing to symmetry,

$$EI\theta_2 = 20 \times 5 = 100$$

With θ_2 known

$$EI\delta_1 = 20 \times EI\theta_2 = 2000$$

Since δ_2 is the deflection of A from the tangent to elastic curve at B'', then

$$EI\delta_2 = 20 \times \tfrac{20}{2} \times \tfrac{2}{3} \times 20 = \tfrac{8000}{3}$$

By symmetry $\qquad\qquad\qquad \delta_4 = \delta_2$

Neglecting changes in length of member BC, joint C'' translates to the left as much as B''. By summarizing all components of displacement, point D'' moves δ to the left in Fig. 16d due to the 1-kip force at D, where

$$EI\delta = -EI(\delta_1 + \delta_2 + \delta_1 + \delta_4)$$
$$EI\delta = -(2000 + \tfrac{8000}{3} + 2000 + \tfrac{8000}{3})$$
$$EI\delta = -\tfrac{28,000}{3}$$

δ is negative since it represents a movement to the left.

The force H_D required to restore D' of Fig. 16b to the required zero deflection position of Fig. 16a may now be obtained. The calculated deflections are related by the equation $\Delta_s + \Delta_i = 0$. In terms of this problem

$$\Delta_s + H_D \cdot \delta = 0$$
$$H_D = -\frac{\Delta_s}{\delta}$$

Since EI is constant it is omitted and the final reaction is

$$H_D = -\frac{40 \times \tfrac{125}{3}}{-\tfrac{28,000}{3}} = +0.179 \text{ kip}$$

where the positive result indicates that H_D acts in the direction assumed.

$$H_A = H_D \text{ by statics}$$

All frames may be solved by such a basic procedure, but as the number of redundants increases the relevant geometry becomes more complex.

3–9. FRAMES—CASTIGLIANO'S THEOREM

The frame of Fig. 17a will be solved by Castigliano's theorem to contrast this method with the moment-area method. In the moment-area

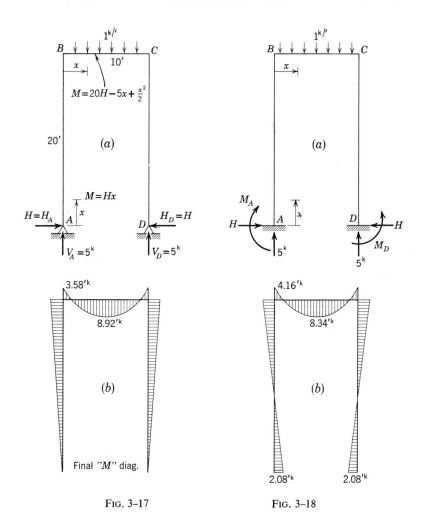

FIG. 3–17   FIG. 3–18

method an analysis of the entire elastic line of the frame was required to obtain the relevant geometry. The Castigliano method, in contrast, is a mathematical approach avoiding the geometric details except for a recognition of the boundary conditions of zero deflection at A or D. The redundant is the horizontal reaction H and the strain energy expression for the entire frame, noting symmetry, is formulated with origins indicated in Fig. 17a, from the basic equation

$$U = \int \frac{M^2 \, dx}{2EI}$$

The sign convention for moments is optional, but once established it must be consistently followed for all members.   In the subsequent analysis the moment in the column has been taken as positive and equal to Hx.   It is to be noted that the positive moment effect of H is retained in the moment equation for the beam.   The integrals are set up to account for U in all members of the frame.   Particularly note that the first integral following has been multiplied by 2 to determine U for both columns.   The basic equation may then be expanded to

$$U = \frac{2}{2EI} \int_0^{20} (Hx)^2 dx + \frac{1}{2EI} \int_0^{10} \left(20H - 5x + \frac{1 \cdot x^2}{2}\right)^2 dx$$

Then, since the horizontal deflection in direction of H is zero, we may differentiate with respect to H under the integral sign and obtain

$$\frac{\partial U}{\partial H} = 2H \int_0^{20} x^2 \, dx + \int_0^{10} \left(20H - 5x + \frac{1 \cdot x^2}{2}\right)(20)dx = 0$$

By integration we obtain

$$2H\left[\frac{x^3}{3}\right]_0^{20} + 400H\left[x\right]_0^{10} - 100\left[\frac{x^2}{2}\right]_0^{10} + 10\left[\frac{x^3}{3}\right]_0^{10} = 0$$

which simplifies to

$$\tfrac{16,000}{3}H + 4000H - 5000 + \tfrac{10,000}{3} = 0$$

and, upon solution, we find

$$H = +0.179 \text{ kip}$$

This value checks that found from the previous solution in Art. 3-8.   The final moment diagram is shown in Fig. 17b.

   If the columns were fixed against rotation at A and D as in Fig. 18a, then two redundant quantities are involved, namely, H_A and M_A.   The total strain energy employing the origins shown in Fig. 18a, is

$$U = \frac{2}{2EI} \int_0^{20} (Hx - M_A)^2 dx + \frac{2}{2EI} \int_0^5 \left(20H - M_A - 5x + \frac{1 \cdot x^2}{2}\right)^2 dx$$

Since the deflection in the direction of H_A is zero, partial differentiation under the integral sign provides the equation

$$\frac{\partial U}{\partial H} = \int_0^{20} (Hx - M_A)x \, dx + \int_0^5 \left(20H - M_A - 5x + \frac{x^2}{2}\right)(20)dx = 0$$

and after integration and simplification the following equation results:

$$28,000H - 1800M_A = 5000 \tag{3-9}$$

A second equation is required and may be obtained, since the rotation at A is zero, by a second application of Castigliano's theorem, resulting in

$$\frac{\partial U}{\partial M_A} = \int_0^{20} (Hx - M_A)(-1)dx + \int_0^5 \left(20H - M_A - 5x + \frac{x^2}{2}\right)(-1)dx = 0$$

or after integration and simplification the second equation is found to be

$$1800H - 150M_A = 250 \tag{3-10}$$

The solution of equations 3-9 and 3-10 will give

$$H = 0.313 \text{ kip}; \quad M_A = 2.08 \text{ ft-kips}$$

The final moment curve is shown in Fig. 18b with moment diagrams plotted on the tension side of members.

3-10. FRAMES—UNIT LOAD-UNIT COUPLE METHOD

The frame of Fig. 19a will now be solved by the use of the unit load-unit couple method expressed as $1 \cdot \varDelta$ or $1 \cdot \theta = \int \dfrac{Mm \, dx}{EI}$. Again, it is

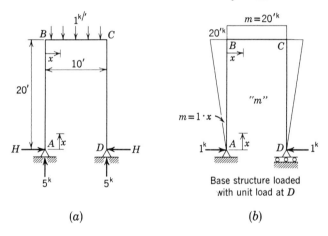

(a)   (b)

FIG. 3–19

important to observe that the structure is first unloaded and made statically determinate, as in Fig. 19b. The unit horizontal load is placed on the base structure at D and a unit reaction is also produced at A. This loading establishes the "m" moments. Now consider that the hinged reaction is restored at D to replace the rollers, fixing D against further translation and also locking in the "m" moments. With the structure so restored to its true conditions of support, the real loading is placed on the structure of Fig. 19*b* and the *M* moments are developed. Taking account of symmetry, and using the origins shown in Figs. 19*a* and 19*b*, the following is obtained.

$$1 \cdot \Delta_D = \frac{2}{EI} \int_0^{20} (Hx)(x) \, dx + \frac{1}{EI} \int_0^{10} \left(20H - 5x + \frac{x^2}{2}\right)(20) \, dx = 0$$

It should be observed that this equation is identical to the equation secured by the Castigliano method in previous article, and the same result, namely

$$H = 0.179 \text{ kip}$$

would be obtained.

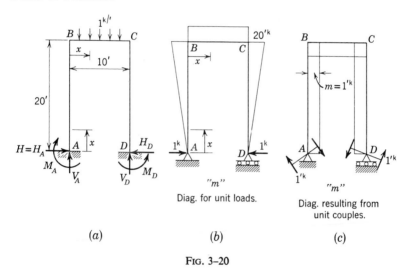

Diag. for unit loads.

Diag. resulting from unit couples.

(a)   (b)   (c)

FIG. 3–20

As a further example of the $\int \dfrac{Mm \, dx}{EI}$ method, the frame of Fig. 20*a* with the base of columns fixed ended will be analyzed. The geometrical condition of zero horizontal deflection at *D* and the fact that the tangents to the elastic curve of columns do not rotate at *A* or *D* form the basis for setting up two equations involving the redundants *H* and M_A. In Fig. 20*b* the unit loading produces the "*m*" moment used in expressing $\Delta_D \cdot 1$ which of course must be zero. Using $\int \dfrac{Mm \, dx}{EI}$, the following equation may be written employing symmetry as

$$1 \cdot \Delta_D = \int_0^{20} (Hx - M_A)(x) \, dx$$

$$+ \int_0^5 \left(20H - M_A - 5x + \frac{x^2}{2}\right)(20) \, dx = 0$$

In Fig. 20c the unit couple loading produces the "m" moment used in $\int \dfrac{Mm\,dx}{EI}$ for expressing θ_A or θ_D, which are both equal to zero. The zero rotation condition is written as

$$1 \cdot \theta_A = \int_0^{20} (Hx - M_A)(-1)\,dx$$
$$+ \int_0^5 \left(20H - M_A - 5x + \frac{x^2}{2}\right)(-1)\,dx = 0$$

Since the two foregoing equations are identically equal before integration to the two equations formulated previously in Art. 3-9 by the Castigliano approach, they need not be further resolved.

3–11. FRAMES—SOLUTION BY MOMENT-AREA METHOD

Although fundamental solutions of indeterminate systems generally provide for diversified choices for the redundants, many analysts prefer moments as redundants. Moments were used as the redundants in the three-moment equation, and frequently moment will be an expeditious choice for frames.

To illustrate the use of moments, once more refer to the already familiar frame of Fig. 21a. The redundant moments are assumed to be M_A and

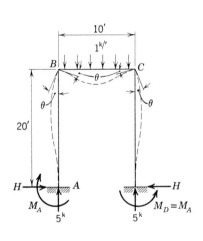

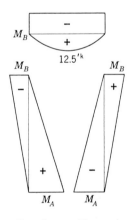

Sign of moment by usual beam convention. Columns viewed from the right.

(a)

(b)

FIG. 3–21

M_B, the moments at joint A and joint B respectively, and the moment diagram by parts, using symmetry, is drawn in Fig. 21b. The geometric relations to form equations expressing continuity of the frame are as follows: First, B has a zero deflection with respect to a vertical tangent to elastic curve at A, and second, the rotation of the tangent to BA at B must equal the rotation of the tangent to BC at B, when B is considered as a rigid joint.

By the moment-area method and the first geometric relation of deflection we may write

$$EI\Delta = \frac{M_A \cdot 20}{2} \times \frac{2}{3} \times 20 - \frac{M_B \cdot 20}{2} \times \frac{1}{3} \times 20 = 0$$

solving,
$$M_A = \frac{M_B}{2}$$

The second condition of equal rotations at B may be formulated by finding θ for the column AB and the beam BC. For the column AB

$$EI\theta = \frac{M_A \cdot 20}{2} - \frac{M_B \cdot 20}{2}$$

and on substituting $M_A = \frac{M_B}{2}$,

$$EI\theta = -5M_B$$

A separate analysis for member BC by the moment-area method provides

$$EI\theta = 5M_B - 12.5 \times 5 \times \tfrac{2}{3}$$

equating the two separate values of $EI\theta$ since they must be equal

$$-5M_B = +5M_B - \tfrac{125}{3}$$
$$10M_B = \tfrac{125}{3}; \qquad M_B = +4.17 \text{ ft-kips}$$

where the positive sign indicates that the assumed direction for M_B was correct. M_A equals one-half of M_B or 2.08 ft kips. The numerical values agree with previous solutions presented.

3–12. FRAMES—WITH SIDESWAY

Many frames deflect horizontally, as a result of unsymmetrical framing, unsymmetrical loading, or horizontal forces due to wind action. For example, it should be quite clear that the structure shown in Fig. 22 will deflect or sway to the right, owing to load P. Joint B will translate through a distance Δ and, neglecting the change in length of member BC, joint C also deflects an amount Δ. The elastic line of the frame is shown dotted with tangents at joints B and C rotating through equal clockwise angles θ. The geometric conditions of translation and rotation are thus

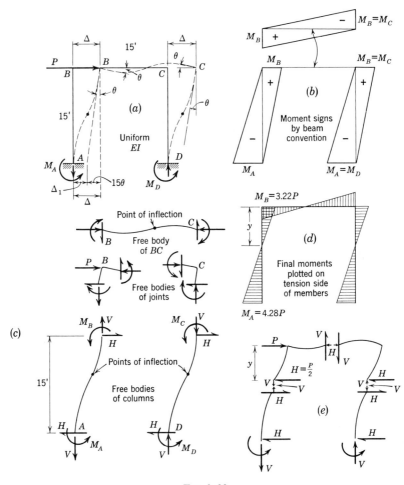

FIG. 3–22

established for defining the elastic configuration of the statically indeterminate structure.

The first solution for the moments of this structure will demonstrate the use of moments as redundants and moment-area principles will be utilized. Recognizing symmetry, the moment diagram by parts in terms of redundant moments at A, B, C, and D is drawn as shown in Fig. 22b. Attention must now be given to formulating the proper geometrical equations. The student will discover that there is some difficulty in maintaining a consistent sign convention unless he possesses an extremely clear understanding of past conventions of the moment-area method. Although it is preferable to draw the bending moment diagrams for frames so that moment ordinates are plotted on tension side of member, it is advisable from the point of view of moment-area analysis to give conventional moment signs to moments. In viewing a vertical member look at it from the right side of the drawing. The horizontal member follows beam practice. Hence, M_A for the column is termed negative since, when the column is viewed from the right, the M_A moment acting as shown in Fig. 22a produces tension on the equivalent of the top side of the member. Conversely, M_B for the column as well as its effect is termed positive.

The analysis may be started by determining θ for the beam BC. By moment-area method we obtain

$$15EI\theta = \frac{M_B \cdot 15}{2} \times \frac{2}{3} \times 15 - \frac{M_B \times 15}{2} \times \frac{1}{3} \times 15$$

or $\qquad EI\theta = 2.5 M_B$

An analysis of the elastic curve for column AB permits the calculation of the deflection of joint B from a vertical tangent at A. This displacement is given by

$$EI\Delta = -\frac{M_A \cdot 15}{2} \times \frac{2}{3} \times 15 + \frac{M_B \cdot 15}{2} \times \frac{1}{3} \times 15$$

or $\qquad EI\Delta = -75 M_A + 37.5 M_B$

θ and Δ must still be related by means of another condition which can be made clear by referring to Fig. 22a. The deflection of A from the tangent to elastic curve at B is Δ_1. Therefore Δ_1 plus 15θ equals Δ, and Δ can be restated as

$$EI\Delta = 15\theta(EI) + \frac{M_B \cdot 15}{2} \times \frac{2}{3} \times 15 - \frac{M_A \cdot 15}{2} \times \frac{1}{3} \times 15$$

$$EI\Delta = 15\theta(EI) + 75 M_B - 37.5 M_A$$

An equating of the two values of Δ, noting the necessary sign correction similar to that encountered in developing the three-moment equation, provides the equation

$$-(-75 M_A + 37.5 M_B) = 15\theta \cdot EI + 75 M_B - 37.5 M_A$$

and with $\theta = 2.5 M_B / EI$ as previously determined

$$M_A = \tfrac{4}{3} M_B \qquad\qquad (3\text{-}11)$$

This is the sole equation obtained from the elastic conditions and merely serves to relate M_A and M_B in order for consistent geometry to exist. An equation of static equilibrium may now be written to obtain the second necessary equation. Referring to Fig. 22c, note that free bodies are drawn of the two columns. On account of symmetry the moments on the separate columns are the same. The end moments M_A and M_B are couples acting in a counterclockwise sense and, neglecting the effect of the axial forces, the resultant couple, having a magnitude of $M_A + M_B$, must be held in equilibrium by another couple. The only possible way for moment equilibrium of the externally unloaded free body to be achieved is by means of the couple formed by the shear forces H, as shown. With symmetry, H must equal $P/2$ to maintain horizontal equilibrium for structure taken as a whole, hence,

$$15 \times \frac{P}{2} - (M_A + M_B) = 0$$

$$M_A + M_B = 7.5P \tag{3-12}$$

Equations 3-11 and 3-12 may now be solved giving

$$M_A = 4.28P$$
$$M_B = 3.22P$$

The results may be summarized in the form of the moment diagram of Fig. 22d. The sections at which zero bending moment occurs are termed the *inflection points*, or points of contraflexure. At such points the center of curvature of elastic curve shifts from one side of the member to the other. The position of these points reflects the structural interaction of column and beam. If the beam were less stiff the points of inflection for the columns would rise, and, conversely, the points of inflection would lower if the beam were stiffer, although in the latter case the inflection points would not fall below the half-height of the column as the beam approached infinite stiffness.

To emphasize the importance of the inflection points, Fig. 22e may be drawn, assuming that the frame is disconnected at the points of inflection. Only shear and axial forces need be shown at the inflection points since they are the equivalent of structural hinges. Thus, the frame is subdivided into structural units with each unit in static equilibrium.

A study of Fig. 22e should indicate to the student how the force P is transmitted to the columns and why, if the location of the points of inflection had been known at start of problem, the solution of moments and other internal forces could have been calculated from principles of static equilibrium alone. This may appear to be only an academic statement of no practical value until it is realized that all indeterminate structures must usually have a preliminary design before the final analysis.   Very often an approximate estimate of the points of inflection enables trial design moments to be found by methods of statics.   Since these points may well be the key to trial design or an approximate solution, they should be carefully observed in all problems.   In this way an excellent insight of structural interaction may be gradually acquired.

3–13. TRUSS SYSTEMS

A system of two-force members could be analyzed by principles of static equilibrium alone if the outer reactions for the system were determinate and if no more members than necessary for stability were employed.   When this situation is not true, the structure so formed may be either statically indeterminate internally or externally or both.   The redundants are either the reactions or stress in selected members or both in some combination.

For beams and frames, slope and deflection were the primary geometric quantities involved, but in a truss system, deflections alone normally become the relevant geometric quantity.   For beams and frames, the axial change in length of members is generally ignored, whereas in a truss system these elastic changes in length are the significant factors.

The fundamental problems of a simple statically indeterminate system may be illustrated by the three-member suspension of load P as shown in Fig. 23a.   The system is coplanar with the cross-sectional area of all members known as well as their modulus of elasticity.   To simplify the discussion L/A of each member is taken as unity.   Since three members meet at A statics alone will not suffice to determine the forces or stresses in the members.   The structure is indeterminate to the first degree, which may be verified by noting that if one of the bars is removed the stresses of the *base structure* can be calculated by methods of static equilibrium.   In this instance, owing to symmetry, member AC is removed as in Fig. 23b, thus establishing the stress S_{AC} as the redundant quantity.   The reduction of structure to statically determinate conditions enables a stress to be computed in members AB and AD equal to $+\frac{5}{8}P$.   Since these members are deformable, point A moves down Δ.   The change in length of AB and AD is

$$\Delta L = \frac{5}{8} \frac{P \cdot L}{AE}$$

and if $L/A = 1$

$$\Delta L = \frac{5P}{8E}$$

hence by geometry and the graphical approach used in constructing Williot diagrams (see Fig. 23c)

$$\varDelta = \frac{25P}{32E}$$

The finding of $\varDelta$ is but a preliminary to establishing a geometrical equation based on the observation that point A of the original structure can only move down the amount that member AC elongates.   Figure 23d

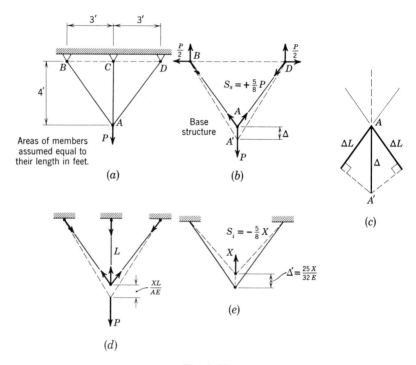

Fig. 3–23

illustrates this final condition with the deflection of A stated in terms of the elongation, XL/AE, or X/E of member AC, where X is the unknown stress in this member.   Figure 23e illustrates that the X stress developed in member AC, if acting alone as a load on statically determinate structure, will produce compressive stress in AB and CD and forces A to rise.   The amount of upward movement is $\varDelta' = 25X/32E$, which may be easily deduced from the already calculated value of $\varDelta$ by substituting X for P.

An equation may now be written consistent with required geometry.

$$\Delta - \frac{25X}{32E} = \frac{X}{E}$$

$$X = +0.438P$$

With the stress known in AC, the final stress in AB and AD may be found by combining the stresses of the statically determinate system of Fig. 23b with the stresses produced by the redundant force X. Thus

$$S_{AB} = S_{AD} = +\tfrac{5}{8}P - 0.418P \times \tfrac{5}{8}$$

$$= +0.351P$$

A further example concerning an indeterminate truss is shown in Fig. 24a, where again, for simplicity of presentation, equal L/A values of unity are assumed for all members.   There is one more member than needed for a stable statically determinate system, hence one redundant member. For solution the stress in member CD, termed X, is selected as the redundant, and the truss system is made determinate by removal of member CD as shown in Fig. 24b.   The stresses for the resulting determinate base truss are computed and the deflection of point C determined from the geometry shown in Fig. 24b.   A simple Williot diagram, Fig. 24c, suffices to find Δ equal to $41P/9E$.   In this instance the diagram was sketched and Δ computed from the relevant geometry.

The one geometrical equation needed is based on the fact that points C and D of truss in Fig. 24a can never be displaced relative to one another by more than the change in length of the member CD.   This change in length may be stated as XL/AE, or simply X/E when L/A equals unity as assumed in this problem.   But in Fig. 24b, C moved down $41P/9E$ relative to itself as well as to D.   It is apparent that the function of the force or stress X is to pull C and D together an amount Δ_x, so that $\Delta - \Delta_x = X/E$ becomes the geometric equation.

To determine Δ_x an analysis is made by the unit load method.   First the stresses in determinate truss system are determined as produced by X, Fig. 24d.   Two unit loads are applied as shown in Fig. 24e and the u stresses determined.   The use of two unit loads will be made clear if it is realized that the Δ_x to be found is the total relative movement of point C or D with respect to the other.   Virtual work principles indicate that the work done by each unit load is equal respectively to the unit load times the movement of the point to which it is attached.   Thus, two unit loads will produce virtual work equal to $1 \cdot \Delta_x$.   Hence

$$1 \cdot \Delta_x = \sum \frac{S_i u L}{AE}$$

where S_i represents the stress due to the X force.   With a summation for all members we obtain

$$1 \cdot \Delta_x = 2 \times \frac{5}{3} \times \frac{5}{3}\frac{X}{E} + 2 \times \frac{4}{3} \times \frac{4}{3}\frac{X}{E} + 1 \cdot \frac{X}{E}$$

which yields

$$\Delta_x = \frac{91X}{9E}$$

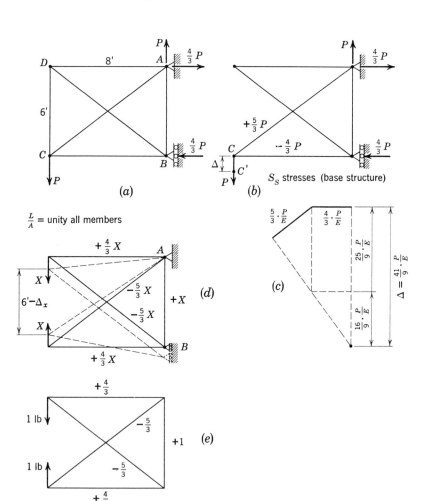

FIG. 3–24

The geometric equation may now be utilized

$$\Delta - \Delta_x = \frac{X}{E}$$

$$\frac{41P}{9E} - \frac{91X}{9E} = \frac{X}{E}$$

$$X = 0.41P$$

Once the stress in CD is known as $0.41P$, the final stress in all members may be computed by superimposing the stresses produced by the redundant member and the stresses of the statically determinate system.   The equivalent of this superimposed condition for final stress in general algebraic terms is

$$S = S_s + S_i = S_s + uX$$

where S_s = stress in a given member of base or statically determinate structure.

S_i = stress in a given member of structure due to action of the redundants, where ($S_i = uX$ in example problem).

Then $S_{AC} = +\frac{5}{3}P - \frac{5}{3}(0.41P)$

$\qquad = +0.99P$

$S_{BD} = 0 - \frac{5}{3}(0.41P) = -0.68P$

REFERENCES

1. Cross, H. and N. D. Morgan, *Continuous Frames of Reinforced Concrete*, Chapter 2, New York: John Wiley, 1932.
2. Laurson, P. G., and W. J. Cox, *Mechanics of Materials*, second edition, Chapter 11, New York: John Wiley, 1948.
3. Parcel, J. I., and R. B. B. Moorman, *Analysis of Statically Indeterminate Structures*, New York: John Wiley, 1955.
4. Popov, E. P., *Mechanics of Materials*, Chapter 12, New York: Prentice-Hall, 1952.
5. Seely, F. B., and J. D. Smith, *Advanced Mechanics of Materials*, Chapters 13, 14, 15, New York: John Wiley, 1952.
6. Wang, C. K., *Statically Indeterminate Structures*, Chapters 1, 3, 4, 5, and 6, New York: McGraw-Hill, 1953.

PROBLEMS

3–1. Determine the reaction at B by the moment-area method. EI is constant.
Ans. $V_B = 3wL/8$.

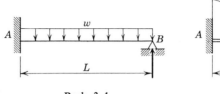

Prob. 3–1   Prob. 3–2

3–2. Determine the reaction at B by the moment-area method. EI is constant.
Ans. $V_B = kL^2/10$.

3–3. (a) Find V_A.   (b) Determine M_B by treating it as the redundant.
Ans. $V_A = 16.88$ kips; $M_B = 7.5$ ft-kips.

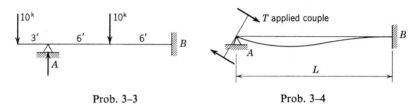

Prob. 3–3   Prob. 3–4

3–4. Determine M_B directly by the moment-area method.
Ans. $M_B = T/2$.

3–5. Determine V_B by (a) using your results from problem 3–4, (b) by treating
V_B as the redundant.   Ans. $V_B = 3T/L$.

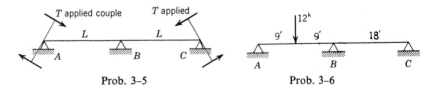

Prob. 3–5   Prob. 3–6

3–6. Solve this problem first using V_B as the redundant, and secondly by
using M_B as the redundant.   Draw the final shear and bending moment diagrams.

3–7. Determine V_B and draw the shear and bending moment diagrams.

Ans. $V_B = 7wa/4$.

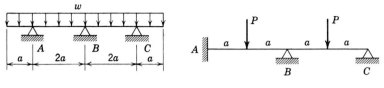

Prob. 3–7   Prob. 3–8

3–8. Determine the reactions.

3–9. Determine the fixed-end moments by the moment-area method.

Ans. $3Pa/4$.

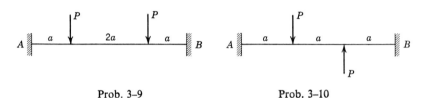

Prob. 3–9   Prob. 3–10

3–10. Determine the fixed-end moments by moment-area method and draw the shear and bending moment diagrams.   *Ans. $2Pa/9$.*

3–11. Determine the fixed-end moments by making use of the general case as described for problem of Fig. 7 in text.   *Ans. $5wL^2/96$.*

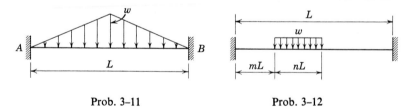

Prob. 3–11   Prob. 3–12

3–12. Same instructions as for problem 3–11.   Check your answers against known results for a fully loaded span by letting $m \rightarrow O$ and $n \rightarrow L$.

3–13. Determine the fixed-end moments by making use of the general case as described for problem of Fig. 7 in text.

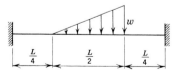

Prob. 3–13

3–14. Solve problem 3–6 by the three moment equation.

3–15. Solve problem 3–7 by the three moment equation.

3–16. Solve problem 3–8 by the three moment equation.

3–17. Solve problem 3–9 by the three moment equation.

3–18. Solve for the moment at B by the three moment equation.

Ans. $wL^2/30$.

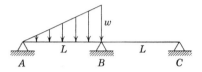

Prob. 3–18

3–19. Solve problem 3–2 by applying the Castigliano theorem.

3–20. Solve problem 3–7 by applying the Castigliano theorem.

3–21. Solve problem 3–8 by the Castigliano theorem, using M_A and V_A as the redundants.

3–22. Solve by the Castigliano theorem, using M_A as the redundant.

Ans. $M_A = Pa/2$.

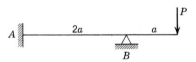

Prob. 3–22

3–23. Solve problem 3–5 by the $\int \dfrac{Mm\,dx}{EI}$ method.

3–24. Solve problem 3–6 by the $\int \dfrac{Mm\,dx}{EI}$ method.

3–25. Solve problem 3–10 by the $\int \dfrac{Mm\,dx}{EI}$ method.

3–26. Determine the horizontal reaction at A by each of the following methods: (*a*) moment-area analysis; (*b*) Castigliano's theorem; (*c*) by use of $\int \dfrac{Mm\,dx}{EI}$. Uniform EI.

Ans. $H_A = 0.05P$.

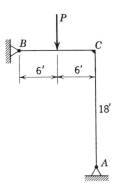

Prob. 3–26

3–27. Solve problem 3–26 directly for the moment at C by each of the three methods given in problem 3–26.

3–28. Solve for the horizontal reaction by each of the following methods: (*a*) moment-area analysis; (*b*) Castigliano's theorem; (*c*) by the $\int \dfrac{Mm\,dx}{EI}$ method. EI is constant.   *Ans. H = P/6.*

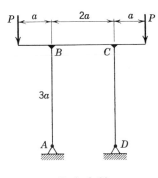

Prob. 3–28

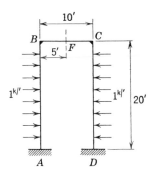

Prob. 3–29

3–29. Solve for the redundant reactions at A or D, first by the Castigliano method and then check your solution by the $\int \dfrac{Mm\,dx}{EI}$ method. Constant EI.

Ans. $H_A = 11.26$ kips; $M_A = 41.7$ ft-kips.

3–30. For the frame of problem 3–29, solve, using the redundants at the center line of member BC. Solve by a method of your choice.

3–31. Find the total reaction at A by an analysis and procedure of your choice. EI is constant.   *Ans.* 4.36 kips.

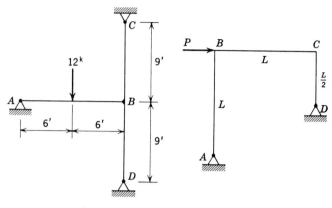

Prob. 3–31   Prob. 3–32

3–32. Solve for the reactions by a method of your choice.   Constant EI.
Ans. $H_A = 0.217P$; $V_A = 0.608P$.

3–33. Solve for the reactions by a method of your choice and draw the bending moment diagram.   Constant EI.

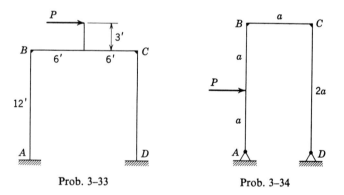

Prob. 3–33   Prob. 3–34

3–34. Solve for the reactions and draw the moment diagram.   Constant EI.

3–35. Noting that A is on rollers, make an analysis and appropriate computations and then draw the moment diagram.

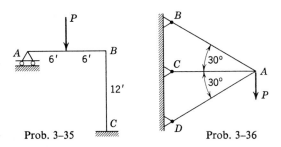

Prob. 3-35   Prob. 3-36

3-36. Determine the stress in all members.   Use $L/A = 5$ for all members.

3-37. Determine the stress in member CD.   Use $L/A = 10$ for all members.
Ans. $S_{CD} = +7.96$ kips.

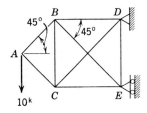

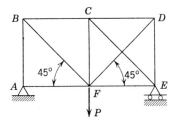

Prob. 3-37   Prob. 3-38

3-38. Determine the stress in CE.   Use $L/A = 4$ for all members.
Ans. $S_{CE} = -0.354P$.

3-39. Determine the stress in AC. $L/A = 1$ for all members.

3-40. Determine the stress in AC. $L/A = 1$ for all members.

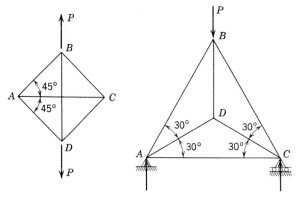

Prob. 3-39   Prob. 3-40

Moment Distribution

4

4–1. INTRODUCTION

The moment distribution method as conceived by Hardy Cross is one of the most significant contributions in several decades to the analysis of indeterminate structures. Although many classical methods preceded the concepts to be described here, and in most textbooks these classical methods are considered first, the author is of the opinion that moment distribution should be presented first. The reasons for this are quite simple and are based chiefly on a belief that the student can obtain a firmer grasp of the importance of structural geometry and the factors that affect this geometry. It may be further stated that once the method is thoroughly understood it will serve as a quick check on subsequent classical methods, a check so often needed by the student.

The method of moment distribution is better adapted to analysis of beams and frames than to analysis of arches and trusses. Thus, in the latter situations classical analysis or straight geometric approaches are usually employed, although not necessarily so. In many instances the classical methods will yield a more descriptive solution or a formulated equation as well as furnishing a basis for electronic computation. An engineer can not be properly instructed without possessing all viewpoints.

4–2. BASIC TERMS

A high premium is placed on a physical understanding of several factors. It is best to consider first these factors separately from their ultimate application.

Let a beam AB of uniform EI be supported at A and fixed against rotation at B, as in Fig. 1a. Now apply a moment M_A to beam at support A. Physically this causes the beam to deflect and the tangent to the elastic curve at A to rotate clockwise by θ_A. It should be noted that the tangent to elastic curve at B is restrained or fixed against rotation. This restraint is in the nature of an end moment M_B. Thus, it can be said that M_B is the result of applying M_A while the end B is restrained. This natural circumstance raises the next question, How is M_B related to M_A. To ascertain the answer to this question recourse must be made to deflection theory. Geometrically point A has no vertical deflection relative to the horizontal tangent to elastic curve at B. With separate moment

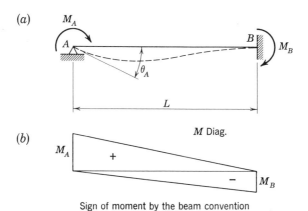

FIG. 4–1

diagrams due to M_A and M_B, and by the moment-area method, the zero deflection condition leads to

$$\frac{1}{EI}\left(\frac{M_A \cdot L}{2} \times \frac{L}{3} - \frac{M_B \cdot L}{2} \times \frac{2}{3}L\right) = 0$$

Then it may be determined from this equation that

$$M_B = \tfrac{1}{2}M_A \qquad (4\text{-}1)$$

This result is very significant and means that M_B will always be equal to one-half of M_A for condition of full restraint at B, shown in Fig. 1a for a member of uniform EI. The one-half factor is given the name "carry-over factor," and is usually denoted as C.O.F. For the time being the factor can be written as C. The term *carry over* is descriptive of one of the mathematical steps of moment distribution, as will be described subsequently.

With M_B taken as one-half of M_A, the angle θ_A of Fig. 1a may be computed by moment-area principles in terms of M_A. The rotation is

$$EI\theta_A = \frac{M_A \cdot L}{2} - \frac{M_A}{2} \cdot \frac{L}{2} = \frac{M_A L}{4}$$

which simplifies to

$$M_A = \frac{4EI}{L} \cdot \theta_A \qquad (4\text{-}2)$$

Thus it may be seen that M_A is directly proportional to θ_A for a given beam. If θ_A is considered to be one radian, $M_A = \frac{4EI}{L}$. This moment is defined as *absolute stiffness* and will be denoted as K_A. Equation 4-2 establishes the concept of relative stiffness by which different beams of unlike I and L, but of uniform cross-section, may be related by comparing their I/L values if E is also equal. The greater the value of I/L the greater the required value of M_A for rotating the tangent at A through an equal angle θ_A.

Enough basic data is now at hand to explore the first application of moment distribution. Refer to Fig. 2a, where a continuous beam is supported at A and fixed at ends B and C. The beam has a constant moment of inertia I. A clockwise couple T of 30 ft-lb is applied externally to beam at A. The question is, how much will the tangent to the elastic curve at A rotate and what will be the moments produced in the beam. Physically the tangent will rotate until the internal moments in beam at A equal the external moment of 30 ft-lb. Figure 2b indicates this physical situation.

To understand the full significance of the equilibrium condition of Fig. 2b, the student should refer to Fig. 2c. This is an isolated free body of the beam showing static moment equilibrium at point A, where M_{AC} is the counterclockwise resisting moment of beam AC and M_{AB} the counterclockwise resisting moment of beam AB. The equation of moment equilibrium at A is

$$M_{AB} + M_{AC} = 30 \qquad (4\text{-}3)$$

To find explicit values of M_{AB} or M_{AC} it is essential to find θ_A. It is known from $M = \frac{4EI}{L} \cdot \theta_A$, since A and C are fixed, that for the right-hand span

$$M_{AB} = \frac{4EI}{20} \cdot \theta_A$$

and for the left-hand span

$$M_{AC} = \frac{4EI}{10} \cdot \theta_A$$

Substitution of these values in the moment equilibrium equation 4-3 follows, resulting in the equation

$$\frac{4EI}{20} \cdot \theta_A + \frac{4EI}{10} \cdot \theta_A = 30$$

$$\theta_A = \frac{30}{4E\left(\dfrac{I}{10} + \dfrac{I}{20}\right)}$$

a substitution of θ_A back into $M = \dfrac{4EI}{L} \theta_A$ provides

$$M_{AB} = \frac{4EI}{20} \cdot \theta_A = \frac{4EI}{20} \times \frac{30}{4E\left(\dfrac{I}{10} + \dfrac{I}{20}\right)} = 10 \text{ ft-lb}$$

$$M_{AC} = \frac{4EI}{10} \cdot \theta_A = \frac{4EI}{10} \times \frac{30}{4E\left(\dfrac{I}{10} + \dfrac{I}{20}\right)} = 20 \text{ ft-lb}$$

For a complete solution of the problem the moments in the right and left spans at B and C may be determined as equal to one-half of M_{AB} and M_{BA}, or $M_{BA} = 5$ ft-lb and $M_{CA} = 10$ ft-lb. This is the practical significance of the carry-over factor. When moments were induced in the beam at A, the far ends of the respective spans were fixed. Such restraint automatically establishes the moment at B or C as one-half of the moment at the A end of these spans. The moment is said to have been "carried over" from A to B or to C.

The foregoing illustrates the basic physical concepts of moment distribution. It was demonstrated that the internal resistance to the external moment was shared by the two adjacent spans at A as depicted in Fig. 2c. In a sense the adjacent members divided or distributed the task of resisting the 30 ft-lb couple on the basis of the stiffness of each span. The ability of each span to resist moment depends on its stiffness or the moment required to rotate its tangent at A through angle θ_A.

A well-understood sign convention is needed for extensive application or discussion of the moment distribution method. The most universally adopted convention defines clockwise rotation of the tangent to the members at a joint as positive and counterclockwise rotation as negative. θ_A in Fig. 2b is positive. The sign convention for moments refers to the direction of the moments acting on a joint. If the moment acts clockwise on the joint it is taken as positive, and negative if counterclockwise. In Fig. 2b the 30 ft-lb external moment is positive, but M_{AB} and M_{AC}, the internal moments, are negative since they act counterclockwise on joint *A*. From the standpoint of rotation of joint *A*, since the applied external moment is positive, it rotates the joint or tangent to elastic curve at this joint in a positive direction.   As was also pointed out, the algebraic sum

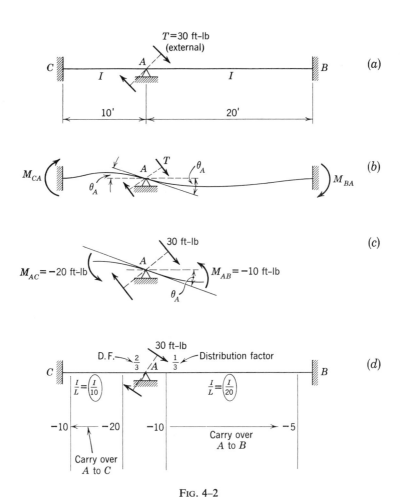

Fig. 4-2

of the moments acting on any joint must be equal to zero to satisfy moment equilibrium of a free body of that joint.

All of the basic concepts are now at hand to permit the solution of the problem of Fig. 2a to be systematized.   Figure 2d should be referred to. The encircled figure on the right and left spans is the I/L value of these spans.   A reference to $M_A = \dfrac{4EI}{L} \theta_A$ indicates that I/L of a given member made of a given material rates the relative stiffness of the member.   The sum of the relative stiffnesses of the two members meeting at A is $I/20 + I/10$.   The ratio of the relative stiffness of a single member to the sum of the relative stiffnesses of the two members, or in the case of member AC

$$\frac{I/10}{I/20 + I/10} = \frac{2}{3}$$

is called the *distribution ratio* or *factor*, which will be denoted as D.F. in the example problems.

For member AB the distribution factor at A is

$$\text{D.F.} = \frac{I/20}{I/20 + I/10} = \frac{1}{3}$$

The general formulation of the distribution factor, for a given member at a given rigid joint, is equal to the relative stiffness of that member divided by the sum of the relative stiffnesses of all members meeting at that joint.   Each member must have a constant cross section throughout its length and must also be restrained against rotation at its far end.   All members must be of the same material.   This definition of the distribution factor is mathematically stated as

$$\text{D.F.} = \frac{\dfrac{I}{L}}{\sum \dfrac{I}{L}}$$

It is a self-evident fact that the sum of all distribution factors at a given joint must equal unity.

A return to the example, and a review of the previous presentation, should make it clear that the unbalanced external moment at A must be resisted by internal moments and that these moments are equal to the appropriate distribution factor times the unbalanced moment.   Then,

$$M_{AB} = \tfrac{1}{3} \times 30 = 10 \text{ ft-lb}$$
$$M_{AC} = \tfrac{2}{3} \times 30 = 20 \text{ ft-lb}$$
$$\text{Total} \qquad\qquad \overline{30 \text{ ft-lb}}$$

The question of sign is logically taken care of by assuming M_{AB} and M_{BC} to be positive when writing the equation of moment equilibrium for joint A where the external couple is $+30$ ft-lb. The equilibrium equation is

$$M_{AB} + M_{AC} + 30 = 0$$
$$M_{AB} + M_{AC} = -30$$

This equation indicates that the induced moments must be negative or opposite in sign to the unbalanced moment to achieve moment equilibrium at joint A.

The foregoing values of M_{AB} and M_{AC} are now tabulated as shown in Fig. 2d. The remaining concept is the carrying over of moment from A to B, and from A to C. This has been indicated by arrows. The carried-over moments have the same sign as the distributed moments, as may be determined from the counterclockwise action of the moments of members on joints B and C.

Assume the spans of Fig. 3a where span AB is subjected to a load P. The final elastic curve must assume a configuration consistent with equilibrium and geometry. If the final θ_A were known, the moments in the span could be derived from deflection principles. The student should recognize that if the two-span beam actually existed and load P were applied to span AB, the elastic curve would automatically adjust itself geometrically to satisfy static equilibrium and boundary restraints. We are now dealing with a method of analysis that permits the determination of this final configuration along with the moments accompanying the configuration by a mathematical relaxation procedure.

The importance of physical visualization can not be overstressed. A mechanical mock-up of the problem is provided in Fig. 3b. To permit a full description of the processes involved, imagine that the beam is supported at points A, B, and C on screws passing through the beam, with the screws tightened securely and holding the beam clamped to a vertical board. This locked condition will be termed the fixed or locked joint condition of Stage I. To complete Stage I, place the load P on the span AB. Since all joints are locked against rotation, fixed-ended moments will be developed in beam AB at A and B, with the clamping action of the screws at A and B furnishing the external moments resisting these internal fixed-ended moments. A mechanical step must now be taken to remove the fictitious external clamping restraint at A. This is accomplished in Stage II by loosening the clamping action of the screw at A so that the external clamping moment is completely released. The beam immediately assumes its final geometric form accompanied by a rotation of the tangent to curve at A. The rotation at A has induced additional moments in the two spans consistent with θ_A and the stiffness of each span. These additional moments may be thought of as corrective moments or balancing

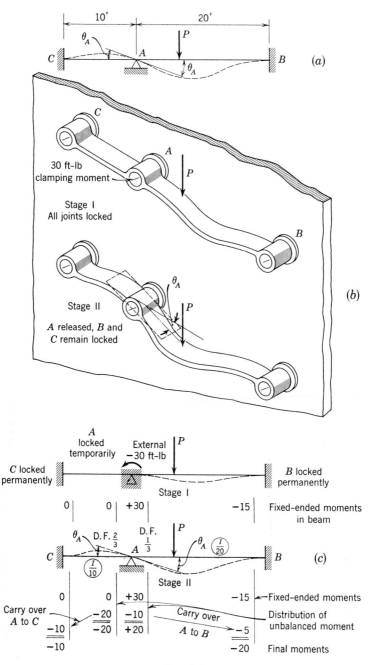

FIG. 4–3

moments superimposed on the moments of Stage I. Joints B and C remain locked, as in this example they are fixed and never rotate.

Analytically Stages I and II may be recorded as in Fig. 3c. The fixed-ended moments are first computed (note signs, moment acting on joint A is positive, moment acting on joint B is negative) with all joints fixed against rotation. This is Stage I. To accomplish Stage II in a mathematical manner, recourse must be made to two facts, first that joint A must be placed in moment equilibrium and second, to the relation of induced moments in the left- and right-hand spans to each other when tangent to joint A rotates. The latter relation was identified by the distribution factors previously determined. Stage II may now be executed. The fixed-ended moment of $+30$ ft-lb acting on A necessitated an external clamping moment of -30 ft-lb for moment equilibrium at A. When A is unclamped or unlocked to permit its rotation, the burden of providing the balancing moment is transferred from the clamping action to the members meeting at A. Thus, the right- and left-hand spans must collectively provide a balancing moment of -30 ft-lb, with each span furnishing its share in accordance with the predetermined distribution ratios. Hence, the right-hand span assumes a -10 ft-lb moment, and the left-hand span a -20 ft-lb moment. These moments are immediately responsible for producing moments at the far ends B and C. This is, in effect, the carry-over concept. Mathematically, the distributed moments and the carried-over moments are corrections to the moments of Stage I, and accordingly an algebraic summation gives the final moments shown in Stage II.

A recapitulation of the physical events is valuable at this time since the student should avoid being satisfied with a knowledge of the routine steps without thoroughly visualizing the flexure of the structure and its innate desire to seek its own equilibrium. Such a recapitulation follows.

(*a*) First, the structure was imagined to have all joints locked against rotation by an external clamping action.

(*b*) Second, with joints locked as in (*a*), the loads are placed on the structure and the fixed-ended moments generated. The external clamping moments act opposite to the fixed-ended moments, thus providing moment equilibrium at joint A. (Note: for the analytical solution these fixed-ended moments are computed values.)

(*c*) Third, the external clamping action was released at A, and the tangent to the joint rotated until static moment equilibrium was established at A. (Note: The analytical solution involved a prior knowledge of stiffness and distribution factors and the mathematical distribution of the released clamping moment. The carry-over moments are obtained after distribution.)

4–3. BEAMS

To further explain the principles of moment distribution, refer to Fig. 4a. The beam is the same as that of Figs. 2 and 3, but the loading differs.

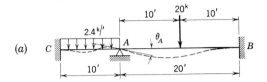

(a)

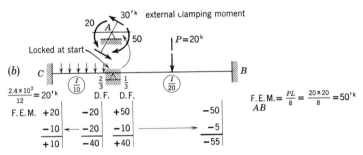

(b)

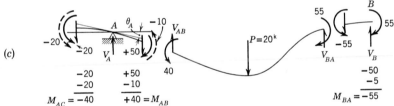

(c)

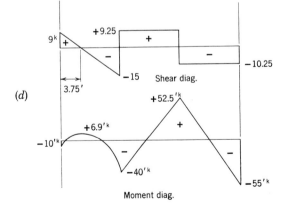

(d)

FIG. 4–4

Figure 4b combines on one sketch the relative stiffness or I/L values, the distribution factors, the fixed-ended moments due to the loading, and once more emphasizes that joint A was locked before loading with an external clamping moment of -30 ft-kips required for moment equilibrium. This is the total unbalanced moment at A. The above data basically illustrates the situation of Stage I. Particular note should again be paid to the sign of moment. Stage II, the unlocking of joint A for rotation, may now be accomplished. The unbalanced moment at joint A is the algebraic sum of the fixed-ended moments of the members meeting at A but with opposite sign. ($+50 - 20 = +30$ ft-kips is the algebraic sum of the fixed-ended moments, and the clamping moment is -30 ft-kips). Upon release of A for rotation, -30 ft-kips must be furnished by the separate beams instead of by the external clamping action. The distribution factors determine the resisting moment each span contributes as the tangent to joint A rotates through θ_A, and these basic factors may be computed following the procedure explained in Figs. 2 and 3. After the balancing of moments at A the carry-over moments are calculated and the final moments determined as the beam is now in its final configuration. It is again to be noted that the final moments at A provide moment equilibrium. This entire process of moment distribution is often termed the balancing or relaxation of the unbalanced fixed-ended moments.

It is further instructive to draw the free bodies of span AB as well as joints A and B, showing the separate moment components representing the fixed-ended moments and the moments derived by moment distribution which lead to the final moment values. This is illustrated in Fig. 4c.

The end shears in the beam are shown and the derived shear and bending moment diagrams are drawn in Fig. 4d. The final moment diagram uses the beam convention for sign of moment. This necessitates a translation of the signs of the moment distribution convention wherein the sign of moment represented the direction of moment on the joint into the beam convention signs.

The beam problem of Fig. 5a differs from that of Fig. 4a, in that the beam is hinged at C instead of being fixed against rotation. The tangent to beam at C will rotate through an angle θ_C. This new situation involves a temporary locking of all joints against rotation including joint C, so that when loads are applied to the spans, the fixed-ended moments will be generated at all supports equal to those shown in Fig. 4b.

To recognize the hinged-end condition at C and the fact that the external moment at C as well as the internal moment in the beam must accordingly be equal to zero, joint C must be unlocked or released. This means that the moments at the joint must be balanced to zero and when joint C is unlocked, a moment of -20 ft-kips must be given or distributed

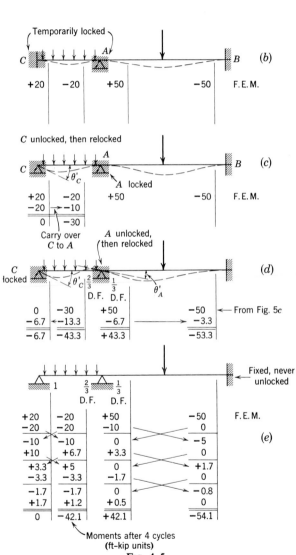

Moments after 4 cycles
(ft-kip units)

FIG. 4-5

to the beam.   The moments at C then total zero and the tangent at C has rotated counterclockwise through an angle θ'_C.   *It should now be noted that C will immediately be locked again against further rotation.*   It must also be recalled that joint A was locked at the start and, up to this point in the solution, has never been unlocked.   Consequently at the A end of beam CA an additional moment, the carry-over moment, appeared when joint C was first released.   This carry-over moment is $\frac{1}{2}(-20) = -10$ ft-kips. All of the preceding operations have been recorded in Fig. 5c.

The next step is shown in Fig. 5d.   The starting moments shown in Fig. 5d are the final moments after the step shown in Fig. 5c.   The joint A is now unlocked and the unbalanced moment of $+20$ ft-kips distributed to spans AB and AC.   It should also be noted that joint A is immediately relocked.   Since C was relocked in a previous step a moment is generated at C equal to the carry-over moment.   A carry-over moment also appears at B, a truly fixed end.   Also note that tangent to A has rotated through an angle θ'_A clockwise, owing to $+20$ ft-kips of unbalanced moment. Moments are summarized as shown.   It is significant to note that the -6.7 ft-kips of moment appearing at C does not comply with the condition that $M_{CA} = 0$.   This result implies that the unlocking-relocking procedures and resultant distributions and carry-over operations of Figs. 5c and 5d must be repeated as often as necessary until no unbalanced moment appears at C.   However, if the student understands the physical happenings as explained, a more expeditious procedure may be used, consisting of orderly cycles of distribution and carry-over.   The new procedure is represented in Fig. 5e.

First, fixed-ended moments are recorded in Fig. 5e.   Each joint in order is unlocked for distribution of unbalanced moment *and then relocked.* The distributed moments of proper sign for moment equilibrium of joints are written below the fixed-ended moments.   A line drawn below the distributed values signifies the balancing of the joint.   The next operation is to carry over all distributed moments to the far ends of the respective members.   The carry-over moments are balanced by the external clamping action, and again internal moment equilibrium at each joint is achieved by repeating the unlocking of joints and making the required distributions of the unbalanced moment.   The solution is then continued by these repetitive cycles until the structure has all joints in moment balance or until the carry-over moments become too small to be significant. It should be noted that B is a true fixed end and is never unlocked.   The solution shown has four complete cycles.

It should be reiterated that moment distribution is a simple method of permitting the structure to be relaxed gradually to its true final configuration shown in Fig. 5a, after starting with an assumed condition of all joints fixed against rotation. The balancing of joints has also enabled the computer to visualize at each step the rotation of each joint.

4-4. MODIFIED STIFFNESS—HINGED-ENDED MEMBER

When a member has its far end hinged instead of fixed, less moment is required to rotate the tangent through a given angle. In Fig. 6, a moment M_A, applied at A when C is hinged, produced a rotation θ_A. By the conjugate-beam method

$$EI\theta_A = \frac{M_A \cdot L}{2} \times \frac{2}{3}L \div L$$

$$EI\theta_A = \frac{M_A L}{3}$$

or $$M_A = \frac{3EI}{L} \cdot \theta_A \qquad (4\text{-}4)$$

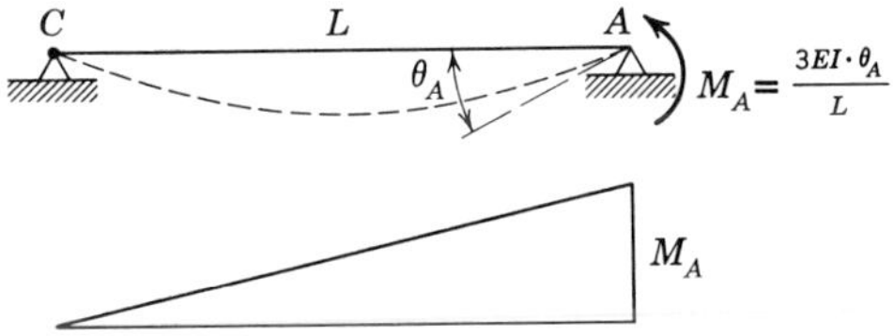

FIG. 4–6

This means that M_A is three-fourths of the moment required to rotate the tangent through an equal angle when C is fixed by comparison with equation 4-2. That is, if C were fixed against rotation,

$$M'_A = \frac{4EI}{L} \cdot \theta_A$$

hence the ratio of end moments for the two types of end restraint for equal rotation of the tangent is

$$\frac{M_A}{M'_A} = \frac{(3EI/L) \cdot \theta_A}{(4EI/L) \cdot \theta_A} = \frac{3}{4}$$

The theoretical significance of this ratio is that it permits the establishment of distribution factors consistent with true end conditions. This can best be illustrated by the problem of Fig. 7a, which is the same problem as in Fig. 5a. The starting point of the solution follows the familiar procedure of fixing or externally clamping all joints against rotation and calculating fixed-ended moments (F.E.M.'s). These are recorded in Fig. 7b. Since C is a hinged end, as a first step unlock C and leave it unlocked, at the same time balancing the joint by a − 20 ft-kip moment to provide a total moment at C equal to zero. This procedure

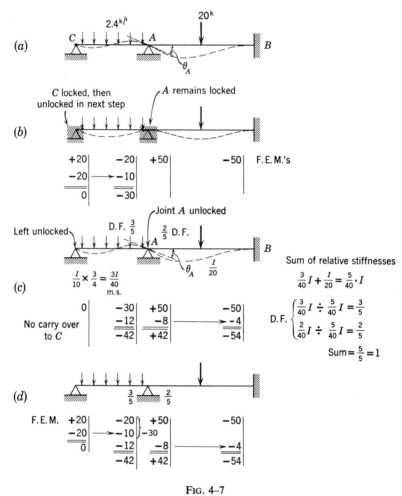

FIG. 4-7

reduces C to its true support condition. Since A was fixed when C rotated, a − 10 ft-kip moment is carried over to A. In Fig. 7b, the joint A is now unbalanced by a + 20 ft-kip moment. To explain the balancing of moments at joint A, refer to Fig. 7c. The first moments listed are the moments from Fig. 7b. It remains for joint A to be unlocked to permit rotation and distribution of the unbalanced moment.   It must be recalled that C was left free to rotate, hence the stiffness factor for span AC must be modified.   A direct multiplication of the I/L value by the $\frac{3}{4}$ factor will suffice to rate this new modified stiffness factor relative to the unmodified stiffness of the adjacent span AB.   Calculations for distribution factors are shown opposite Fig. 7c.

The distribution follows at A and there is a carry-over moment to B but not to C since C is free to rotate at this stage.   For this problem all unbalanced moments have been distributed and carried over, and no further cycles of distribution are necessary.

The problem can now be done, following standard procedures, in Fig. 7d by combining the concepts of Figs. 7b and 7c.   The results of this solution should be compared with those of Fig. 5e, which took 4 cycles of distribution.   The results of Fig. 7d are exact, whereas the results of Fig. 5e are nearly exact but can only approach the true values as the number of cycles is increased.   The shorter analysis of Fig. 7d is to be used when not confusing, but the longer version of Fig. 5e is often termed the most general approach.

Example

Figure 8a represents a beam similar to Fig. 7a, except that a loaded cantilever arm is added at left.   The relative stiffness factors and the distribution factors have been determined in Fig. 8a.   Three new considerations require discussion.   First, the relative stiffness of the cantilever span AB equals zero.   This may be proved by taking out AB as a free body and noting that if there were no load on the cantilever the moment at B would be zero, but that if a load is on the cantilever AB the bending moment at B will be statically determinable and independent of loads on adjacent spans.   Insofar as moment distribution is concerned, the span AB must be treated as having zero stiffness, as statics rules the moment at B and not joint rotation.   Therefore the distribution factor at A is zero for span AB and the factor for span BC is one.   The second new consideration requiring clarification is the modification of relative stiffness for span BC. The analysis takes into account that joint B, after its initial unlocking, will be left unlocked.   The moment M_{BC} must be equal to $+30$ ft-kips, owing to static considerations, and will be unchanged by subsequent moment distributions in the event that B remains free to rotate.   The third new consideration deals with distribution factors at the fixed end D.   Since it is actually fixed by the given conditions, D will never be unlocked for moment distribution, but for purposes of this explanation assume that it is. Then the relative stiffness of the beam DC would be $I/20$.   Treating the rigid support as a finite member would reveal its infinite resistance to

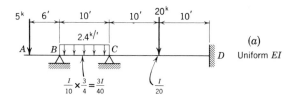

(a) Uniform EI

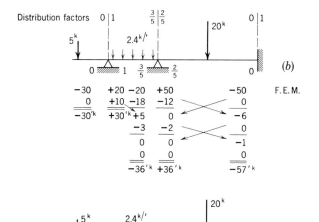

Distribution factors

(b)

F.E.M.

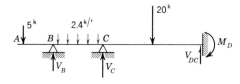

(c)

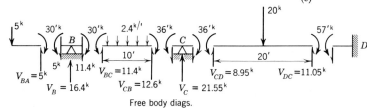

Free body diags.

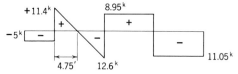

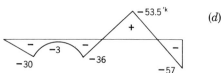

(d)

FIG. 4–8

rotation. Consequently, on a routine basis the ratio of the stiffness of the beam to the sum of the stiffness of the beam and the joint at joint D would be zero. Hence, the distribution factor for the beam is zero; for the rigid support it is one. Thus, since no changes would take place in beam moments, no distribution need be made at D.

To follow normal procedures, lock all joints against rotation and compute the fixed-ended moments. These are recorded in Fig. 8b. The moment distribution was started by unlocking B first and leaving it unlocked. Distribution leaves the cantilever moment unchanged and the moment unbalance must be assumed by member BC for moment equilibrium of joint B. From this point the solution follows in cycles as previously demonstrated.

For completeness the shear and moment diagrams for this beam are developed in Figs. 8c and 8d. A systematic removal of each member from the structure as a free body is shown with moment and shear forces. Principles of static equilibrium permit the shears to be computed and then the vertical reactions. The routine construction of the shear diagram and moment diagram follows Fig. 8d.

4–5. SIMPLE RIGID FRAMES

Simple frames will be defined as those where no sidesway or lateral deflection occurs. In that event every joint of the frame that is free to rotate will do so without translation. To comply fully with the requirement of zero translation it may be assumed further that the change in length of members on account of their axial load is negligible.

Figure 9a represents a one-story frame loaded as shown. For purposes of this problem assume that EI is constant for both beam and columns. The relative I/L for each member is computed and the distribution ratios determined.   To prevent B and C from rotating, consider them temporarily locked against rotation and then place the uniform loading on beam.   This establishes the F.E.M.'s of $\dfrac{wL^2}{12} = \dfrac{1 \times 10^2}{12} = 8.33$ ft-kips at ends of beam. No F.E.M.'s are generated in columns since columns are free of any transverse load.

The solution by moment distribution is carried out in Fig. 9b. The relaxing of structure towards its final configuration and final moments proceeds by the familiar cycles of distribution and carry-over. Figure 9c and 9d are free body diagrams. The moment diagram has been plotted with ordinates on tension side of members. Compare final results with the solutions in Chapter 3 for the same frame.

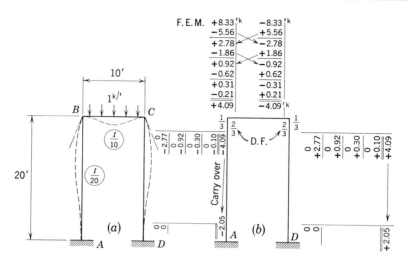

F.E.M. +8.33'ᵏ   −8.33'ᵏ
        −5.56   +5.56
        +2.78   −2.78
        −1.86   +1.86
        +0.92   −0.92
        −0.62   +0.62
        +0.31   −0.31
        −0.21   +0.21
        +4.09   −4.09'ᵏ

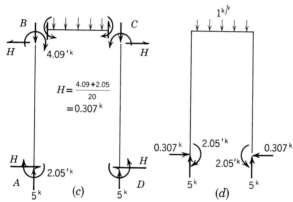

$$H = \frac{4.09 + 2.05}{20} = 0.307^k$$

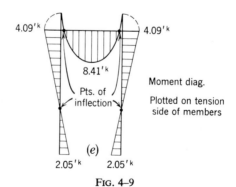

Moment diag.

Plotted on tension
side of members

FIG. 4–9

Example

Heretofore only two members have met at a joint, but in general any number of members may meet at a joint in a rigid frame. The same basic procedures of moment distribution apply. Figure 10a represents a situation where three members meet at A. The relative moments of inertia are indicated from which the relative I/L values are computed. The sum of the I/L values for members meeting at A is $3I/10$, hence the distribution factor to each member at A is one-third. Joint A is held in

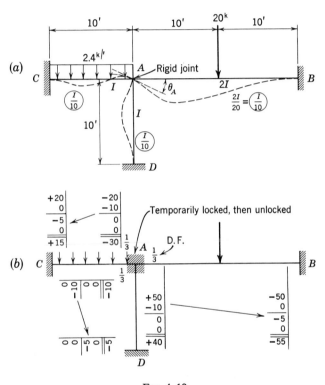

Fig. 4–10

equilibrium by an external clamping moment of -30 ft-kips. If A is unlocked the tangent to the elastic curve at A will rotate an amount θ_A until joint A is in moment equilibrium. The joint is balanced by providing for -30 ft-kips of internal moment in the three members meeting at A. The routine of distribution and carry over is then performed in Fig. 10b. It is to be noted that D is a fixed end. If D had been hinged the I/L of $I/10$ would be modified by a $\frac{3}{4}$ factor to rate its reduced stiffness and new distribution factors computed. Then no moment would be carried over to D from A.

4–6. TRANSLATION OF JOINTS

Until now, in discussing moment distribution the joints have been restricted against translation or deflection and rotation only has been permitted. Although this situation is ideal from the standpoint of simple analysis, it does not represent the general problem. A structure may be forced to deflect by lateral loadings or the joints may translate, on account of unsymmetrical framing or loading.

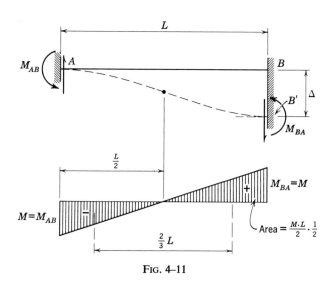

Fig. 4–11

The translation effects can first be considered by referring to Fig. 11. The beam AB is fixed against rotation at A and B. The end B is deflected downward from B to B' an amount Δ while both ends are restrained against rotation. This restraint induces end moments M which are a function of Δ, L, and EI. Moment-area relations suffice to determine theoretical value of M. Thus we obtain

$$EI\Delta = \frac{ML}{2} \times \frac{1}{2} \times \frac{2}{3}L = \frac{ML^2}{6}$$

or

$$M = \frac{6EI\Delta}{L^2} \tag{4-5}$$

M will be termed a fixed-ended moment caused by the translation Δ. This fixed-ended moment will now be introduced into the moment distribution procedures by considering the beam of Fig. 12a. Although the problem can be expeditiously solved by other methods, it is desired to determine the fixed-ended moments acting on beam of Fig. 12a by moment-distribution procedures through an introduction of deflection or translation concepts. The point C under the load will deflect an unknown amount Δ_C. To view this deflection as the controlling element in the following solution, point C will be temporarily considered as a real joint subdividing the total span into individual spans of 5 and 10 ft as shown. The 10-kip load, for the present, is considered to be removed.

The next step in analysis is to recognize that C can temporarily be termed a joint and that it also deflects downward. Figure 12b indicates C being translated downward an arbitrary amount Δ while rotation of C is prevented. The point C is pinned in this translated position to permit subsequent rotation and is held temporarily in this fictitious condition by an external clamping couple and an external downward force. The translation has induced fixed-ended moments in the beam segments AC and CB which may be expressed respectively as M_1 and M_2. These moments are related to the deflection Δ as follows, by applying equation 4-5.

$$M_1 = \frac{6EI\Delta}{5^2}$$

$$M_2 = \frac{6EI\Delta}{10^2}$$

or, in terms of a ratio

$$\frac{M_1}{M_2} = \frac{10^2}{5^2} = 4$$

The ratio of M_1/M_2 is the important relation and is true for any given Δ. Hence, if M_1 is established arbitrarily as 100 ft-kips, M_2 should be taken as 25 ft-kips. If one were fortunate enough to know the real and final Δ_C in advance, it could be used to determine the initial fixed-ended moments. The moment distribution sign convention must be followed, hence for the left beam segment the 100 ft-kip fixed-ended moments are positive since they act on joints A and C in a clockwise direction. The end moments in the beam segment CB are negative.

The solution is continued by unlocking joint C, permitting the tangent to rotate clockwise through an angle θ_C until moment equilibrium is reached at joint C. This is accomplished by the usual moment distribution procedures. The calculation of relative stiffness is shown and the distribution factors are indicated at C.   One cycle of distribution suffices, as shown in Fig. 12c, to determine the moments.   It must be understood that these moments are still arbitrary, but are consistently proportional to

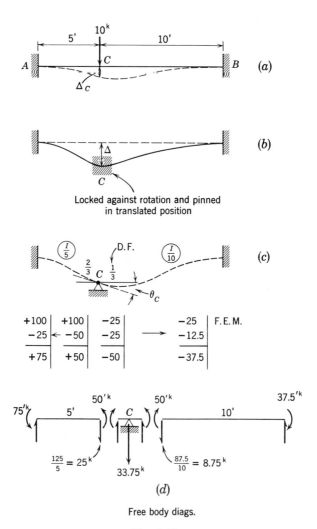

(a)

(b)

Locked against rotation and pinned
in translated position

(c)

+100	+100	−25		−25	F. E. M.
−25	← −50	−25		−12.5	
+75	+50	−50	→	−37.5	

(d)

Free body diags.

FIG. 4–12

the actual moments of Fig. 12a.   To determine final moments it is necessary to calculate the vertical force at C acting through the pin, Fig. 12c.   Free bodies of the beam segments are drawn in Fig. 12d and the vertical end shears calculated.   Vertical equilibrium principles applied to joint C

indicate that the required downward force is 33.75 kips. The final moments of the original problem shown in Fig. 12a may be determined by correcting the moments of Fig. 12c. The correction factor, since the real load at C is 10 kips instead of 33.75 kips, is $\frac{10}{33.75} = 0.296$.

Final moments using the derived correction factor are

$$M_{AB} = 0.296 \times 75 \quad = 22.2 \text{ ft-kips}$$
$$M_{BA} = 0.296 \times 37.5 = 11.1 \text{ ft-kips}$$

Although easier methods exist to determine the above moments, the example indicates the basic analysis to be introduced later into more complex problems.

4–7. FRAMES WITH SIDESWAY

The horizontal deflection of a frame is spoken of as sidesway. In Fig. 13a, sidesway is produced by the horizontal load P. The elastic curve indicates that joints B and C translate by Δ while the tangents to the elastic curve at B and C rotate through equal angles of θ. If Δ and θ were known, moments could be determined immediately. The value of the moment distribution procedure lies in the fact that these geometrical requirements are considered together while satisfying static equilibrium.

To emphasize the geometric aspects of the method, consider that the frame has holes drilled at the joints B and C, with screws or pins inserted, to permit the pinning as well as the locking of these joints to a vertical backboard. See Fig. 13b. The frame has been translated Δ' and joints B and C have been fixed against rotation. The frame is held in this position by clamping and pinning to the vertical backboard. Thus, fixed-ended moments are produced in each column proportional to Δ', whereas no moments are produced in the beam BC. The screws furnish the physical force and external clamping moment necessary to hold the structure in this position.

The solution may now proceed by assuming, for the arbitrary deflection of Δ', that fixed-ended moments of $+10P$ have been generated in the columns. Moment distribution may now be performed by unlocking joints B and C to permit rotation about the pins. This follows the familiar procedure of balancing the unbalanced moments at B and C. The initial moments at A and D are modified by the carry-over moments. The distribution is shown in Fig. 13c. It must be recognized once again that the structure by this procedure is permitted to seek its true configuration in conformity with the arbitrary translation assumed and hence achieves moment equilibrium at each joint.

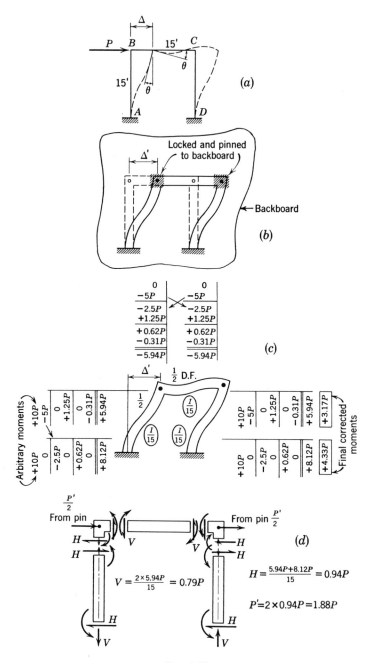

FIG. 4–13

Figure 13*d* represents the free bodies of all members and the joints. Particular attention should be given to the forces exerted by the pins on the structure. These forces are real, but their mode of application is unreal, as no such pins exist in actual structure. For a full physical understanding, now consider that a force P' is added at B to replace the two horizontal pin forces and withdraw the pins. The structure will remain in its distorted configuration with moments unchanged. P' of course is not equal to the actual P applied in 13*a* since Δ' was arbitrarily assumed, but the moments in Fig. 13*c* are proportional to the actual moments.

The ratio of P/P' may be thought of as a correction factor, and in this case its value is $P/1.88P$ or 0.531.

The final moments are

$$M_{AB} = M_{DC} = 0.531 \times 8.12P = 4.33P$$
$$M_{BA} = M_{CD} = 0.531 \times 5.94P = 3.17P$$

It is well for the student to restudy the physical aspects of this problem. He should pay particular attention to the latitude provided in the assuming of initial fixed-ended moments. It is important to note that the joints of the structure may be physically locked against rotation while translation is induced and that the pins, although fictitious, hold the structure in the translated position while rotation takes place.

Example

A further case of sidesway is illustrated in Fig. 14*a*. The columns are of unequal length but are assumed to have the same moment of inertia. Again an arbitrary deflection of Δ' must be assumed for joints B and C while B and C are locked against rotation, and fixed-ended moments assigned to the columns. These must be assigned consistent with the properties of the columns. Since this is likely to be a general problem, refer to Fig. 14*b*, where two columns of unlike length and moment of inertia are deflected an amount Δ' while both ends are restrained against rotation. From equation 4-5,

$$M_1 = \frac{6EI_1\Delta'}{L_1^2}$$

and

$$M_2 = \frac{6EI_2\Delta'}{L_2^2}$$

or, in terms of a ratio,

$$\frac{M_1}{M_2} = \frac{I_1 \cdot L_2^2}{I_2 \cdot L_1^2} \tag{4-6}$$

For the current problem, since $I_1 = I_2$

$$\frac{M_1}{M_2} = \frac{L_2^2}{L_1^2} = \frac{20^2}{10^2} = 4$$

$$M_2 = \tfrac{1}{4}M_1$$

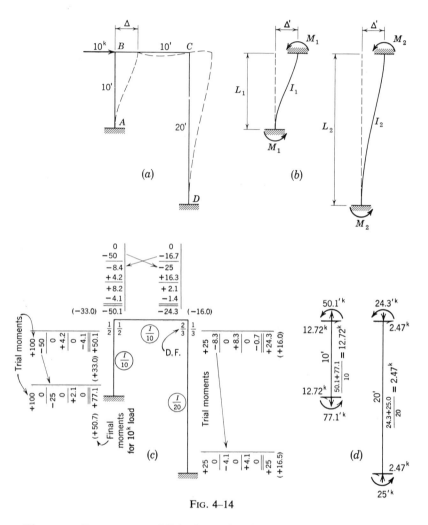

FIG. 4–14

The preceding steps establish the ratio that must exist between fixed-ended column moments. If M_1 is assumed to be 100 ft-kips, M_2 is 25 ft-kips. These moments become the starting moments and are shown in Fig. 14c. The moment distribution then follows. It should be particularly noted that the final moments in the right-hand column approach the value of the starting moments. The only physical fact that can justify this result is that joint C has no rotation in the final configuration, although joint C was unlocked for distribution and is free to rotate during the cycles of distribution. The importance of moment distribution as a means of interpreting the physical results is thus aptly illustrated.

The remainder of the problem is routine and is shown in Fig. 14d. Free bodies reveal that the total horizontal shear, the sum of the horizontal reactions at A and D, is 15.19 kips. For static equilibrium this also represents the lateral load at B directed to the right, to produce the final moments of Fig. 14c; however, the true moments due to the 10-kip load are computed by correcting these moments by the ratio of 10 to 15.19 kips. The corrected moments are given in parentheses.

4–8. SIDESWAY—UNSYMMETRICALLY LOADED FRAME

The frame of Fig. 15a will sway horizontally, as a result of the un-symmetrical vertical loading. The visualization of this deflection is important, and though intuition may suggest that the frame deflects toward the left it will be shown that frame deflects to the right instead. As a matter of fact, this very uncertainty establishes the first step of the solution, as it is realized that sidesway could be prevented if a temporary external support is provided, as shown in Fig. 15b.

If the moments are determined for the condition of Fig. 15b, the moments are due to two forces, namely the actual load P and the force F_1 exerted through the temporary support. F_1 is a real force, owing to the action of temporary support, and as it pushes on the structure to the left it prevents sidesway to the right. Since the moments of Fig. 15b include the effect of F_1, the final moments due to P alone can be determined by subtracting the moments due to F_1. This is a superposition point of view and the intent is shown by Fig. 15c. Note that F_1 in Fig. 15c is directed opposite to F_1 in Fig. 15b, which is as it should be to create moments of opposite sign to those produced by the temporary support force. Super-position of Fig. 15c on top of Fig. 15b will result in cancelation of the F_1 forces and provide the moments of Fig. 15d. These moments are the final moments for structure. A restatement of this form of solution would be: first, prevent sidesway by the temporary support of Fig. 15b; second, allow the structure to sway, as in Fig. 15c, under the action of a force opposite to force at temporary support; third, combine the effects of Figs. 15b and 15c, to provide a solution for the structure free to sway.

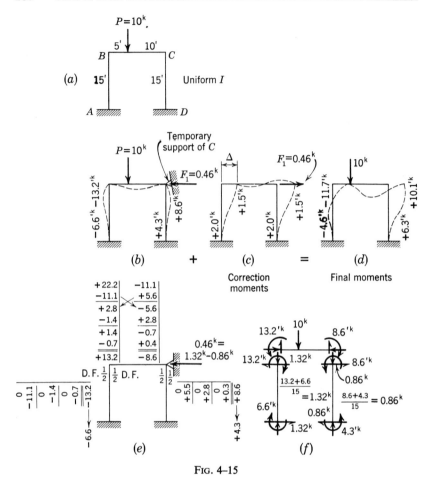

FIG. 4-15

The detailed numerical solution must start with all joints fixed against rotation and with translation prevented. The fixed-ended moments for beam are computed or taken from the previous example.

$$M_{BC}^F = \frac{Pab^2}{L^2} = \frac{10 \times 5 \times 10^2}{15^2} = 22.2 \text{ ft-kips}$$

$$M_{CB}^F = \frac{Pba^2}{L^2} = \frac{10 \times 10 \times 5^2}{15^2} = 11.1 \text{ ft-kips}$$

Figure 15e represents the moment distribution, and in Fig. 15f the horizontal shears in columns are computed. To comply with horizontal static equilibrium, the temporary support force F_1 equals 0.46 kip to the left. Since this external force pushes on frame, it is evident that it

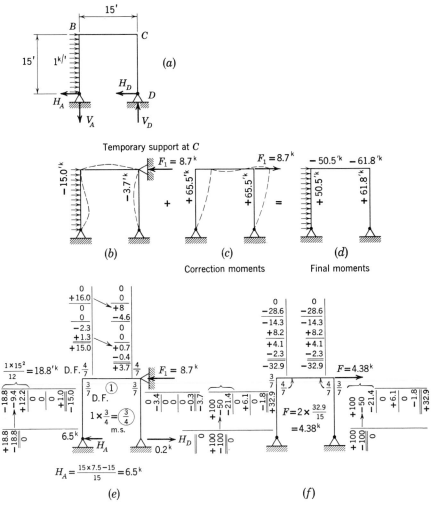

FIG. 4–16

prevents translation of the frame to the right.   The correction moments of Fig. 15c are derived from the results of Fig. 13c by determining the moments for a horizontal force of $P = 0.46$ kip.   Final moments are summarized in Fig. 15d.   A check on the horizontal shears in columns will disclose that H_A equals H_D, and since they act in opposite directions the structure is in horizontal static equilibrium.

Example

As a further example consider the frame in Fig. 16a under action of a uniform load acting on the left-hand side of a structure. As can easily be noted, the structure will deflect to the right while all joints rotate. The bottoms of the columns are pinned.

The solution is similar to the previously discussed problem and the steps in the solution are shown in Figs. 16b, c, and d. First, a temporary support prevents sidesway and exerts a force of F_1 on the frame. Second, the moment effect of this force is removed by finding the moments due to an equal but oppositely directed force, Fig. 16c. Final moments are obtained by superposition and are given in Fig. 16d.

After the analysis comes the detailed moment distributions. These are shown in Figs. 16e and 16f. Figure 16e determines the moments for conditions of Fig. 16b. Figure 16f represents a distribution of assumed moments for an arbitrary horizontal deflection of points B and C. It must be recalled that all joints are fixed against rotation during this arbitrary translation, hence, fixed-ended moments are generated at A and D. The moment distribution takes the true hinged condition into account. The author finds it expeditious to consistently assume fixed-ended conditions as a starting condition, although many others prefer to take account of the actual hinge condition and modify starting moments. Either basis is acceptable. The final moments of Fig. 16f form the basis of determining the horizontal force necessary for establishing these moments. Here the force is 4.38 kips and the correction factor to modify the moments of Fig. 16f to determine the moments of Fig. 16c is $F_1/4.38 = 1.99$.

4–9. TWO-STORY FRAMES—SIDESWAY

One of the most common horizontal forces acting on buildings is caused by the wind. In multistory buildings with many lines of columns, the stress analysis is complex. Figure 17a represents a two-story wind bent with two lines of columns where, for purposes of illustration, the moment of inertia of columns and beams is assumed to be equal. The wind loads are arbitrarily taken as 10 kips each at the roof and second floor level.

The structure deflects about as shown in Fig. 17a, with B deflecting Δ_2 and C a distance Δ_1. This structure may be said to have two degrees of freedom with respect to translation, whereas those dealt with previously had only one degree of translational freedom. For a given loading Δ_1 and Δ_2 must adjust themselves in amount and relativeness to one another, so that the final configuration of structure will be one that also provides for static equilibrium of structure. If final values of Δ_1 and Δ_2 were known in advance the structure could be deflected into this position while all joints were fixed against rotation. A relaxation of the joints by means of the

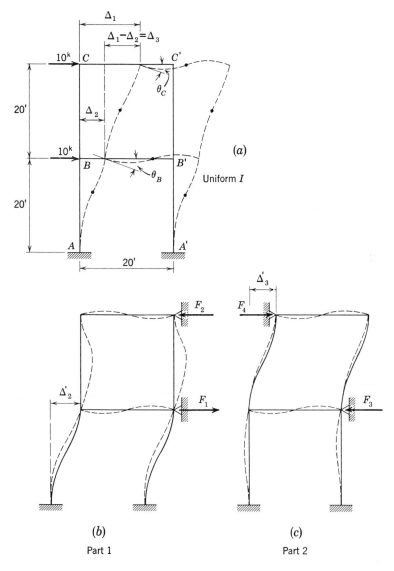

FIG. 4–17

moment distribution method would immediately provide all final moments and also satisfy the loading requirement of 10 kips at roof and second floor level. This cannot be accomplished since it would only be through fortuitous circumstances that the final deflections were known in advance. The only value in recognizing that it could be done under these ideal conditions is the insight provided by fully visualizing the deflections of structure.

The principle of superposition may be applied in reaching a final solution if the elements or components to be superimposed each represent a true geometric condition.   To achieve this result here the problem may be viewed in two parts.   Part one, Fig. 17b, represents a translation of structure to introduce the deflection of the second floor while making the relative deflection of the roof with respect to the second floor equal to zero.   This initial translation is arbitrary and is indicated as Δ'_2.

While translation is taking place all joints are fixed against rotation. The solid outlines indicate that under initial conditions fixed-ended moments are developed in the lower story only.   While held in this translated position by temporary external supports, the joints may be permitted to rotate, providing the dotted configuration of elastic curve shown in Fig. 17b.   The moments in columns in this relaxed condition will determine column shears and forces F_1 and F_2.

Part two, Fig. 17c, represents translation of the structure to introduce the relative deflection of the roof level with respect to the second floor level. Δ'_3 is an arbitrary choice for this deflection.   Note that the second floor is held against translation and all joints are fixed against rotation.   A relaxation of structure by rotation of the joints permits the structure to adjust to the dotted configuration.   Forces F_3 and F_4 are determined after calculation of shears in columns.

With the separate trial solutions, parts 1 and 2, a final solution can be formulated.   The forces on given structures were 10 kips to the right at roof and second floor level, but for the solutions of part 1 and part 2, forces different than these are developed through the temporary external supports.   The basic problem is to determine how F_1, F_2, F_3, and F_4, should be modified so that the actual loading is satisfied.   If the forces in Fig. 17b are multiplied by a factor f_1 and the forces of Fig. 17c multiplied by a factor of f_2, two separate force equations may be formulated to relate the forces F_1, F_2, F_3, and F_4, to the actual loading.

At the second floor level $f_1 \cdot F_1 - f_2 \cdot F_3 = 10$ kips

and at the roof level $-f_1 \cdot F_2 + f_2 \cdot F_4 = 10$ kips

A solution of these two simultaneous equations gives the values of f_1 and f_2, representing proportional parts of the trial solutions.   The moments of part 1 and part 2 are superimposable, after they are modified by the above factors, to provide the final solution.

Figure 18 represents the solution of the problem of Fig. 17 by moment distribution.   All work is shown in the figure.   A large amount of free

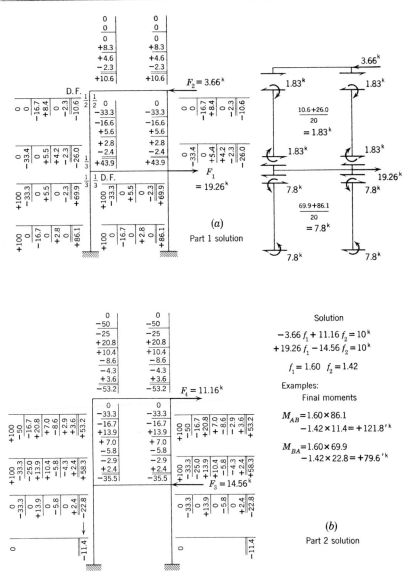

FIG. 4–18

body work is encountered in establishing the outer reaction forces at the temporary supports. Once again the physical events should be thoroughly understood. The distortions are real, the forces are finite, and the arbitrariness of the starting conditions is eventually resolved by correction factors to satisfy actual conditions.   Figure 18 indicates two examples of computation for final moments.

4–10. SPLIT-LEVEL FRAME SIDESWAY

A frequent case of structural framing wherein the roof is at separate levels is shown in Fig. 19.   Although horizontal loads are shown acting

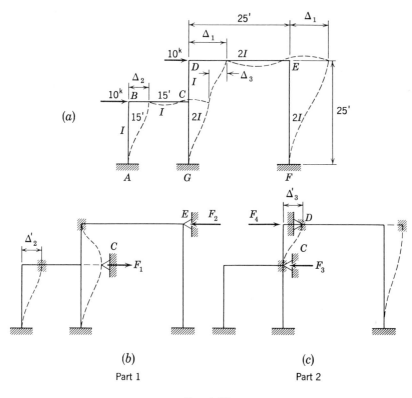

(a)

(b)   (c)

Part 1   Part 2

FIG. 4–19

on structure to produce sidesway, it must be realized that, owing to unsymmetrical framing, vertical loads on roof would also cause structure to sway.

The analysis of the frame of Fig. 19 indicates that a solution bringing in two degrees of freedom relative to horizontal deflection is required.   The lateral deflection of the two roof levels is completely specified by Δ_1 and

Δ_2. Since Δ_1 and Δ_2 are not known, the solution must proceed by means of two separate solutions wherein arbitrary deflections are imposed as in Figs. 19b and 19c, with temporary supports. With the forces F_1, F_2, F_3, and F_4, known at the temporary supports, the two modifying factors f_1 and f_2 may be determined by writing two force equations similar to those

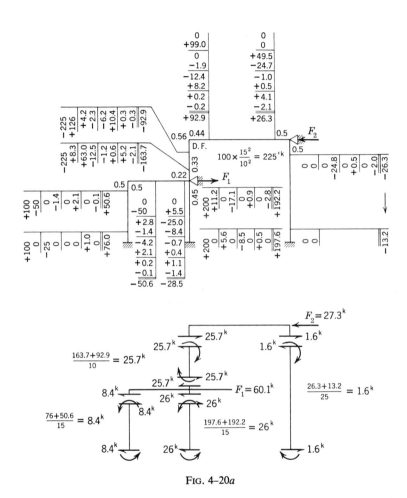

Fig. 4–20a

explained in the previous article. Physically it must be realized that if through some power of foresight Δ'_2 and Δ'_3 had been estimated as exactly equal to the final deflections Δ_2 and Δ_3, then the combination of F_4 and F_2 would equal 10 kips and F_1 and F_3 would combine to equal 10 kips without modification or correction. Inasmuch as this is not the case, the final situation must be a combination of a calculated percentage of each solution as defined by the factors f_1 and f_2.

Although the analysis is complete the routine of moment distribution must still be performed. Figures 20a and 20b represent the detailed solution of the problem.

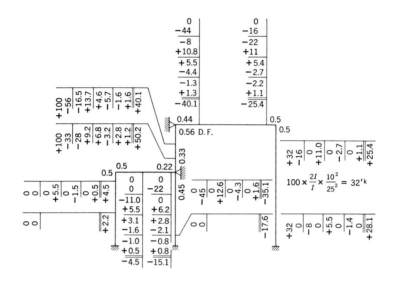

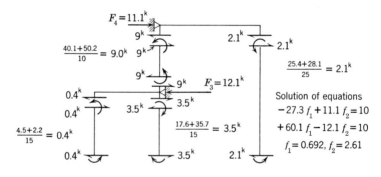

$$\frac{40.1+50.2}{10} = 9.0^k$$

$$\frac{4.5+2.2}{15} = 0.4^k$$

$$\frac{25.4+28.1}{25} = 2.1^k$$

$$\frac{17.6+35.7}{15} = 3.5^k$$

Solution of equations

$$-27.3 f_1 + 11.1 f_2 = 10$$

$$+60.1 f_1 - 12.1 f_2 = 10$$

$$f_1 = 0.692, \ f_2 = 2.61$$

FIG. 4–20b

4–11. FRAMES WITH SLOPING MEMBERS

Figure 21a depicts a rigid frame with battered columns loaded and framed symmetrically. A solution for moments in this frame by moment distribution involves no new situations not already encountered. The solution for moments is completely given in Fig. 21*b*.   Particular attention must be given to constructing complete free bodies to determine the reaction components.

The loading of Fig. 22*a* produces a translation of joints *B* and *C*, with this translation having both horizontal and vertical components.   Assuming that the members have no axial changes in length, joints *B* and *C* will

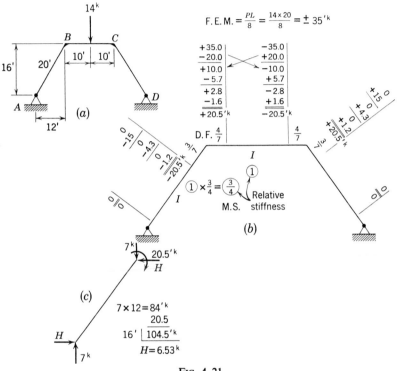

$$\text{F. E. M.} = \frac{PL}{8} = \frac{14 \times 20}{8} = \pm\ 35\,'^k$$

Fig. 4–21

move along the arc of a circle, which, as in the Williot diagram construction, may be taken as a straight line perpendicular to the initial position of radius.   The deflection of *B'* from a tangent to elastic curve may be taken as Δ.   Then the vertical downward translation of *B'* by similar triangles is $\frac{3}{5}\Delta$.   Joint *C'* is similarly located and the translation upward of *C'* with respect to *B'* is $\frac{6}{5}\Delta$.   It should be particularly noted that the movements have been computed with respect to the perpendicular to the original direction of the member.   This deflection is necessary in order to rate the fixed-ended moments in the analysis to follow.

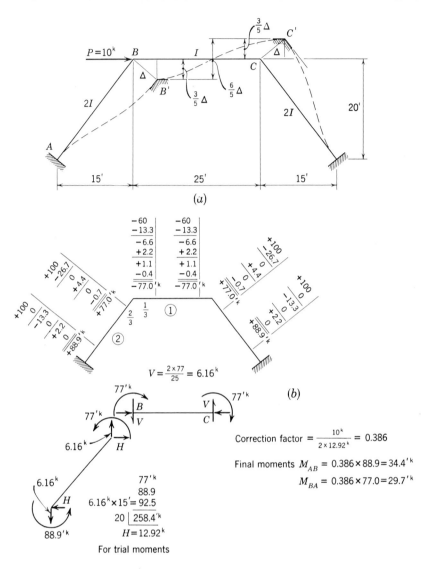

FIG. 4–22

We may now revert to the common approach for all solutions by the moment distribution method, namely, let the translation take place with all joints fixed against rotation and establish a set of consistent arbitrary fixed-ended moments in all members. Letting M_1 equal fixed-ended moments in columns and M_2 the similar quantity for the beam, we have

$$M_1 = \frac{6E \cdot 2I \cdot \Delta}{25^2}$$

and

$$M_2 = \frac{6E \cdot I \cdot \frac{6}{5}\Delta}{25^2}$$

or, in the form of a ratio,

$$\frac{M_1}{M_2} = \frac{10}{6}$$

In Fig. 22b, in order to provide proportional moments, M_1 has been chosen as $+100$ ft-kips and M_2 as -60 ft-kips. (Note: the sign of moment is as required by the deflected configuration.) Distribution follows and equilibrium studies of free bodies of the columns establish the horizontal force consistent with assumed moments. Particularly observe that the free body of the beam is essential in order to determine V. The correction factor is computed and final moments are established. The solution is given in full in Fig. 22.

4–12. GABLED FRAME

A gabled rigid roof framing as shown in Fig. 23 will be analyzed for the effect of one vertical load applied at the ridge. The deflected configuration indicates that joints B and D deflect outward by Δ while C moves down Δ_C. The roof framing may be considered as two separate members BC and CD, with C considered as a common joint. Since there are no loads acting directly on the members, initial fixed-ended moments depend on the relative deflections of the members. In Fig. 23b, temporarily disjoint the frame at C and let B and D move outward by an amount Δ. The ends of the rafters C' and C'' must be reconnected at point C''' by rotating the radii $B'C'$ and $D'C''$. (Note the Williot diagram technique of erecting perpendiculars.) The point C has moved down Δ_C equal to $\frac{4}{3}\Delta$, while the translation of C with reference to a tangent to the elastic curve at B is $C'C'''$ or $\frac{5}{3}\Delta$. This is the important quantity since all joints must be assumed fixed against rotation during these initial translations and fixed-ended moments established in terms of the deflection of one end of the member relative to the tangent to the other end. It should also be realized that structure may be held in this deflected configuration by securing point C''' to a temporary support. If the student prefers, the analysis could also start by pushing C down while permitting B and D to move outward.

Figure 23c indicates the details of the moment distribution after writing

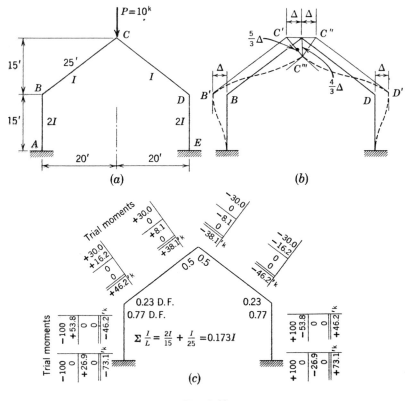

Fig. 4–23

correctly proportioned relative fixed-ended moments in the columns and rafters. Letting M_1 equal fixed-ended moments for the columns and M_2 the similar moments for the rafters, we find

$$M_1 = \frac{6E(2I)\Delta}{15^2}$$

$$M_2 = \frac{6E(I)\frac{5}{3}\Delta}{25^2}$$

or, in the form of a ratio,

$$\frac{M_1}{M_2} = \frac{2 \times 25^2}{15^2 \times \frac{5}{3}} = \frac{50}{15}$$

In Fig. 24, the results of the moment distribution using trial moments are used to finalize the reactions and moments due to the 10-kip load. Two free bodies are necessary, as shown in Fig. 24, one to secure the horizontal reaction and the other to find $P/2$ agreeing with trial moments. P then equals 20.35 kips for the trial moments in Fig. 23c. The trial moments of Fig. 23c are multiplied by the ratio 10:20.35, to determine final moments. The corrected moments for the actual 10-kip load are given in Fig. 24b, along with the final moment diagram.

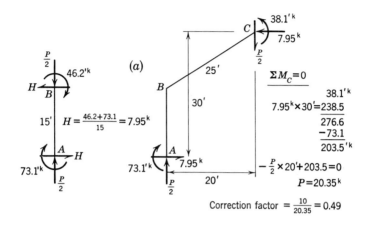

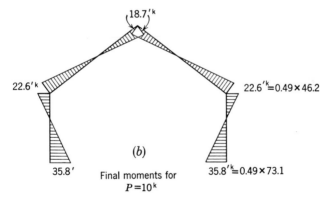

(b) Final moments for $P=10^k$

FIG. 4–24

Figure 25a is a gabled frame loaded laterally at B by a 10-kip force. This is a situation wherein two degrees of freedom must be considered since Δ_B and Δ_D are not equal. The analysis will therefore be performed in two easily visualized steps. The first step is indicated in Fig. 25b, where, with all joints fixed against rotation, the frame is distorted by arbitrary Δ_1 deflections and pinned and clamped in this position through

the action of temporary supports.   In Fig. 25c, the frame is distorted by
the arbitrary deflections Δ_2 and pinned and clamped in this position
through the action of temporary supports.   The objective in making the
original distortions in this manner is to recognize the inequality of Δ_B and
Δ_D and to prepare the way for the final solution for the single force of
10 kips at B only.

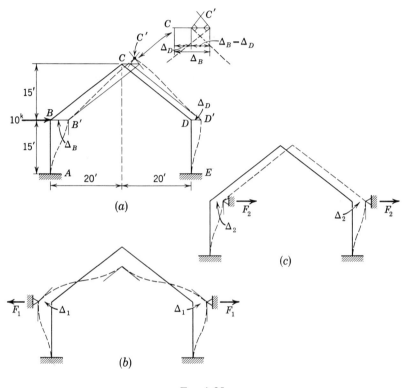

FIG. 4–25

Figures 26 and 27 present the relevant calculations for the structure of
Fig. 25a.   A new moment distribution is not needed for the determination
of force F_1, since a previous solution for the problem of Fig. 23 determined
a set of distributed moments for an arbitrary outward movement of B and
D.   With the moments of Fig. 23c, the free body diagram of Fig. 26b is
drawn.   The Δ_1 displacements are produced by two equal and opposite
forces at B and D, consequently there are no vertical forces at A and E.
Static equilibrium conditions applied to the free body of Fig. 26b provides
the basis for the calculation of F_1 equal to 13.53 kips.

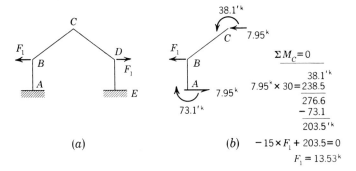

FIG. 4–26

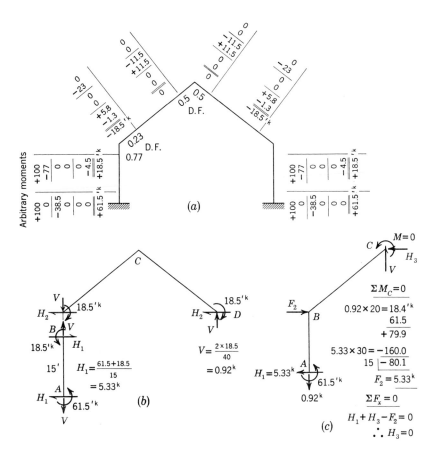

FIG. 4–27

A new moment distribution must be performed to determine F_2. In Fig. 27a, arbitrary moments are distributed for an arbitrary displacement of Δ_2. (See Fig. 25c.) It should be particularly noted that no fixed-end moments are developed in the rafter members. Figure 27b and 27c represent the essential free bodies and the applications of equilibrium principles to determine the value of F_2.

With F_1 and F_2 known, for arbitrary displacements Δ_1 and Δ_2, the data is available for computing the final solution of this two degree of freedom problem. Identifying the solution of Figs. 23c and 26 as the f_1 solution, and the solution of Fig. 27a as the f_2 solution, then the equations expressing the total actual forces at B and D may be written where f_1 and f_2 represent proportional parts of the two solutions.

At B, since the actual force is 10 kips (see Fig. 25a),

$$-13.53f_1 + 5.33f_2 = 10 \text{ kips}$$

and at D, since the external force is actually zero,

$$+13.53f_1 + 5.33f_2 = 0$$

The two equations may then be solved simultaneously

$$-13.53f_1 + 5.33f_2 = 10$$
$$+13.53f_1 + 5.33f_2 = 0$$
$$\overline{\phantom{+13.53f_1 + {}} 10.66f_2 = 10}$$
$$f_2 = +0.937$$
$$f_1 = -0.370$$

With f_1 and f_2 known the final moments for the single 10-kip force at B will be obtained by combining the solutions of Fig. 23c and Fig. 27a. For example, at A

$$M_{AB} = (-0.370)(-73.1) + (0.937)(+61.5)$$
$$= +27.0 + 57.7 = +84.7 \text{ ft-kips}$$

At B $\qquad M_{BA} = (-370)(-46.2) + (0.937)(+18.5)$
$$= +17.1 + 17.3 = +34.4 \text{ ft-kips}$$

Figure 28 indicates an alternate displacement procedure which may be substituted for the displacements of Fig. 25b. Point B is displaced and point D is held against translation. The geometry of the relative displacements for the rafter members are shown. A distribution using an arbitrary set of consistent fixed-end moments will lead to the calculations of the forces to be applied at B and D to maintain this configuration. This solution may then be combined with the solution of Figs. 25c and Fig. 27a by calculating new values of f_1 and f_2.

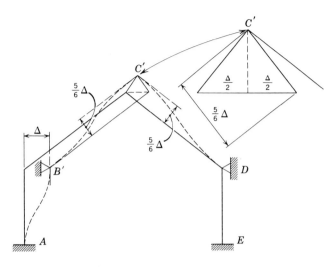

FIG. 4–28

4-13. MOMENT DISTRIBUTION—ADDITIONAL TOPICS

Moment distribution principles for structures with variable moment of inertia are developed in Chapter 9. The modifications of moment distribution principles for flexible members axially loaded are presented in Chapter 13.

REFERENCES

1. Cross, Hardy, "Analysis of Continuous Frames by Distributing Fixed-End Moments," *Trans. Am. Soc. Civil Engineers*, **96**, 1–10 (1932).
2. Cross, Hardy, and N. D. Morgan, *Continuous Frames of Reinforced Concrete*, Chapter 4, New York: John Wiley, 1932.
3. Griffiths, J. D., "Multiple-Span Gabled Frames," *Trans. Am. Soc. Civil Engineers*, **121**, 1288–1317 (1956).
4. Kinney, J. S., *Indeterminate Structural Analysis*, Chapters 8 and 9, Reading, Mass.: Addison Wesley, 1957.
5. Maugh, L. C., *Statically Indeterminate Structures*, Chapter 3, New York: John Wiley, 1946.
6. Morris, C. T., and S. T. Carpenter, *Structural Frameworks*, Chapter 6, New York: John Wiley, 1943.
7. Parcel, J. I., and R. B. B. Moorman, *Analysis of Statically Indeterminate Structures*, Chapter 6, New York: John Wiley, 1955.
8. Wilbur, J. B., and C. H. Norris, *Elementary Structural Analysis*, Chapter 14, New York: McGraw-Hill, 1948.

PROBLEMS

Note.   All problems are to be solved by the moment distribution method.

4–1. Determine the moment at C by moment distribution. T is a couple applied at A in vertical plane.   Draw the moment diagram.

Ans. $M_C = 0.143T$.

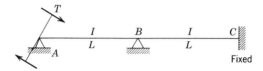

Prob. 4–1

4–2. If the end C of the beam of problem 4–1 is pinned instead of fixed, what is the moment at B?   *Ans. $M_B = 0.25T$.*

4–3. If T is an external couple applied at C in vertical plane, determine the final moments and draw the moment diagram.

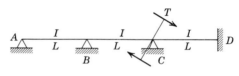

Prob. 4–3

4–4. Determine the final moments by moment distribution.   Draw the bending moment diagram.

Ans. $M_A = 15$ ft-kips; $M_B = 30$ ft-kips; $M_C = 45$ ft-kips.

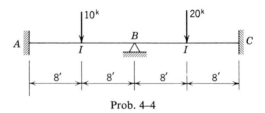

Prob. 4–4

4–5. Assume that end A of beam of problem 4–4 is pinned and determine the reactions and final moments.

4–6. Assume that both A and C are pinned in problem 4–4 and solve for final moments.   *Ans. $M_B = 45$ ft-kips.*

4–7. Determine the reactions and final moments and draw the bending moment diagram.   *Ans.* $M_B = M_C = wL^2/10$.

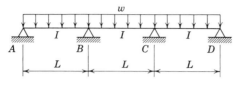

Prob. 4–7

4–8. Determine the final moments and draw the bending moment diagram.

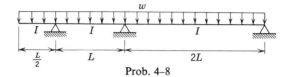

Prob. 4–8

4–9. Determine the final moments.   *Ans.* $M_B = Pa/2$.

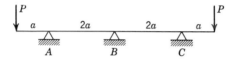

Prob. 4–9

4–10. Determine the final moment at *B*.

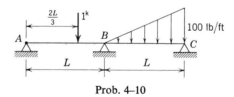

Prob. 4–10

4–11. Determine the final moment at *B*.

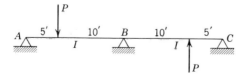

Prob. 4–11

4-12. Noting that the moment of inertia of span AB is twice that of span BC, determine the final moment at B.   *Ans. $M_B = 36.43$ ft-kips.*

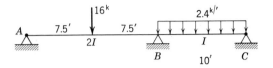

Prob. 4–12

4-13. Determine the reactions and final moments, noting the variation in moments of inertia.   *Ans. $M_A = 19.43$ ft-kips; $M_B = 9.14$ ft-kips.*

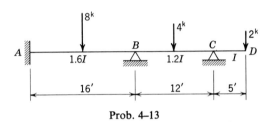

Prob. 4–13

4-14. Determine the reactions.   *Ans. $H_C = 0.017P$; $V_C = 0.364P$.*

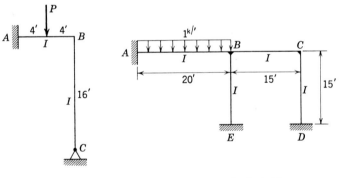

Prob. 4–14   Prob. 4–15

4-15. Determine all reactions and draw the final bending moment diagram.

4–16. Determine all moments.

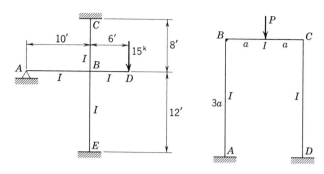

Prob. 4–16   Prob. 4–17

4–17. Determine the final moments and draw the moment diagram.

Ans. $M_A = 0.072Pa$; $M_B = 0.144Pa$.

4–18. Determine the reactions. Each joint is rigidly framed. Change in length of members to be neglected.

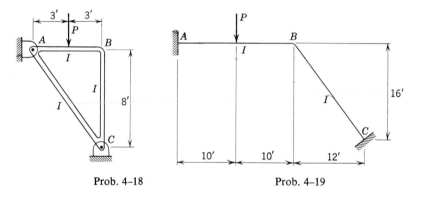

Prob. 4–18   Prob. 4–19

4–19. Determine the reactions and draw the bending moment diagram.

Ans. $M_C = 0.625P$; $H_C = 0.421P$; $V_C = 0.406P$.

4–20. Determine the reactions.   Note differences in the moment of inertia.

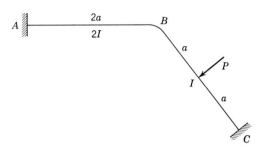

Prob. 4–20

4–21. Determine the bending moment at *B*.

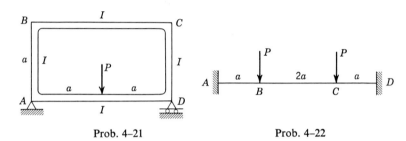

Prob. 4–21   Prob. 4–22

4–22. It is desired to determine the fixed-ended moments for the beam shown. Displace *B* and *C* through an arbitrary vertical distance of *Δ*, with all beam segments considered deflecting as fixed-ended elements, and lock and pin *B* and *C* in this position.   After moment distribution, determine the true fixed-ended moments for the original problem.   *Ans.* 3*Pa*/4.

4–23. Determine the final moments at A and C.   Constant I.

Ans. $M_C = 0$; $M_A = 10$ ft-kips.

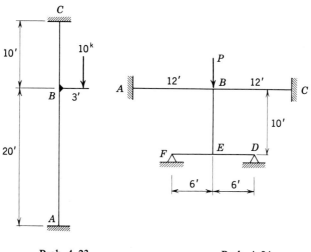

Prob. 4–23   Prob. 4–24

4–24. Calculate the vertical reaction at F.   Constant I.

4–25. Determine all of the final reactions.   Draw the moment diagram.
Ans. $V_A = 8.21$ kips; $H_A = 1.18$ kips; $V_E = 8.21$ kips; $H_E = 8.82$ kips.

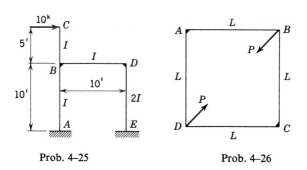

Prob. 4–25   Prob. 4–26

4–26. Determine the moments and draw the moment diagram.   The frame shown is in a horizontal plane.   Constant I.

4–27. A steel frame of the dimensions shown is subjected to an increase in temperature of ΔT degrees. Determine the horizontal reactions in terms of ΔT, L, EI, and the coefficient of expansion C_x.   *Ans.* $3.75 C_x \Delta T EI/L^2$.

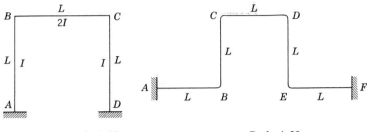

Prob. 4–27   Prob. 4–28

4–28. A steel pipe of uniform size conducts steam from A to F. The configuration for the pipeline has been chosen to minimize moments and thrusts on the mechanical equipment at A and F. Considering that the pipe is fully supported in a horizontal plane, compute the moment and other induced forces at A due to a temperature increase of ΔT degrees. State your answers in terms of L, EI, ΔT, and the coefficient of expansion.

4–29. Determine the moments at A and D. Determine the deflection of B in terms of EI.   *Ans.* $M_A = 14.1$ ft-kips; $M_D = 20.0$ ft-kips; $\Delta_B = 318/EI$.

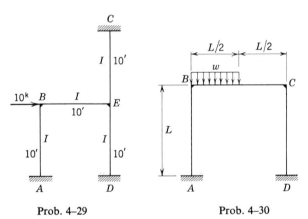

Prob. 4–29   Prob. 4–30

4–30. Determine the final moments corrected for sidesway and draw the moment diagram. Uniform I.

4–31. Determine the final moments corrected for sidesway.   Constant I.

Ans. $M_A = -0.343P$; $M_D = +0.078P$.

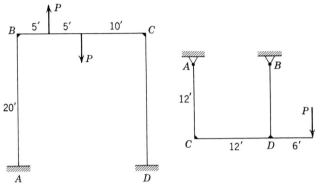

Prob. 4–31   Prob. 4–32

4–32. Determine the final moments corrected for sidesway.   Constant I.

4–33. Determine the final moment, corrected for sidesway, at B.   Constant I.

Ans. $M_{BA} = +5.2$ ft-kips.

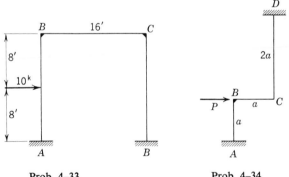

Prob. 4–33   Prob. 4–34

4–34. Determine the final moments corrected for sidesway and draw the moment diagram.   Constant I.   *Ans.* $M_A = +0.573Pa$; $M_D = -0.132Pa$.

4–35. Determine the final moments and all reactions.   Constant I.

Ans. $M_A = +6.4$ ft-kips.

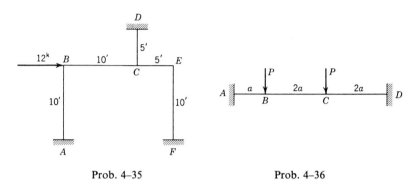

Prob. 4–35   Prob. 4–36

4–36. This problem is similar to problem 4–22, except it must be recognized that it must be treated as a two degrees of freedom problem.   Determine the fixed-ended moments at A and B.

4–37. Solve this two degrees of freedom problem for moments.   Constant I.

Ans. $M_A = +31.6$ ft-kips.

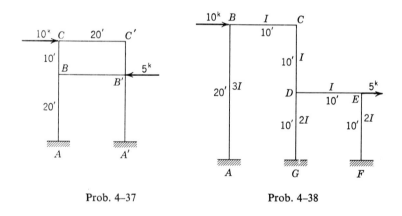

Prob. 4–37   Prob. 4–38

4–38. Solve this two degrees of freedom problem for moments.

4-39. Solve this problem for moments.   Constant *I*.

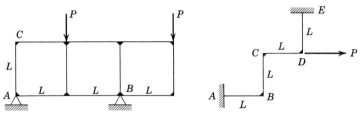

Prob. 4-39   Prob. 4-40

4-40. Solve this two degrees of freedom problem for moments.   Constant *I*.
$$Ans.\ M_A = +0.062PL.$$

4-41. Solve this two degrees of freedom problem for moments.   Assume constant *I*.

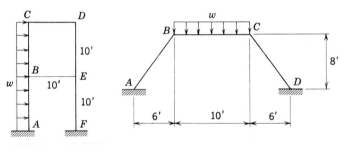

Prob. 4-41   Prob. 4-42

4-42. Determine the horizontal reactions.   Constant *I*.   *Ans.* 4.79*w*.

4-43. Determine the horizontal reactions at *A* and *D*.   Draw the bending moment diagram.   How much does *B* deflect horizontally in terms of *EI*?
$$Ans.\ H_A = 7.10\ \text{kips};\ \varDelta = 473/EI.$$

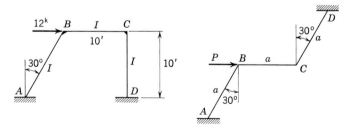

Prob. 4-43   Prob. 4-44

4-44. Determine the moment at *A*.   Constant *I*.

4-45. Determine all moments and draw the moment diagram.   Constant *I*.

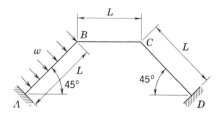

Prob. 4-45

4-46. Determine the reactions and all moments.   Draw the moment diagram.
Ans. $H_A = 0.1685P$.

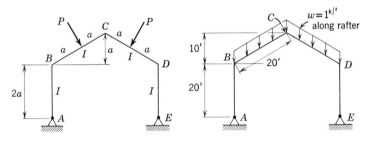

Prob. 4-46   Prob. 4-47

4-47. Determine the reactions and all moments.   Draw the moment diagram for *BC*.   Constant *I*.

4-48. Solve for all moments and reactions.   Constant *I*.

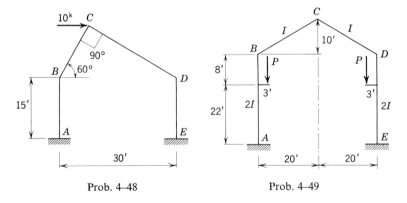

Prob. 4-48   Prob. 4-49

4-49. The columns of a building support a craneway.   The loads *P* are the craneway reaction.   Determine all moments in the frame.

Slope Deflection Method

5

5-1. INTRODUCTION

The method of slope deflection was developed before the moment distribution method; it remains a valuable method of analysis in many situations and is particularly adaptable to electronic computation procedures. By properly associating the rotations of the tangents to the elastic curve and the deflections at the ends of the individual members, standard equations for end moments may be formulated. In this chapter the method will be developed for either fully continuous members, such as a beam, or for members rigidly connected into a framework.

5-2. SLOPE-DEFLECTION EQUATIONS

In Fig. 1a a continuous elastic beam is shown and, owing to the loading, deflections occur and the tangents to the elastic curve at A and B rotate from their horizontal unloaded positions. Let the angular rotation of the tangent at A be noted as θ_A and at B as θ_B in radian units. A sign convention may be adopted as follows: if the tangent rotates clockwise the angle of rotation is positive, and if counterclockwise rotation takes place the angle is termed negative. Thus θ_A and θ_B are both positive in Fig. 1a.

Figure 1b isolates the span AB and indicates the end moments M_{AB} and M_{BA}. A sign convention is also necessary for moment and the same sign convention used in moment distribution is convenient. Accordingly, M_{AB} is positive, since the direction of moment on the joint (not the beam) is clockwise; M_{BA} is shown as negative acting counterclockwise on joint B.

To develop the standard equation, first assume, as in Fig. 1c, that the ends A and B are held momentarily in a fixed-ended condition, that is,

189

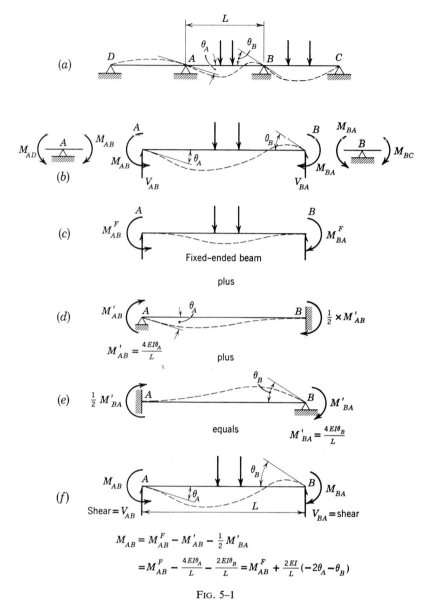

$$M_{AB} = M_{AB}^F - M'_{AB} - \tfrac{1}{2} M'_{BA}$$

$$= M_{AB}^F - \frac{4EI\theta_A}{L} - \frac{2EI\theta_B}{L} = M_{AB}^F + \frac{2EI}{L}(-2\theta_A - \theta_B)$$

FIG. 5–1

with zero rotation of the end tangents.   The loading on span then produces
the fixed-ended moments M_{AB}^F and M_{BA}^F, as illustrated in Fig. 1c.   These
fixed-ended moments must be modified by moments which, if applied
independently, will rotate the tangents at A and B through the angles θ_A

and θ_B as required by the final configuration of Fig. 1a.   The fundamentals of moment distribution assist in modifying the fixed-ended moments. For example, in Fig. 1d, the tangent at A is rotated by M'_{AB} while the B end is fully restrained.   The moment at B is one-half the moment at A by moment distribution principles. M'_{AB} is equal to $(4EI/L)\theta_A$ by previous relations for a beam of uniform moment of inertia as given by equation 4-2.

In Fig. 1e, with end A restrained, M'_{BA} rotates the tangent at B by θ_B in a clockwise direction.   This develops a moment of $\frac{1}{2}M'_{BA}$ at the A end as shown in Fig. 1e.   The separate moments of Figs. 1c, d, and e may be combined or superimposed, as in Fig. 1f, to represent the true end moments for span AB of the continuous beam.   At A we obtain M_{AB} as

$$M_{AB} = M^F_{AB} - M'_{AB} - \tfrac{1}{2}M'_{BA}$$

or in terms of rotation of the tangents at A and B

$$M_{AB} = M^F_{AB} - \frac{4EI\theta_A}{L} - \frac{2EI\theta_B}{L}$$

$$M_{AB} = M^F_{AB} + \frac{2EI}{L}(-2\theta_A - \theta_B) \qquad (5\text{-}1)$$

Similarly at B we obtain M_{BA} by combining the values of moment in Figs. 1c, d, and e.

$$M_{BA} = -M^F_{BA} - \tfrac{1}{2}M'_{AB} - M'_{BA}$$

or in terms of rotation of the tangents at A and B,

$$M_{BA} = -M^F_{BA} - \frac{2EI\theta_A}{L} - \frac{4EI\theta_B}{L}$$

$$M_{BA} = -M^F_{BA} + \frac{2EI}{L}(-2\theta_B - \theta_A) \qquad (5\text{-}2)$$

The angles θ_A and θ_B have been assumed as positive quantities in the derivation of equations 5-1 and 5-2.   Refer to Figs. 1d and 1e.

Equations 5-1 and 5-2 state, if the rotations of the tangents are known at A and B, that the final end moments may be determined.   Furthermore, owing to their similarity, equations 5-1 and 5-2 may be restated in the form of a single standard equation,

$$M_{AB} = \frac{2EI}{L}(-2\theta_A - \theta_B) \pm M^F_{AB} \qquad (5\text{-}3)$$

wherein θ_A is interpreted as the rotation of tangent at the end where moment is desired, and θ_B the rotation of the tangent at the far end of the same member.   The fixed-ended moment is written as plus or minus, depending on the loading and applies to the end of the beam to which the designation A refers.

The student may have attached no importance to the assumption which appears to have arbitrarily taken the rotations as positive in Fig. 1a. This was done to make the sign convention a general one and to insure that any computed value of an angle may be interpreted correctly. For example, a negative value of θ_A or θ_B will imply counterclockwise rotation, since the sign is opposite to the assumed positive direction which denoted clockwise rotation in the original development of the standard equation.

The standard equation must be augmented if a relative deflection as well as rotation occurs. In Fig. 2a, end B is shown displaced vertically downward to B' by an amount Δ, while the tangents to A and B are considered

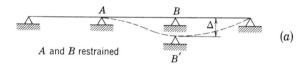

A and B restrained

(a)

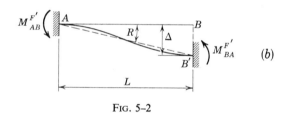

(b)

FIG. 5–2

fixed against rotation. Figure 2b enlarges the view of the beam AB under the above conditions. A line is drawn from A to B' and the angle R is indicated as the rotation of line AB from its original horizontal position. Since R is a small angle its value in radians is Δ/L. The fixed-ended moments $M_{AB}^{F'}$ and $M_{BA}^{F'}$ may be expressed, using equation 4-5, as

$$\frac{6EI\Delta}{L^2} = \frac{6EIR}{L}$$

If Δ is such as to make R a clockwise angle as shown, which is a positive angle by the adopted convention, the fixed-ended moments produced by the translation are positive. If R is counterclockwise, the angle R and moments are negative and B' would be above A.

Assuming that the member AB is also a loaded member and part of a continuous beam, the standard equation may be augmented by adding the moments $M_{AB}^{F'}$ and $M_{BA}^{F'}$ to equations 5-1 and 5-2, giving the general equation

$$M_{AB} = \frac{2EI}{L}(-2\theta_A - \theta_B + 3R) \pm M_{AB}^F \qquad (5\text{-}4)$$

Equation 5-4 is the general standard equation associated with the slope deflection method.   If $R = 0$, equation 5-4 reduces to equation 5-3.

This method will, by nature of the developed equations, consider rotations and deflections as the unknown quantities.

5–3. BEAMS BY SLOPE DEFLECTION METHOD

Figure 3 is a beam of constant moment of inertia of two full spans and an overhanging end.   It is desired to determine the final bending moment

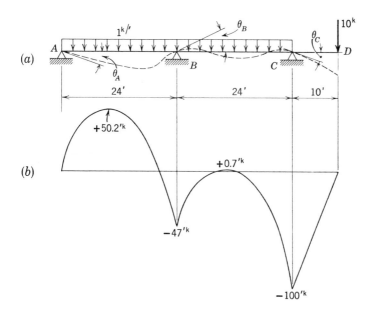

Fig. 5–3

diagram by the slope deflection method.   In this instance all supports are firm and no foundation settlement is expected; consequently, equation 5-3 may be used since R is zero for all spans.

The application of the subject method may begin with an analysis of the unknowns where these unknowns will be the slopes or rotations of the tangents at the ends of the members at each support. In Fig. 3a, these unknown angles are indicated as θ_A, θ_B, and θ_C. No matter what the analyst may surmise concerning the direction of rotation, the original assumption of positive values should be retained. The proper sign of these angles will result from the solution.

To set up the general equations for each member the fixed-ended moments are needed. In this instance they are numerically equal to

$$\frac{wL^2}{12} = \frac{1 \times 24^2}{12} = 48 \text{ ft-kips}$$

Hence
$$M_{AB} = \frac{2EI}{24}(-2\theta_A - \theta_B) + 48 \tag{5-5}$$

where it will be noted that in keeping with general development of the standard equation the fixed moment must be included, although it is known that the final moment at A is zero. Later the equation for M_{AB} set equal to zero will provide a condition equation.

Following the form of the general equation

$$M_{BA} = \frac{2EI}{24}(-2\theta_B - \theta_A) - 48 \tag{5-6}$$

$$M_{BC} = \frac{2EI}{24}(-2\theta_B - \theta_C) + 48 \tag{5-7}$$

$$M_{CB} = \frac{2EI}{24}(-2\theta_C - \theta_B) - 48 \tag{5-8}$$

A slope deflection equation need not be written for the cantilever end CD as statics alone suffices to determine the moments in CD.

By statics with a section to the right of C

$$M_{CD} = +100 \text{ ft-kips} \tag{5-9}$$

The positive sign for moment is required since the moment acts on C in a clockwise direction.

With three unknowns, three independent equations are required, and in this method they must be formulated from equations expressing static equilibrium. The static equations for the beam problem will be referred to as joint equations where the joint equations separately express the static moment equilibrium of the moments in all of the members meeting at a joint. With three supports or joints we have the static condition equations:

at A $\quad M_{AB} = 0$ $\tag{5-10}$

at B $\quad M_{BA} + M_{BC} = 0$ $\tag{5-11}$

at C $\quad M_{CB} + M_{CD} = 0$ $\tag{5-12}$

It is of particular importance to note that no attempt is made to impress a sign on these moments. All moments are assumed as positive. The true sign, whatever it may be, will be used whenever a known value is substituted in the condition equation; however, the sign of all unknown moments is deliberately kept as positive until the true sign, depending on the values of θ_A, θ_B, and θ_C, has been established.

The formulation of the equations in terms of θ_A, θ_B, and θ_C follows.

Substituting the values of moment from equations 5-5, 5-6, 5-7, 5-8, and 5-9 into equations 5-10, 5-11, and 5-12, respectively, equation 5-10 becomes

$$\frac{2EI}{24}(-2\theta_A - \theta_B) + 48 = 0 \qquad (5\text{-}13)$$

Equation 5-11 becomes

$$\frac{2EI}{24}(-2\theta_B - \theta_A) - 48 + \frac{2EI}{24}(-2\theta_B - \theta_C) + 48 = 0 \qquad (5\text{-}14)$$

Equation 5-12 becomes

$$\frac{2EI}{24}(-2\theta_C - \theta_B) - 48 + 100 = 0 \qquad (5\text{-}15)$$

The simultaneous equations 5-13, 5-14, and 5-15 may now be solved by a number of approaches too well known to discuss here. Solution yields the values of the angular rotations

$$\theta_A = +\frac{388}{EI}$$

$$\theta_B = -\frac{200}{EI}$$

$$\theta_C = +\frac{400}{EI}$$

The only moment that is unknown, although by this method three unknown angles were involved, is the moment over the support at B. Either M_{BA} or M_{BC} may be computed by substituting the values of the rotations in the proper equations. By equation 5-6

$$M_{BA} = \frac{2EI}{24}\left(+\frac{2 \times 200}{EI} - \frac{388}{EI}\right) - 48$$

$$= +1 - 48 = -47 \text{ ft-kips}$$

M_{BC} would be $+47$ ft-kips by equation 5-7.

5–4. FRAMES WITHOUT SIDESWAY

Rigid frames that do not involve sidesway may be solved using the slope deflection standard equation with the $3R$ term eliminated. The frame of Fig. 4 will be taken as an example. This has been solved previously by other methods in the preceding chapters.

The solution of the frame of Fig. 4 will be expedited by first sketching the deflected outline. This forces a recognition of the angles of rotation of each joint and identifies the quantities that must be selected as unknowns. With the end conditions at A and D taken as fixed against rotation, it is

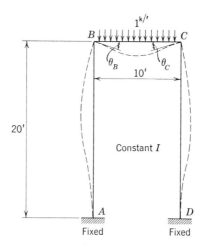

FIG. 5–4

clear that θ_A and θ_D equal zero. At B and C the joints rotate through angles of θ_B and θ_C; however, symmetry of framing and loading permits a statement to the effect that $\theta_B = -\theta_C$, and, consequently, there is only one unknown rotation to be found for a complete solution.

As a preliminary calculation the fixed-ended moments are found as

$$M_{BC}^F = \frac{1 \times 10^2}{12} = +8.33 \text{ ft-kips}$$

and $$M_{CB}^F = -8.33 \text{ ft-kips}$$

With only one unknown, only one static equilibrium condition equation need be formulated, such as the equilibrium of moments at joint B, or

$$M_{BA} + M_{BC} = 0 \tag{5-16}$$

The equations for M_{BA} and M_{BC} based on the general equation 5-3 may now be substituted in equation 5-16, providing a basis for determining θ_B. By substitution we obtain

$$\frac{2EI}{20}(-2\theta_B) + \frac{2EI}{10}(-2\theta_B + \theta_B) + 8.33 = 0 \qquad (5\text{-}17)$$

and
$$\theta_B = +\frac{83.3}{4EI}$$

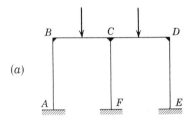

(a)

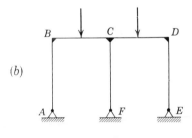

(b)

Fig. 5–5

The end moments in members of the frame may now be determined from equations based on the general equation 5-3.

$$M_{AB} = \frac{2EI}{20}\left(-\frac{83.3}{4EI}\right) = -2.08 \text{ ft-kips}$$

$$M_{BA} = \frac{2EI}{20}\left(-\frac{2 \times 83.3}{4EI}\right) = -4.16 \text{ ft-kips}$$

$$M_{BC} = \frac{2EI}{10}\left(-\frac{2 \times 83.3}{4EI} + \frac{83.3}{4EI}\right) + 8.33 = +4.16$$

All of the above signs are stated according to the slope deflection convention for signs and it will be recalled that the signs establish the direction of the moment acting on the joints.

Figure 5 represents other symmetrically loaded and framed structures which do not involve sidesway. In Fig. 5a, taking full advantage of symmetry, only one unknown θ_B would enter the solution. In Fig. 5b, the hinging of the bottom of columns would introduce an unknown rotation at A and E, as well as at B and D. The determination of two unknown rotations θ_A and θ_B would suffice for full solution.

Although there appears to be no advantage of the subject method for problems of this type, an advantage would disclose itself if the solutions were written out in terms of algebraic expressions assigning L_1, L_2, L_n, and I_1, I_2, I_n to the lengths and moments of inertia. Equations for moments may then be established for repetitive types of frame problems. The method also has great merit when digital computers are available for solving the system of simultaneous equations.

5-5. FRAMES WITH SIDESWAY

Frames with laterally applied forces or frames unsymmetrically framed or loaded will sway. The effect of such translation is to cause the ends of many members to deflect relative to initial unloaded positions. The angle R is a measure of the relative movement of the far end of a member with respect to the near end.

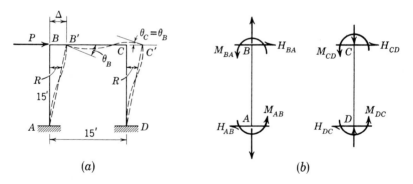

(a)   (b)

FIG. 5–6

Figure 6a represents a two-column rigid frame with a lateral load P applied at B. The frame will deflect to the right while joints B and C rotate clockwise. The deflection Δ of B relative to A is unknown. If A and B' are joined by a straight line, the angle that this line makes with the original unloaded position of AB may be termed R. It is to be noted that the sign of R is positive since AB' turns clockwise with respect to AB.

The analysis may now proceed with an identification of the unknowns θ_B, θ_C, and R. Symmetry demands that $\theta_B = \theta_C$, hence only two unknowns enter the solution, namely θ_B and R. Thus, with two unknowns, two condition equations based on static equilibrium are required. The first equation is easily formulated from moment equilibrium at joint B.

$$M_{BA} + M_{BC} = 0 \qquad (5\text{-}18)$$

The second condition equation can be formulated by considering the horizontal static equilibrium of the structure. If the columns are removed from the frame as free bodies, as in Fig. 6b, the end moments or end couples may be shown as well as the axial and shearing forces. The moment equilibrium of column AB requires the two end couples to be held in equilibrium by the single couple formed by the horizontal shear forces. (Note: $H_{AB} = H_{BA}$ to provide horizontal force equilibrium.) Thus

$$H_{AB} = \frac{M_{AB} + M_{BA}}{15} \qquad (5\text{-}19)$$

The right-hand column in this case is identical to the left-hand column, and a condition equation expressing the static equilibrium of horizontal forces for entire frame is

$$\frac{2(M_{AB} + M_{BA})}{15} - P = 0 \qquad (5\text{-}20)$$

Equations 5-18 and 5-20 must now be restated in terms of the unknown rotation and angle R. The joint equation derived from the general equations 5-3 and 5-4 becomes

$$\frac{2EI}{15}(-2\theta_B + 3R) + \frac{2EI}{15}(-2\theta_B - \theta_B) = 0 \qquad (5\text{-}21)$$

and the horizontal equilibrium equation gives

$$\frac{2\left[\frac{2EI}{15}(-\theta_B + 3R) + \frac{2EI}{15}(-2\theta_B + 3R)\right]}{15} - P = 0 \qquad (5\text{-}22)$$

Equations 5-21 and 5-22 must be solved simultaneously. The solution is left to the student who may check his results with the values found by moment distribution in Chapter 4 for the same problem.

Although the first example of sidesway did not involve fixed-ended moments, caused by direct loading on the members, the problem of Fig. 7a has a wind loading imposed on column AB. An analysis of rotation and deflection indicates that θ_B is different than θ_C, whereas the angle R is the same for column CD as for column AB assuming that the change in length of BC may be ignored.

The three condition equations from statics follow (see Fig. 7b), where the first two come from joint equilibrium considerations and the third from the static equilibrium of horizontal forces. *The student should again note the consistent assumption of positive moment.*

$$M_{BC} + M_{BA} = 0$$

$$M_{CB} + M_{CD} = 0$$

$$H_{AB} + H_{DC} - 40 = 0$$

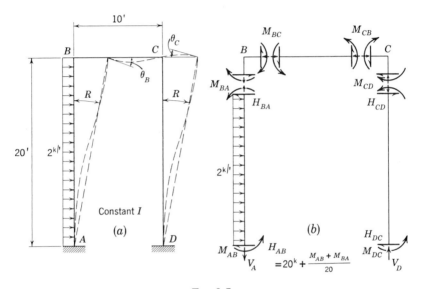

FIG. 5–7

The last equation must be restated in terms of end moments. The application of the principles of static equilibrium to the free bodies of the columns permits its expansion to

$$20 + \frac{M_{AB} + M_{BA}}{20} + \frac{M_{CD} + M_{DC}}{20} - 40 = 0$$

or

$$M_{AB} + M_{BA} + M_{CD} + M_{DC} = 400$$

The standard slope deflection equations may now be substituted in the above static condition equations as follows:

$$\frac{2EI}{10}(-2\theta_B - \theta_C) + \frac{2EI}{20}(-2\theta_B + 3R) - \frac{2 \times 20^2}{12} = 0 \qquad (5\text{-}23)$$

$$\frac{2EI}{10}(-2\theta_C - \theta_B) + \frac{2EI}{20}(-2\theta_C + 3R) = 0 \qquad (5\text{-}24)$$

$$\frac{2EI}{20}(-\theta_B + 3R) + \frac{2 \times 20^2}{12} + \frac{2EI}{20}(-2\theta_B + 3R) - \frac{2 \times 20^2}{12}$$

$$+ \frac{2EI}{20}(-2\theta_C + 3R) + \frac{2EI}{20}(-\theta_C + 3R) = 400 \quad (5\text{-}25)$$

The three equations may be simplified for simultaneous solution and, in order, become

$$-6\theta_B - 2\theta_C + 3R = \frac{1333.3}{2EI} \qquad (5\text{-}26)$$

$$-2\theta_B - 6\theta_C + 3R = 0 \qquad (5\text{-}27)$$

$$-3\theta_B - 3\theta_C + 12R = \frac{8000}{2EI} \qquad (5\text{-}28)$$

To demonstrate a method of solution, the three equations will be solved in the following tabular form arranged for computing machine operation:

Equation	\multicolumn{3}{c}{Coefficients of}	Constant
	θ_B	θ_C	R	$\dfrac{}{2EI}$
1	−6	−2	+3	1333.3
2	−2	−6	+3	0
3	−3	−3	+12	8000.0
1′	1	0.333	−0.500	−222.2
2′	1	3.000	−1.500	0
3′	1	1.000	−4.000	−2667.0
1′−2′ = a		−2.667	+1.000	−222.2
3′−1′ = b		0.667	−3.500	−2444.8
a'		−1.000	0.374	−83.1
b'		+1.000	−5.250	−3670.0
$a' + b'$			−4.876	−3753.1
R			1	+768.0
θ_C		1		+370
θ_B	1			+35

The unknowns become

$$\theta_B = +\frac{35}{2EI}; \qquad \theta_C = +\frac{370}{2EI}; \qquad R = +\frac{768}{2EI}$$

The end moments of the members in frame may now be evaluated.

$$M_{AB} = \frac{2EI}{20}(-\theta_B + 3R) + 66.7 = \frac{1}{20}(-35 + 3 \times 768) + 66.7$$
$$= +183.7 \text{ ft-kips}$$

$$M_{BA} = \frac{2EI}{20}(-2\theta_B + 3R) - 66.7 = \frac{1}{20}(-2 \times 35 + 3 \times 768) - 66.7$$
$$= +44.0 \text{ ft-kips}$$

$$M_{BC} = \frac{2EI}{10}(-2\theta_B - \theta_C) = \frac{1}{10}(-2 \times 35 - 370) = -44.0 \text{ ft-kips}$$

$$M_{CB} = \frac{2EI}{10}(-2\theta_C - \theta_B) = \frac{1}{10}(-2 \times 370 - 35) = -77.5 \text{ ft-kips}$$

$$M_{CD} = \frac{2EI}{20}(-2\theta_C + 3R) = \frac{1}{20}(-2 \times 370 + 3 \times 768) = +77.5 \text{ ft-kips}$$

$$M_{DC} = \frac{2EI}{20}(-\theta_C + 3R) = \frac{1}{20}(-370 + 3 \times 768) = +96.7 \text{ ft-kips}$$

Modifications of the slope-deflection equations for members of non-uniform cross-section are developed in Art. 9-7 of Chapter 9.

REFERENCES

1. Sutherland, H., and H. L. Bowman, *Structural Theory*, fourth edition, Chapter 8, New York: John Wiley, 1950.
2. Kinney, J. S., *Indeterminate Structural Analysis*, Chapter 11, Reading, Mass.: Addison-Wesley, 1957.
3. Maney, G. A., *Studies in Engineering*, No. 1, University of Minnesota Press, 1915.
4. Morris, C. T., and S. T. Carpenter, *Structural Frameworks*, New York: John Wiley, 1943.
5. Parcel, J. I., and R. B. B. Moorman, *Analysis of Statically Indeterminate Structures*, Chapter 5, New York: John Wiley, 1955.
6. Wilbur, J. B., and C. H. Norris, *Elementary Structural Analysis*, Chapter 14, New York: McGraw-Hill, 1948.
7. Wilbur, J. B., "Successive Elimination of Unknowns in the Slope Deflection Method," *Trans. Am. Soc. Civil Engineers*, **102**, 346 (1937).
8. Wang, C. K., *Statically Indeterminate Structures*, Chapter 7, New York: McGraw-Hill 1953.

PROBLEMS

5–1. The beam is originally fixed at A and B, after which the support at B gradually rotates until the moment at B is equal to one-half the original fixed-end moment. Determine the rotation at B to produce this condition in terms of P, EI, and L. What is the value of this angle if $P = 12$ kips, $L = 15$ ft, $E = 30 \times 10^6$ psi, and $I = 500$ in.^{4}.   *Ans.* $\theta_B = -4.05 \times 10^{-4}$ radians.

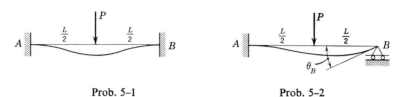

Prob. 5–1   Prob. 5–2

5–2. Determine θ_B by slope deflection principles in terms of P, EI, and L.
Ans. $\theta_B = -PL^2/32EI$.

5–3. Determine the moment at B by the slope deflection method. Constant *I.* $L = 12$ ft.   *Ans.* $M_{BC} = +36$ ft-kips.

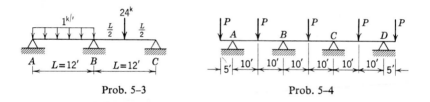

Prob. 5–3   Prob. 5–4

5–4. Determine the moments at B and C by the slope deflection method. Constant *I.*   *Ans.* $M_B = M_C = 2.0P$.

5–5. Do problem 4–13 by the slope deflection method.

5–6. Do problem 4–14 by the slope deflection method.

5–7. Do problem 4–15 by the slope deflection method.

5–8. Do problem 4–16 by the slope deflection method.

5–9. Do problem 4–17 by the slope deflection method.

5–10. Do problem 4–21 by the slope deflection method.

5–11. Do problem 4–24 by the slope deflection method.

5–12. Do problem 4–26 by the slope deflection method.

5–13. Do problem 4–29 by the slope deflection method.

5–14. Do problem 4–33 by the slope deflection method.

5–15. Solve for the deflection of B, and the rotation of B and C.   Constant I. Sketch the final elastic curve.

$$Ans.\ \theta_B = +8.3P/EI;\ \theta_C = 0;\ \Delta_B = +111P/EI;\ M_{AB} = +5P.$$

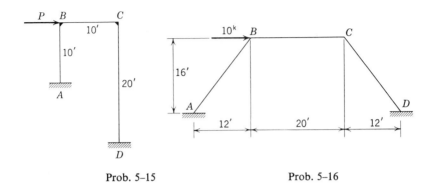

Prob. 5–15   Prob. 5–16

5–16. Solve for the moments in this frame and draw the bending moment diagram.   Constant I.

5–17. Determine all moments by the slope deflection method.   Sketch the final elastic curve.

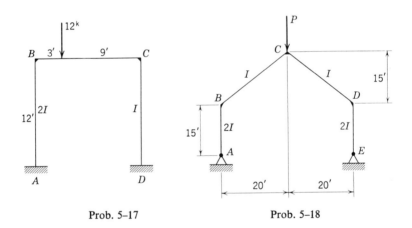

Prob. 5–17   Prob. 5–18

5–18. Solve for the reactions by the slope deflection method.

$$Ans.\ H_A = 0.228P.$$

5–19. A section of concrete culvert 1 ft long is subjected to the soil pressures shown.   Assuming uniform thickness of concrete, calculate the bending moments in culvert structure.

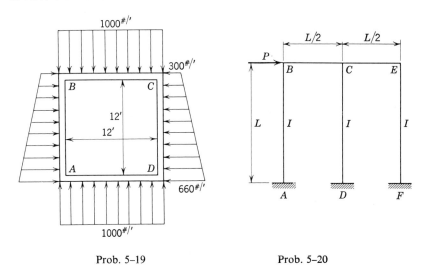

Prob. 5–19   Prob. 5–20

5–20. Determine the horizontal reactions at *A*, *D*, and *F* by assuming that (*a*) *BC* and *CE* are infinitely rigid, (*b*) *BC* and *CE* have a moment of inertia equal to *I*.

Additional problems for solution by the slope deflection method may be selected from the list of problems for Chapter 4.   Problems suggested are 4–4, 4–25, 4–32, and 4–46.

Elastic Center and Column Analogy Methods

6

6-1. INTRODUCTION

The elastic center and the column analogy methods are special methods of solving indeterminate structures of the one-loop form. Many structures may be thought of as consisting of one continuous member forming a completely closed loop. A round pipe is such a structure, as well as all closed quadrangular frames. A two-column portal rigid frame is a closed loop since the soil, through the action of the foundations, closes the loop. A single-span arch is another form of closed loop structure which will be dealt with in a later chapter.

The methods will be found to be formalized approaches involving theory previously presented. The elastic center method will be discussed first and the column analogy method generalized from the elastic center equations.

6-2. ELASTIC CENTER METHOD—BASIC CONSIDERATIONS

Basic considerations of elastic frame distortion due to flexure may begin by referring to the determinate frame of Fig. 1a. In Fig. 1a at a section C, referenced to A by x and y coordinates, a local angular kink of $d\theta$ is introduced. Owing to this local rotation the frame to the left of the section rotates as a rigid structure with A moving along an arc to A'. Since $d\theta$ is to be defined as a small angle AA' equals $r\,d\theta$. The original vertical tangent to A has rotated through $d\theta$, and A has translated $d(\Delta_x)$ and $d(\Delta_y)$ in the x and y directions respectively.

Since triangles $A'AD$ and ACE are similar

$$d(\Delta_x) = -y\,d\theta$$
$$d(\Delta_y) = x\,d\theta$$
and
$$d(\theta_A) = d\theta$$

206

It is to be noted that A has translated to the left in the negative x direction and upward in the positive y direction.

The local kinking of $d\theta$ would in actuality be produced by a bending moment at C. If M is the local moment caused by a real load system, then $d\theta$ may be considered as the differential rotation of a differential length ds. From slope and deflection principles $d\theta = M\,ds/EI$. In Fig. 1a, M has been considered as a positive moment producing tension on the inside of the frame. For a frame under the influence of actual forces a similarly calculated angle change $d\theta$ will occur at many sections. Then the total rotation and displacements of A will be obtained by summarizing

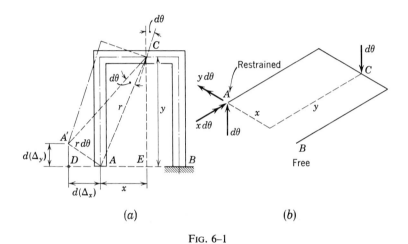

(a)   (b)

FIG. 6–1

all differential effects. In terms of integral calculus, with integration performed over all relevant lengths, the summation of all angle changes is

$$\theta_A = \int \frac{M\,ds}{EI}$$

and the summations of all differential displacements give

$$\Delta_x = -\int \frac{My\,ds}{EI}$$

$$\Delta_y = \int \frac{Mx\,ds}{EI}$$

These expressions relating geometric effects at A to physical angle changes throughout the frame are basic in the development to follow; however, they will also be rederived in terms of a general system of M_s and

M_i moments by using the $\int \dfrac{Mm\ ds}{EI}$ method. It is desirable for the student to reconcile mathematical approaches with their physical meaning at all times.

As was the case in the development of the conjugate-beam method, (see Fig. 10, Chapter 2), the angle $d\theta$ may be considered as an elastic load on a conjugate-space frame with reactions again analogous to rotations and moments analogous to deflections. Figure 1b indicates such a loading condition wherein the load acts downward on the frame shown in a horizontal plane. It is to be noted that A of the conjugate frame is now restrained and B is free to meet the requirements of the analogy as explained in Chapter 2. The reaction at A is $d\theta$ and represents the rotation of the tangent at A. The moment of $d\theta$ about the x axis through A is $y\ d\theta$. The moment of $d\theta$ about the y axis through A is $x\ d\theta$. These moments represent the magnitude of the translations of A of $d(\Delta_x)$ and $d(\Delta_y)$ in the x and y directions. The direction of the moment vectors representing these translations indicates the direction of the translations. Owing to the assumed directions of positive x and positive y with the origin at A, these moment vectors must be drawn in Fig. 1b as shown in agreement with the usual right-hand rule, in order to reconcile all sign conventions and to have the moment vector pointing in the direction of translation. The resultant displacement of A is the vector sum.

Figure 2a represents a closed rectangular frame with A and B fixed. The frame is acted upon by any general form of external loading, and bending moments are generally caused in all members of the frame. Let M signify the final and total moment at any section. Figure 2a also indicates the final values of horizontal and vertical reactions and moment at A, denoted as H_A, V_A, and M_A. The frame, fully cut off at A, but still restrained at B, would be stable and statically determinate. Then H_A, V_A, and M_A would be termed the redundant reactions and the structure is indeterminate to the third degree. This is common for nearly all structures having a closed loop or circuit. The structure with the redundants removed is spoken of as the determinate base structure and is shown in Fig. 2b. The moment in general at any section of this statically determinate structure will be termed M_s. In Fig. 2c, the redundants alone are shown applied in their assumed positive directions. The moments in frame due to these redundants will be termed M_i. Accepting superposition, the final moment, in general for any section of any member of the frame is the sum of M_s and M_i, or in equation form

$$M = M_s + M_i$$

A convenient sign convention for moment has been adopted by denoting moment as positive when the tension side of the member is to the inside of the loop and negative when the tension side is the outside of the loop. Thus, the assumed direction of M_A is in the direction of positive moment.

In common with all other approaches to indeterminate structures, an analysis of slopes and deflections must be made. In Fig. 2b, point A deflects horizontally and vertically, while the tangent at A rotates. Since the structure is determinate in this condition the moments are equal to

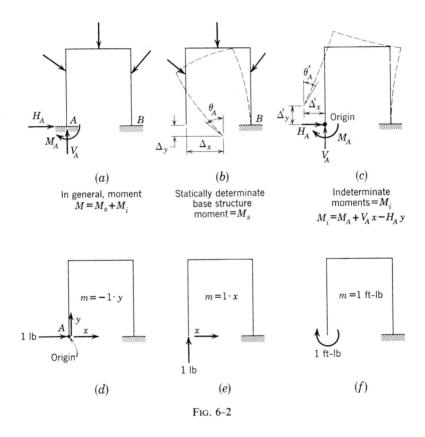

(a)

In general, moment
$M = M_s + M_i$

(b)

Statically determinate
base structure
moment $= M_s$

(c)

Indeterminate
moments $= M_i$

$M_i = M_A + V_A x - H_A y$

(d)

(e)

(f)

Fig. 6–2

M_s, and Δ_y, Δ_x, and θ_A may be determined. In Fig. 2c, the redundants must be of such a value to make $\Delta'_x = -\Delta_x$, $\Delta'_y = -\Delta_y$, and $\theta'_A = -\theta_A$, since A is fixed.

These geometric conditions must be restated in terms of M_s and M_i. For the first developments to follow, the origin of coordinates will be taken at A, letting x be positive to the right and y positive upward. From Fig. 2c, M_i at any section can be stated in terms of the redundants by using a moment equilibrium equation, thus,

$$M_i = M_A + V_A x - H_A y$$

where x and y are the coordinates of any general section of the frame and where the moment sign convention is observed.

Figures 2d, e, and f represent the frame subjected to three different unit loadings to prepare the way for stating deflections and rotations by the general deflection or rotation equation $\int \dfrac{Mm\,ds}{EI}$. It should be particularly noted that ds has been substituted for dx the differential length along a given member. The ds notation is more general and applies for a differential length of horizontal, vertical, or curved members.

In Fig. 2f, the moment m is unity for all sections of the frame due to the unit couple at A. Thus θ_A, of the statically determinate structure Fig. 2b, may be written as

$$\theta_A = \int \frac{M_s\,ds}{EI}$$

where all integrations are to be performed for all members of the frame.

The redundants produce θ'_A, which may be written as $\int \dfrac{M_i\,ds}{EI}$. Hence, the following geometric condition must be satisfied since A does not rotate.

$$\theta_A + \theta'_A = 0$$

This condition may be restated as

$$\int \frac{M_s\,ds}{EI} + \int \frac{M_i\,ds}{EI} = 0$$

and since

$$M_i = M_A + V_A x - H_A y$$

the final mathematical form of the rotation condition is

$$\int \frac{M_s\,ds}{EI} + M_A \int \frac{ds}{EI} + V_A \int \frac{x\,ds}{EI} - H_A \int \frac{y\,ds}{EI} = 0 \qquad (6\text{-}1)$$

The geometric condition expressing zero translation of A in the x direction is written as

$$\Delta_x + \Delta'_x = 0$$

and may be restated in terms of M_s and M_i, by making use of the unit loading given in Fig. 2d, where m, the moment at any section due to the unit horizontal load at A, is equal to $-1 \cdot y$. The results are

$$\Delta_x = \int \frac{M_s(-1 \cdot y) \, ds}{EI} = -\int \frac{M_s y \, ds}{EI}$$

$$\Delta'_x = \int \frac{M_i(-1 \cdot y) \, ds}{EI} = -\int \frac{M_i y \, ds}{EI}$$

hence

$$-\int \frac{M_s y \, ds}{EI} - \int \frac{M_i y \, ds}{EI} = 0$$

and with M_i stated in terms of H_A, V_A, and M_A, this equation becomes

$$-\int \frac{M_s y \, ds}{EI} - M_A \int \frac{y \, ds}{EI} - V_A \int \frac{xy \, ds}{EI} + H_A \int \frac{y^2 \, ds}{EI} = 0 \quad (6\text{-}2)$$

The condition $\Delta_y + \Delta'_y = 0$ at A is similarly expanded, using the unit loading of Fig. 2e, and $m = 1 \cdot x$, to

$$\int \frac{M_s x \, ds}{EI} + M_A \int \frac{x \, ds}{EI} + V_A \int \frac{x^2 \, ds}{EI} - H_A \int \frac{xy \, ds}{EI} = 0 \quad (6\text{-}3)$$

Equations 6-1, 6-2, and 6-3 are sufficient for solving for the three redundants and would be solved by simultaneous equation procedures. Although this is entirely feasible, and the labor is often warranted, a much simplified solution results if the origin of coordinates is established at the so-called elastic center. The concept of the elastic center is credited to Carl Culmann.

To establish such a center we define the elastic center as either the center of gravity of the elastic weights or the centroid of the elastic areas. It may be noted that ds/EI appears in every term of equations 6-1, 6-2, and 6-3. If ds/EI is resolved into its factors ds and $1/EI$, and these factors each considered as linear dimensions, ds can be interpreted as usual, as a differential length along the axis of a member, and $1/EI$ as the elastic width of that member. The product of length and width is termed *area*, and in this instance it is a fictitious area. Figure 3 illustrates the meaning of $ds \times 1/EI$. The term *elastic weight* is usually applied to this area, although the term *elastic area* would be more appropriate. The total elastic weight or area of the member of Fig. 3b of length L and constant moment of inertia is

$$L \times \frac{1}{EI} = \frac{L}{EI}$$

If I had been variable the total elastic weight would be equal to a summation of $ds \times \frac{1}{EI}$ or $\int \frac{ds}{EI}$.

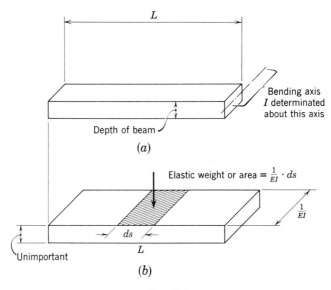

Fig. 6–3

Accepting the definition of elastic weight or area as stated, a frame such as Fig. 2a may have its elastic weight or elastic area variation portrayed as in Fig. 4a. If the separate elastic weights of each member are considered as forces, then a center of gravity exists. If the student also considers that elastic area is a better term, then a centroid of the elastic areas exists. Under either definition, of weight or area, the principle of moments would determine the location of the center of gravity of the forces or centroid of the areas. This center or centroid is called the elastic center

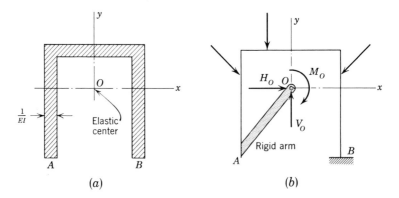

Fig. 6–4

and denoted as O.   In Fig. 4a it is also shown as the origin of new x and y coordinates.   As will be further demonstrated, the use of the elastic center as the origin of coordinates will greatly simplify numerical work. With the definition of elastic areas as described, then the integrals such as $\int \frac{x \, ds}{EI}$ and $\int \frac{y \, ds}{EI}$ may be interpreted as moments of the elastic areas about the coordinate axes.   These integrals will be equal to zero if the coordinate axes are centroidal axes passing through the elastic center.

The argument leading to the simplification of the equations is further based on the concept of an infinitely rigid arm joining A and the elastic center.   Figure 4b shows such an arm rigidly connected to A.   The arm is thought of as rigid so that it will introduce no additional distortions, but will respond as a rigid body to the rotation of A and the end of the arm at O will translate in the x and y directions.   This permits three new redundants to be substituted at O in place of H_A, V_A, and M_A at A.   These redundants are H_O, V_O, and M_O, and their magnitude will be such as to keep the rigid arm from rotating and point O from translating, as would be necessary since A is a fixed end.

The geometric analysis is shown in Fig. 5, where Fig. 5a illustrates the given structure and Fig. 5b the same structure but with the rigid arm attached to A extending to the elastic center O.   The redundant reactions are shown as M_O, H_O, and V_O, and they must hold the rigid arm at O without rotation or translation.

A general solution may be formulated by reducing structure to a statically determinate base structure by removing M_O, H_O, and V_O, as in Fig. 5c.   The statically determinate moments may be represented by M_s. Note that as the frame deflects, the rigid arm rotates by θ_O and its end at O' translates Δ_x and Δ_y.   Figure 5d shows the unloaded structure subjected to the restoring action of the redundants.   Moments created by the redundants are noted as M_i and the rotation and displacement of O denoted as θ'_O, Δ'_x, and Δ'_y.   Geometrical conditions as demanded by the supporting conditions of the original structure are

$$\theta_O + \theta'_O = 0; \qquad \Delta_x + \Delta'_x = 0; \qquad \Delta_y + \Delta'_y = 0$$

Taking the condition $\theta_O + \theta'_O = 0$, and referring to Fig. 5e, where a unit couple is applied to rigid arm at O, the following equation is formulated from equation 6-1 by substituting M_O, V_O, and H_O for M_A, V_A, and H_A, respectively.   The new equation is

$$\int \frac{M_s \, ds}{EI} + M_O \int \frac{ds}{EI} + V_O \int \frac{x \, ds}{EI} - H_O \int \frac{y \, ds}{EI} = 0$$

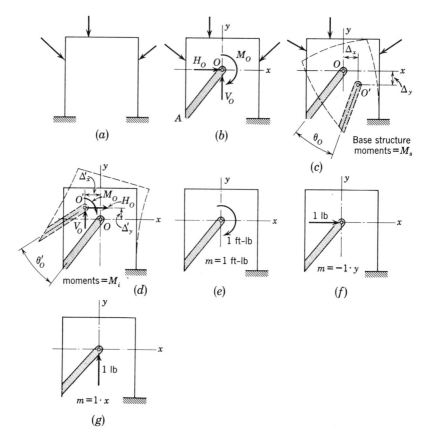

FIG. 6–5

The virtue of the elastic center may now be demonstrated. Since by definition of the elastic center $\int \frac{x\,ds}{EI}$ and $\int \frac{y\,ds}{EI}$ equal zero, the equation reduces to

$$M_O = -\frac{\int \dfrac{M_s\,ds}{EI}}{\int \dfrac{ds}{EI}} \tag{6-4}$$

Thus, the elastic center concept permits direct determination of M_O.

A similar consideration exists with respect to the remaining geometric conditions. For $\Delta_x + \Delta'_x = 0$ and using the unit horizontal force of

Fig. 5*f* and equation 6-2 we obtain

$$-\int\frac{M_s y\,ds}{EI} - M_O\int\frac{y\,ds}{EI} - V_O\int\frac{xy\,ds}{EI} + H_O\int\frac{y^2\,ds}{EI} = 0$$

Substitution of zero for the integrals $\int\frac{y\,ds}{EI}$ and $\int\frac{x\,ds}{EI}$ is again possible.

Additionally, $\int\frac{xy\,ds}{EI}$ equals zero if the x and y axes are principal axes* (or as in this case the y axis is an axis of symmetry of the elastic areas).

The above equation then simplifies to

$$H_O = \frac{\int\dfrac{M_s y\,ds}{EI}}{\int\dfrac{y^2\,ds}{EI}} \qquad (6\text{-}5)$$

which permits a direct evaluation of H_O.

From the geometric condition $\Delta_y + \Delta'_y = 0$, and the unit vertical force of Fig. 5*g* and equation 6-3, it follows that

$$V_O = -\frac{\int\dfrac{M_s x\,ds}{EI}}{\int\dfrac{x^2\,ds}{EI}} \qquad (6\text{-}6)$$

which permits a direct evaluation of V_O.

Equations 6-4, 6-5, and 6-6 are the usual equations of the elastic center method for quadrangular frames possessing an axis of structural symmetry.

The physical significance of the terms of these equations should be appropriately noted. For example, in equation 6-4, $\int\frac{M_s\,ds}{EI}$ equals the rotation of the rigid arm for the statically determinate structure due to the M_s moments, and $\int\frac{ds}{EI}$ represents the rotation of the rigid arm due to a unit couple at O. The indicated division of equation 6-4 may then be interpreted as finding out how many units of the redundant moment, or M_O, are necessary to satisfy $\theta_O + \theta'_O = 0$. Similar interpretations may be placed on the integrals of equations 6-5 and 6-6, wherein the integrals in the numerator represent deflections of O due to M_s, and the integrals in the denominator represent the deflections of O due to the unit loads, in the x direction for equation 6-5 and the y direction for equation 6-6.

* The general case where the x and y axes are not the principal axes is considered in Art. 10–8. This may be taken up now at the option of the instructor and student.

Example

Figure 6a represents a frame that may be easily solved by the elastic center method. For demonstration purposes the frame is assumed to be of uniform moment of inertia.

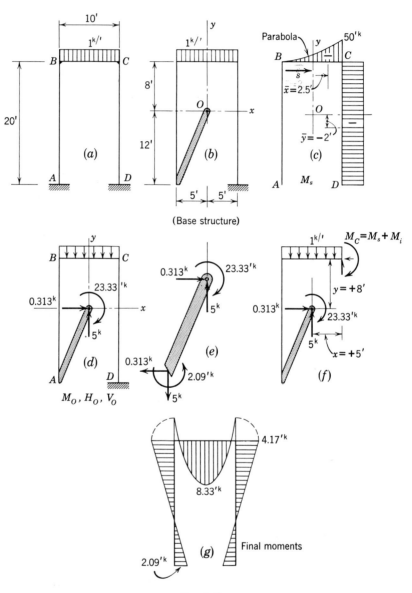

(Base structure)

$M_C = M_s + M_i$

(g) Final moments

FIG. 6–6

The elastic center is found by finding the centroid of the elastic areas. When the moment of inertia is constant for all members, the integrals are easily evaluated by routine methods, such as

$$A = \text{elastic area} = \int \frac{ds}{EI} = \frac{20}{EI} + \frac{10}{EI} + \frac{20}{EI} = \frac{50}{EI}$$

and taking moments of the elastic areas about line BC,

$$\bar{y} = \frac{2 \times (20/EI) \times 10}{(50/EI)} = 8 \text{ ft}$$

The centroid is also located on the vertical axis of symmetry.

The integrals $\int \frac{y^2 \, ds}{EI}$ and $\int \frac{x^2 \, ds}{EI}$ may be interpreted as the moment of inertia of the elastic areas about the x and y axes respectively. (Boldface type is used to differentiate this moment of inertia from that of individual members.) Thus

$$\mathbf{I}_x = \int \frac{y^2 \, ds}{EI}$$

and

$$\mathbf{I}_y = \int \frac{x^2 \, ds}{EI}$$

For the example problem and using standard formulas for moment of inertia of rectangular areas and the transfer formula, we obtain

$$\mathbf{I}_x = \frac{1}{EI} \left(2 \times \frac{1 \times 20^3}{12} + 2 \times 1 \times 20 \times 2^2 + 1 \times 10 \times 8^2 \right) = \frac{2133}{EI}$$

$$\mathbf{I}_y = \frac{1}{EI} \left(1 \times \frac{1 \times 10^3}{12} + 2 \times 1 \times 20 \times 5^2 \right) = \frac{1083}{EI}$$

The values designated above for area and moment of inertia are basic properties of the frame. The next steps involve the loading of the structure and the determination of moments M_s for the statically determinate base structure. Figure 6b indicates the rigid arm from A to O, with the load in place, and Fig. 6c the variation of the M_s moments. The evaluation of the numerator of equations 6-4, 6-5, and 6-6 can now be accomplished by either integration or numerical methods. Each method is illustrated.

INTEGRATION METHOD

To evaluate $\int \frac{M_s \, ds}{EI}$ for the horizontal member an advantage is gained by using the origin for s at B.

Then for BC (note tension produced on outside of member calling for $-$ sign for moment)

$$\int \frac{M_s \, ds}{EI} = -\frac{1}{EI} \int_0^{10} \frac{1s^2}{2} \, ds = -\frac{166.7}{EI}$$

For CD the origin for s can be taken at D, then

$$\int_0^{20} \frac{M_s \, ds}{EI} = -\frac{1}{EI} \int_0^{20} 50 ds = -\frac{1000}{EI}$$

$\int \frac{M_s \, ds}{EI}$ for the frame is the sum of the separate values, or

$$\int \frac{M_s \, ds}{EI} \text{ over all members} = -\frac{1166.7}{EI}$$

Then by equation 6-4

$$M_O = \frac{-\int \frac{M_s \, ds}{EI}}{\int \frac{ds}{EI}} = \frac{-\left(\frac{-1166.7}{EI}\right)}{\frac{50}{EI}} = +23.33 \text{ ft-kips (clockwise)}$$

The next redundant to be computed is H_O, and it is necessary to evaluate $\int \frac{M_s y \, ds}{EI}$ for BC and CD. For BC, y is constant at a value of 8 ft, hence the easiest integral will be secured by again choosing the origin for s at B. Then for BC

$$\int \frac{M_s y \, ds}{EI} = -\frac{1}{EI} \int_0^{10} \frac{1s^2}{2} (8) \, ds = -\frac{4000}{3EI}$$

For the evaluation of the same general integral for CD an advantage is obtained using a variable y, where y is the distance from elastic center vertically to a given ds, and $dy = ds$. This results in

$$\int \frac{M_s y \, ds}{EI} = -\frac{50}{EI} \int_{-12}^{+8} y \, dy = +\frac{2000}{EI}$$

Then by equation 6-5

$$H_O = \frac{-\frac{4000}{3EI} + \frac{2000}{EI}}{\frac{2133}{EI}} = +0.313 \text{ kip (to the right)}$$

Although unnecessary for this illustrative problem, owing to symmetry of loading, the value of V_O will be determined by the integration method. To evaluate $\int \dfrac{M_s x \, ds}{EI}$, x is measured from the elastic center, hence for member BC, let $dx = ds$, then for BC

$$\int \frac{M_s x \, dx}{EI} = -\frac{1}{EI} \int_{-5}^{+5} \frac{(5+x)^2}{2} \cdot (1) \cdot x \, dx = -\frac{416}{EI}$$

and for CD, since $x = +5$ ft for all differential elements

$$\int \frac{M_s x \, ds}{EI} = -\frac{50}{EI} \int_0^{20} 5 dy = -\frac{5000}{EI}$$

The summation of $\int \dfrac{M_s x \, ds}{EI}$ for BC and CD is $-\dfrac{5416}{EI}$. Then by equation 6-6 we may compute the vertical reaction as

$$V_O = -\frac{\int \dfrac{M_s x \, ds}{EI}}{\int \dfrac{x^2 \, ds}{EI}} = -\frac{\left(\dfrac{-5416}{EI}\right)}{\dfrac{1083}{EI}} = +5 \text{ kips (up)}$$

The elastic center redundants are placed on Fig. 6d. Free body methods may now be used to determine the final moments in the frame, or the general equation $M = M_s + M_i$ may be used. To illustrate the free body approach, sever the arm at A, at the end of the column, and draw the free body of Fig. 6e. The equations of static equilibrium establish first that H_A and V_A equal 0.313 and 5 kips respectively. For moment equilibrium

$$M_A = 23.33 + 0.313 \times 12 - 5 \times 5 = +2.09 \text{ ft-kips}$$

directed as shown. The direction is reversed on the column at A. To further illustrate, the free body of Fig. 6f is used to find the moment at C.

$$M_C = -50 - 0.313 \times 8 + 23.33 + 5 \times 5 = -4.17 \text{ ft-kips}$$

We may recompute the moment at C to illustrate the general approach, where

$$M = M_s + M_i = M_s + M_O + V_O x - H_O y$$

The coordinates of C are

$$x = +5 \text{ ft}, \ y = +8 \text{ ft, and } M_s = -50 \text{ ft-kips}$$

Hence

$$M_C = -50 + 23.33 + 5 \times 5 - 0.313 \times 8 = -4.17 \text{ ft-kips}$$

Either procedure will lead to the final moment diagram of Fig. 6g. In general, the moments are computed at the supports and at the joints, and the moment diagram is completed by applying basic principles for construction of such diagrams.

NUMERICAL METHOD

The example of Fig. 6 may be solved by a numerical approach, which depends basically on an interpretation of the integrals in terms of areas and moments of areas.

$\int \frac{M_s \, ds}{EI}$, where $M_s \, ds$ is the area of a differential element of the moment diagram, may be interpreted as

$$\sum \frac{\text{Areas of statically determinate moment diagram}}{EI}$$

$\int \frac{M_s y \, ds}{EI}$, where $y \, (M_s \, ds)$ is the moment of the area of the differential element of the moment diagram, may be interpreted as

$$\sum \frac{\text{Areas of statically determinate moment diagram times } \bar{y}}{EI}$$

where $\bar{y}$ is the vertical coordinate of the centroid of the area from the elastic center measured to, or along, the axis of the member concerned.

Finally, $\int \frac{M_s x \, ds}{EI}$ may be interpreted as

$$\sum \frac{\text{Areas of statically determinate moment diagram times } \bar{x}}{EI}$$

where $\bar{x}$ is the horizontal coordinate of the centroid of the area from the elastic center measured to, or along, the axis of the member concerned.

From Fig. 6c, the areas divided by EI are as follows:

$$\text{For } BC = -\frac{\frac{1}{3} \times 10 \times 50}{EI} = -\frac{500}{3EI}$$

$$\text{For } CD = -\frac{20 \times 50}{EI} = -\frac{1000}{EI}$$

$$\text{summation} = -\frac{3500}{3EI}$$

and $\int \dfrac{M_s y \, ds}{EI}$

$$\text{for } BC = \frac{(-\frac{500}{3})(+8 \text{ ft})}{EI} = -\frac{4000}{3EI}$$

$$\text{for } CD = \frac{(-1000)(-2 \text{ ft})}{EI} = +\frac{2000}{EI}$$

$$\text{summation} \qquad = +\frac{2000}{3EI}$$

and $\int \dfrac{M_s x \, ds}{EI}$

$$\text{for } BC = \frac{(-\frac{500}{3})(+2.5 \text{ ft})}{EI} = -\frac{1250}{3EI}$$

$$\text{for } CD = \frac{(-1000)(+5 \text{ ft})}{EI} = -\frac{5000}{EI}$$

$$\text{summation} \qquad = -\frac{16250}{3EI}$$

Since these summations by the numerical method are identical to those secured by integration, it is obvious that H_O, V_O, and M_O have the same value as before.

6–3. COLUMN ANALOGY

The numerical approach of the elastic center method leads to a generalized procedure known as the column analogy method. Conversely, if the column analogy method had been developed first the elastic center procedures could have been generalized. Principles remain the same; and, as the name suggests, a column stress concept is involved.

The frame of Fig. 6a is reproduced in Fig. 7a. As in every general approach the statically indeterminate structure must be reduced to a statically determinate structure first. Figure 7b shows the moments M_s with the redundants removed at A. To make the analogy, represent the frame as a cross-section of a short column, spoken of as the analogous column, as in Fig. 7c, where the wall thickness of the column is equal to $1/EI$. The cross-sectional area of this column, owing to uniform EI for all members of the original frame, equals $50/EI$. This area is equal to $\int \dfrac{ds}{EI}$. The centroid of the cross-sectional area is found by routine methods, and the principal axes, the x and y axes, are located. *The student can again recognize the physical interpretation of* $\int \frac{ds}{EI}$ *as the rotation of the rigid arm, or the tangent at A, due to the unit couple at O. Physically this rotation represents the total angle changes throughout the frame that are due to this unit couple.*

If x and y are the principal axes of the cross-section, then moments of inertia of the cross-section of the analogous column may be computed about these axes. *The student can also recognize that these moments of inertia physically represent the displacements of the elastic center in the x and y directions respectively, produced by the unit forces at O.* Using a boldface **I** to represent these moments of inertia, it should be clear that

$$\mathbf{I}_x = \int \frac{y^2 \, ds}{EI}$$

$$\mathbf{I}_y = \int \frac{x^2 \, ds}{EI}$$

and as shown before, area is

$$\mathbf{A} = \int \frac{ds}{EI}$$

The above will be immediately recognized as standard quantities employed in the analysis of unit stress for eccentrically loaded short columns.

The next step in the procedure is to consider the M_s diagrams as representing the unit stress variation of an applied loading acting on the top cross-section. This is shown graphically in Fig. 7d. The justification for visualizing the value of M_s as a unit stress intensity will be clearer if it is recalled that differential areas of the analogous column cross-section are equal to ds/EI. If an area is multiplied by a unit stress, a force results. *The student should also recognize, by reference to the basic principles of Fig. 1a, that the total force physically represents the summation of all angle changes produced in frame by M_s.* Hence, an integral of the form $\int \frac{M_s \, ds}{EI}$ may be interpreted as total force. For member BC the total force acting at a centroid of moment diagram is

$$P_1 = -\frac{\frac{1}{3} \times 10 \times 50}{EI} = -\frac{500}{3EI}$$

To provide ultimately an automatic sign convention, P_1 is taken as an upward force (see Fig. 7d), since M_s represented tension on the outside of the frame in Fig. 7b. If the reverse had been true, P_1 would act downward.

P_2 follows as being equal to $-20 \times 50/EI = -1000/EI$. It should be noted that the values agree with those determined previously by the elastic center method. The forces are analogous to concentrated angle loads.

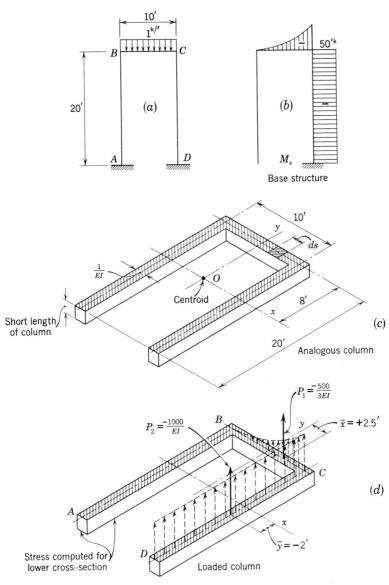

Base structure

Centroid

Analogous column

$P_1 = \dfrac{-500}{3EI}$

$P_2 = \dfrac{-1000}{EI}$

$\bar{x} = +2.5'$

$\bar{y} = -2'$

Stress computed for lower cross-section

Loaded column

FIG. 6–7

After the analogous column loads are computed, then the statical moment of these loads may be found about the x and y axes. *Again the student should recognize, by reference to Fig. 1a, that these moments physically represent the deflection of the elastic center in the direction of the y and x axes respectively.* It is also possible to generalize still further to the concept of moment of the angular loading on the conjugate frame, as depicted in Fig. 1b, wherein moment is equal to the deflections as defined. Letting a boldface **M** represent these moments, then about the x axis by direct analogy

$$\mathbf{M}_x = \int \frac{M_s y \, ds}{EI}$$

and about the y axis

$$\mathbf{M}_y = \int \frac{M_s x \, ds}{EI}$$

We may now substitute the new notation in the general equations of the elastic center solution and obtain

$$M_O = -\frac{\int \dfrac{M_s \, ds}{EI}}{\int \dfrac{ds}{EI}} = -\frac{\Sigma \, \mathbf{P}}{\mathbf{A}}$$

$$H_O = \frac{\int \dfrac{M_s y \, ds}{EI}}{\int \dfrac{y^2 \, ds}{EI}} = \frac{\mathbf{M}_x}{\mathbf{I}_x}$$

$$V_O = -\frac{\int \dfrac{M_s x \, ds}{EI}}{\int \dfrac{x^2 \, ds}{EI}} = -\frac{\mathbf{M}_y}{\mathbf{I}_y}$$

Now, since M_i can be written in its general form as

$$M_i = M_O + V_O x - H_O y$$

then, by substitution

$$M_i = -\frac{\Sigma \, \mathbf{P}}{\mathbf{A}} - \frac{\mathbf{M}_y(x)}{\mathbf{I}_y} - \frac{\mathbf{M}_x(y)}{\mathbf{I}_x} \qquad (6\text{-}7)$$

The equation for M_i, consisting of three terms, permits the analogy to be completed. The first term of equation 6-7 represents the uniform intensity of direct stress on the internal cross-section of the analogous column shown in Fig. 7d, and in this instance represents tension on that cross-section to oppose P_1 and P_2 which are acting upward. The sign of this direct tensile stress will work out algebraically as positive. The second and third terms of the equation represent a flexural unit stress variation on the interior cross-section. The sign of this stress will be established automatically, on an algebraic basis of positive for tension and negative for compression, if the sign convention for moments and co-ordinates is followed.

The foregoing discussion should make it clear that the combined unit stress at any point on the interior column cross-section is equal to M_i, the indeterminate component of the total moment at this section of the frame. It also follows that the final bending moment in actual indeterminate structure is equal to $M_s + M_i$. The sign of stress + for tension and − for compression is identically equal to the original convention where positive moment in the frame was defined as producing tension on the inside of the loop and negative moment tension on the outside of the loop.

The student should never lose sight of the geometric premises from which this analogy stems. The final moments are those that are compatible with the geometric conditions as set forth in the elastic center method.

Following a complete algebraic procedure and employing the data recorded on Fig. 7d, we obtain

$$\Sigma\,\mathbf{P} = -\frac{166.7}{EI} - \frac{1000}{EI} = -\frac{1166.7}{EI}$$

$$\mathbf{M}_x = -\frac{500}{3EI}(8) - \frac{1000}{EI}(-2)$$

$$= +\frac{667.3}{EI}$$

$$\mathbf{M}_y = -\frac{500}{3EI}(2.5) - \frac{1000}{EI}(5)$$

$$= -\frac{16250}{3EI}$$

At A

$$x = -5, \quad y = -12$$

Then by equation 6-7,

$$\underset{\text{at } A}{M_i} = -\frac{(-1166.7/EI)}{50/EI} - \frac{(-16250/3EI)(-5)}{1083/EI} - \frac{(+667.3/EI)(-12)}{2133/EI}$$

$$M_{i(A)} = +23.33 - 25 + 3.76 = +2.09 \text{ ft-kips}$$

and since M_s at A equals zero, M_i represents the total moment.

At C

$$x = +5, \quad y = +8$$

By equation 6-7

$$\underset{\text{at } C}{M_i} = -\frac{(-1166.7/EI)}{50/EI} - \frac{(-16250/3EI)(+5)}{1083/EI} - \frac{(+667.3/EI)(+8)}{2133/EI}$$

$$M_{i(C)} = +23.33 + 25 - 2.50 = +45.83 \text{ ft-kips}$$

Finally the moment at C may be written as

$$M_C = M_s + M_i = -50 + 45.83 = -4.17 \text{ ft-kips}$$

Although it has been pointed out that an analogy exists between M_i and unit stress in an eccentrically loaded column, full realization of the analogy is obtained only when the complex sign convention is avoided. Knowing that the analogy exists, complete freedom from the stated sign convention may be gained by interpretation of unit stresses as for an eccentrically loaded column. This is equivalent to stating

$$M_i = \pm\frac{\Sigma \mathbf{P}}{A} \pm \frac{\mathbf{M}_y x}{I_y} \pm \frac{\mathbf{M}_x y}{I_x}$$

When EI is constant the work can be greatly simplified by calling EI unity.

Hence, for Figs. 7c and 7d

$$\mathbf{A} = 50; \quad \mathbf{I}_x = 2133; \quad \mathbf{I}_y = 1083$$
$$\Sigma \mathbf{P} = 1166.7$$
$$\mathbf{M}_y = \tfrac{16250}{3}$$
$$\mathbf{M}_x = 667.3$$

A stress or moment calculation at A with due recognition of a positive sign for tension and negative sign for compression is

$$M_{i(A)} = \frac{1166.7}{50} - \frac{16250 \times 5}{3 \times 1083} + \frac{667.3 \times 12}{2133}$$
$$= +23.33 - 25 + 3.76 = +2.09 \text{ ft-kips}$$

where the plus sign indicates tension stress at A. Since the interior cross-section of the analogous column is likened to inside of a loop, the tension indicates the presence of a moment at A producing tension on the inside of the frame.

Thus, it may be summarized that the frame can be treated as a case of column stress calculation. There is no particular advantage of the column analogy method over the elastic center method unless the method is

reduced to the routine method last shown. Otherwise, the sign convention remains unsimplified and the two methods are equivalent in time and
effort.

An analysis by column analogy may also be simplified by choosing a
statically determinate structure that will provide a less complicated M_s
diagram.   Figure 8a is the same structure as Fig. 7a; however, in Fig. 8b,

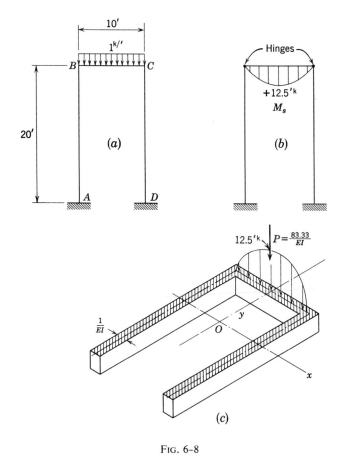

FIG. 6–8

taking full advantage of symmetry of loading, a different statically determinate structure is chosen. The rigid connections at B and C are
replaced with hinges and the M_s diagram is a parabola since BC now acts
as a simple span. The analysis now goes forward by choosing the
analogous column (the same as in Fig. 7, and data are not repeated), and
this column cross-section is loaded with the M_s diagram. The ordinates

of the M_s diagram are directed downward since loading produced tension on inside of frame (contrast with Figs. 7b and 7d).   The relevant quantities are

$$\mathbf{P} = \tfrac{2}{3} \times 10 \times 12.5 = \tfrac{250}{3} = 83.33 \text{ (down)}$$

$$\mathbf{M}_x = 83.33 \times 8 = 666.7$$

$$\mathbf{M}_y = 83.33 \times (0) = 0$$

Then
$$\underset{\text{at } A}{M_i} = -\frac{83.33}{50} + \frac{666.7}{2133} \times 12$$

$$= -1.67 + 3.76 = +2.09 \text{ ft-kips}$$

which checks the previous result at A.

$$\underset{\text{at } C}{M_i} = -\frac{83.33}{50} - \frac{666.7}{2133} \times 8$$

$$= -1.67 - 2.50 = -4.17 \text{ ft-kips}$$

and, since M_s equals zero at C, -4.17 ft-kips is the sum of $M_s + M_i$ and represents the moment at C.

6–4. INDETERMINATE BEAMS BY COLUMN ANALOGY

A single-span beam restrained at the ends may be considered as a structure of one loop if the rigid foundation and beam is thought of as forming a closed loop.   In Fig. 9a, a beam fixed at both ends may be made statically determinate in a number of ways.   To illustrate the method, the support at B is removed and the M_s diagram of Fig. 9b drawn.   The analogous column is drawn in Fig. 9c and loaded correctly with the M_s diagram.

Owing to the uniform EI of the beam, the width of the column may be taken as proportional to unity.   All calculations for M_A and M_B are given in Fig. 9, and they should be verified by the student.   As an exercise the student should rework this example with the beam made statically determinate by hinging A and B.

The student should realize that the column analogy method is a routine method and that his ability to obtain answers by this technique adds little to his basic knowledge of structural action.   This technique is in contrast to the moment distribution method, or elastic center approach, which permits full recognition of structural distortion.   The method, however, is in common use and should be understood.

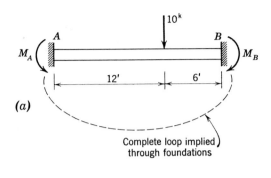

(a)

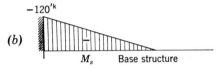

(b)

$$I_y = \frac{1 \times 18^3}{12} = 486$$

$$A = 1 \times 18 = 18$$

$$M_y = 720 \times 5 = 3600$$

$$M_{i(B)} = + \frac{720}{18} - \frac{3600 \times 9}{486}$$

$$= +40 - 66.7 = -26.7'^k$$

$$M_B = M_s + M_i = -26.7'^k$$

$$M_{i(A)} = +40 + 66.7 = +106.7'^k$$

$$M_A = -120 + 106.7 = -13.3'^k$$

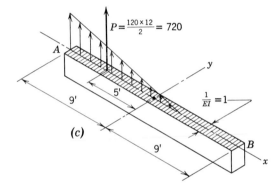

(c)

FIG. 6–9

6–5. STRUCTURES HINGED AT BASE

The elastic center method was demonstrated for symmetrical quad-rangular frames with columns fixed at the base. If the columns had been hinged at the base the elastic center would have to be relocated and new elastic properties computed. The key to the analysis is based on the definition of the elastic area. A differential elastic area was defined as ds/EI, and since $1/EI$ was defined as the analogous width of the member the width of member increases with a decrease in I. This concept of an elastic area for a hinge at A and D leads to the necessity of treating the local width of the member at a hinge as equal to infinity since a hinge possesses zero moment of inertia. Mathematically stated: For a hinge as

$I \to 0$, $1/EI \to \infty$.   With infinite width, the local area representation for a hinge, or ds/EI, must be shown as an area equal to infinity.   This is depicted in Fig. 10c.   In case both A and D are hinged, as in Fig. 10a, two localized areas of infinity are involved.

With both A and D hinged in Fig. 10a and elastic areas as defined, the elastic center would be located at the same level as A and D and half-way between, as shown in Fig. 10b.   The rigid arm then extends from A to this new center.   It must be realized, of course, that if A and D were both hinges, the original structure was indeterminate to the first degree. Hence, in reducing structure to a base structure, A would be placed on rollers instead of remaining absolutely free as in the general case of three redundants.

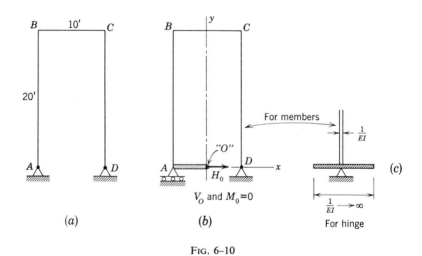

FIG. 6–10

The calculation of elastic center redundants is reduced to the calculation of H_O only.   If the hinges must be treated as possessing infinite area, then $\int \dfrac{ds}{EI}$ for the entire frame is infinite.   With infinite area, equation 6-4 for M_O indicates that M_O equals zero no matter how the frame is loaded. Similarly $\int \dfrac{x^2\,ds}{EI}$ is infinite, hence from equation 6-6, V_O is zero no matter how the frame is loaded.   However, H_O is finite since $\int \dfrac{y^2\,ds}{EI}$ is finite and its value may be computed from equation 6-5.

Figure 11a represents another example of hinged ends at A and C. Following the principles discussed, A and C must be considered as having infinite area as denoted in Fig. 11*b*.   The centroid of the elastic areas is marked in Fig. 11*c* as the elastic center *O*.   The principal *x* and *y* axes are shown and the rigid arm drawn from *A* to "*O*".   Standard procedures and equation 6-5 for any loading will suffice to compute the one redundant reaction H_O.

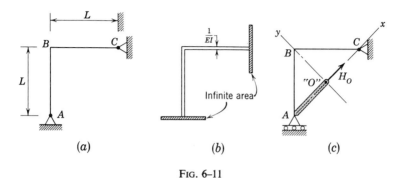

FIG. 6-11

6-6. ELASTIC CENTER METHOD—ADDITIONAL TOPICS

Additional applications of the principles of the elastic center method are given in Chapter 7 and in Chapter 10, dealing with curved members and arches.

REFERENCES

1. Cross, H., and N. D. Morgan, *Continuous Frames of Reinforced Concrete*, Chapter 3, pp. 46–60, New York: John Wiley, 1932.
2. Kinney, J. S., *Indeterminate Structural Analysis*, Reading, Mass.: Addison-Wesley, 1957.
3. Parcel, J. I., and R. B. B. Moorman, *Analysis of Statically Indeterminate Structures*, Chapter 4, pp. 169–206, New York: John Wiley, 1955.

PROBLEMS

6–1. Calculate the displacement of the elastic center due to a unit force in the x direction, the y direction, and the rotation due to a unit clockwise couple, all applied separately at the elastic center.   Constant I.

Ans. $\delta_x = 576/EI$; $\delta_y = 1008/EI$; $\theta = 36/EI$.

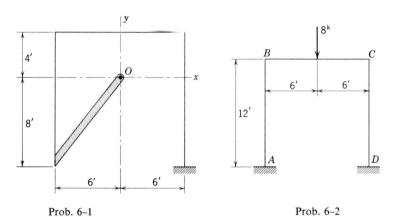

Prob. 6–1   Prob. 6–2

6–2. Determine the moment at A by the elastic center method and check your result by a method of your choice.   Use data from problem 6–1.   Constant I.   *Ans.* $M_A = +4$ ft-kips.

6–3. Determine the moment at A by the elastic center method.   Constant I.

Ans. $M_A = -3.43P$.

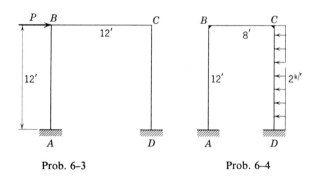

Prob. 6–3   Prob. 6–4

6–4. Determine the moments by the elastic center method and draw the moment diagram.   Constant I.

6–5. Determine the horizontal reactions by the elastic center method. For a hinge $\dfrac{ds}{I}$ approaches infinity and the elastic center is on a level with A and D. Constant I. *Ans.* $H_A = 0.60$ kip.

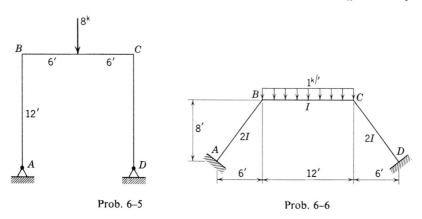

Prob. 6–5   Prob. 6–6

6–6. Determine the moments by the elastic center method and draw the moment diagram. Derive general equations for determining the contribution of an inclined member to the properties of the frame about the elastic center. These general equations may then be used in other problems.

Ans. $M_A = +4.96$ ft-kips; $M_B = -9.91$ ft-kips.

6–7. Determine the moments by the elastic center method.

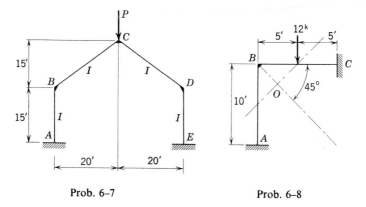

Prob. 6–7   Prob. 6–8

6–8. Determine the moments and the reactions by the elastic center method. Use coordinate axes as indicated by dotted lines. (Why?) Constant I.

Ans. $M_A = +3.75$ ft-kips; $V_A = 4.88$ kips.

6–9. The member BC has its temperature raised by ΔT degrees. Determine the moment at A by the elastic center method in terms of ΔT, EI, L, and the coefficient of expansion.   *Ans.* $M_A = 2C_x \Delta TEI/L$.

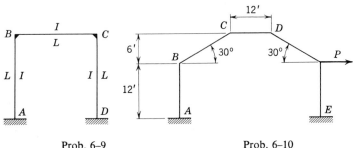

Prob. 6–9   Prob. 6–10

6–10. Determine the reactions and the moments by the elastic center method. Constant I.

Note to Instructor: *Although problems involving the elastic center method and the column analogy method may be assigned for solution by one of the methods, it is advisable to assign some problems to be solved in the same assignment by both methods.*

6–11. Do problem 6–2 by the column analogy method, (*a*) using base structure with redundants removed at A, (*b*) using base structure with hinges at B and C.

6–12. Do problem 6–3 by the column analogy method.

6–13. Do problem 6–6 by the column analogy method.

6–14. Do problem 6–9 by the column analogy method.

6–15. Do problem 6–10 by the column analogy method.

6–16. Determine the fixed-end moments at A and B by the column analogy method, (*a*) using base structure with redundants at A removed, (*b*) using base structure with hinges at A and B.   *Ans.* $M_A = 11wL^2/192$.

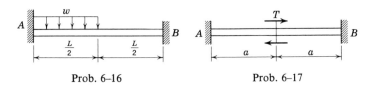

Prob. 6–16   Prob. 6–17

6–17. A couple T is applied in the vertical plane of the beam as shown. Determine the fixed-end moments by the column analogy method.
    Ans. $M_A = T/4$.

6–18. Determine the moments at A and B by the column analogy method.

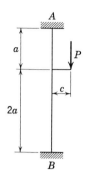

Prob. 6–18

6–19. Determine the moments at A and B by the column analogy method.

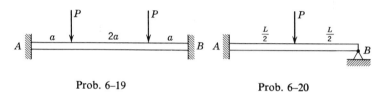

Prob. 6–19   Prob. 6–20

6–20. Determine the moment at A by the column analogy method.

Ans. $M_A = 3PL/16.$

Work and Energy Methods

7

7-1. INTRODUCTION

Work and energy methods have been introduced in earlier chapters on deflections and fundamental analysis of indeterminate structures. This chapter will expand these principles to curved frames and curved beams, complex frames, and indeterminate trusses.

7-2. CURVED FRAMES

Figure 1 illustrates a semicircular frame of uniform EI, loaded with one load P at the center line. A and C are pinned and fixed against displacement. The vertical reactions are determinate, and the prevention of horizontal deflection has caused horizontal redundant reactions of H. As a familiar first step the structure is made determinate by placing a roller at C. Flexure of the structure causes C to move Δ_s to the right as shown in Fig. 1b. If H of the proper magnitude is applied to C', this point would be returned to C.

To determine H first find Δ_s. In Fig. 1c, a unit horizontal force is applied to C and the moment produced in the frame at a section located by angle θ is $m = -R \sin \theta$. From Fig. 1b, M_s, the moment in determinate structure, is $+(P/2) \cdot R(1 - \cos \theta)$. (Note sign convention for moment, negative for tension on outside, positive for tension on inside.)

From the previous theory of Chapter 2 and equation 2-5

$$1 \cdot \Delta_s = \int \frac{M_s m \, ds}{EI}$$

and in terms of this problem, with both quadrants taken into account,

$$1 \cdot \Delta_s = \frac{2}{EI} \int_0^{\pi/2} \frac{P}{2} R(1 - \cos \theta)(- R \sin \theta)R \, d\theta$$

$$1 \cdot \Delta_s = -\frac{PR^3}{EI} \int_0^{\pi/2} (1 - \cos \theta)(\sin \theta)d\theta$$

which results in

$$\Delta_s = -\frac{PR^3}{2EI}$$

(The negative sign indicates that C' moves opposite to the direction of a 1-lb load at C.)

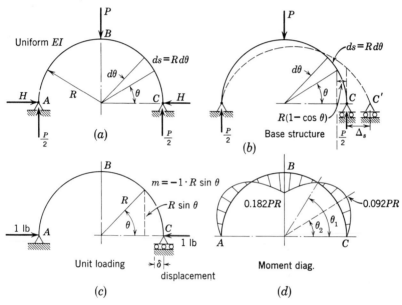

FIG. 7–1

To find H it is necessary to determine the deflection δ of C in Fig. 1c, due to the unit load. If m signifies the moment due to unity load, then

$$1 \cdot \delta = \int \frac{m^2 \, ds}{EI}$$

and in terms of this problem with both quadrants taken into account

$$1 \cdot \delta = \frac{2}{EI} \int_0^{\pi/2} (R \sin \theta)^2 R \, d\theta$$

$$1 \cdot \delta = \frac{2R^3}{EI} \int_0^{\pi/2} \sin^2 \theta \, d\theta$$

$$1 \cdot \delta = \frac{2R^3}{EI} \left[\frac{\theta}{2} - \frac{\sin 2\theta}{4} \right]_0^{\pi/2} = \frac{\pi R^3}{2EI}$$

The force H required to restore C' to point C in Fig. 1b, from the geometric condition $\Delta_s + \delta \cdot H = 0$, is

$$H = -\frac{\Delta_s}{\delta} = -\frac{-PR^3/2EI}{\pi R^3/2EI} = +\frac{P}{\pi}$$

An alternate solution, using $1 \cdot \Delta = \int \frac{Mm \ ds}{EI}$ directly, can be achieved by first drawing Fig. 1c and placing the unit horizontal load on at C. This will induce the m moment in the frame and cause C to move δ. Then, and this is the important step, lock the rollers at C, thus restoring the mode of support to that of the given original structure. With the structure so restored, place the load P on at B, causing the horizontal reactions to be developed. The physical condition in setting up the basic equation uses the fact that the 1-lb load cannot move after restoration of the true end condition at C and hence can do no work, or

$$\int \frac{Mm \ ds}{EI} = 0$$

where M is $[(P/2)R(1 - \cos \theta) - HR \sin \theta]$, and m is $-R \sin \theta$, or

$$\frac{2}{EI} \int_0^{\pi/2} \left[\frac{P}{2} R(1 - \cos \theta) - HR \sin \theta\right][-R \sin \theta]R \ d\theta = 0$$

Integrating and substituting, we again find

$$H = \frac{P}{\pi}$$

Another approach for solving the problem comes from Castigliano's theorem where bending strain energy is used. The strain energy by equation 2-4 is

$$U = \int \frac{M^2 \ ds}{2EI}$$

and in terms of this problem

$$U = \frac{2}{2EI} \int_0^{\pi/2} \left[\frac{P}{2} R(1 - \cos \theta) - HR \sin \theta\right]^2 \cdot R \ d\theta$$

since $\qquad \frac{\partial U}{\partial H} = $ deflection in direction of H

and, since horizontal deflection of supports is zero, we have a condition equation of the form

$$\frac{\partial U}{\partial H} = 0$$

hence, for this problem

$$\frac{\partial U}{\partial H} = \frac{2}{EI} \int_0^{\pi/2} \left[\frac{P}{2} R(1 - \cos \theta) - HR \sin \theta \right] [-R \sin \theta] R \, d\theta = 0$$

which is identical to the final integral of previous approach and will lead to the same result for H.

It may be summarized then that several procedures exist to solve the problem and, although they vary in their methods of analysis, the results are identical. No surprise over this fact is necessary since they are all deflection theories and actually all stem from the first fundamental, namely, $1/\rho = M/EI$.

The bending moment diagram for the curved frame of Fig. $1a$ is shown in Fig. $1d$. The calculations relating to its construction follow:

The bending moment at B is

$$\frac{P}{2} \cdot R - \frac{P}{\pi} \cdot R = 0.182PR$$

The general equation for bending moment at a section located at an angle θ from the horizontal is

$$M = -\frac{P}{\pi} \cdot R \sin \theta + \frac{P}{2} R(1 - \cos \theta)$$

and if the point of inflection is to be located by the angle θ_1, then we set M equal to zero and obtain the equation

$$-\frac{P}{\pi} R \sin \theta_1 + \frac{PR}{2} (1 - \cos \theta_1) = 0$$

$$\frac{\sin \theta_1}{\pi} - \frac{1}{2} + \frac{\cos \theta_1}{2} = 0$$

substituting $\qquad \cos \theta_1 = \sqrt{1 - \sin^2 \theta_1}$

and solving, gives $\qquad \theta_1 = 65°$

To locate the section at which maximum moment occurs between $\theta = 0$ and $\theta = \theta_1$, set the derivative of M with respect to θ equal to 0, or

$$\frac{dM}{d\theta} = -\frac{P}{\pi} R \cos \theta + \frac{PR}{2} \sin \theta = 0$$

$$\tan \theta_2 = \frac{2}{\pi}$$

and $\qquad \theta_2 = 32.5°$

Then M at θ_2 equals $-0.092PR$ by substitution of the functions of θ_2 in the general moment equation.

7–3. CURVED FRAME—RESTRAINED SUPPORTS

Figure 2a represents a semicircular frame of uniform moment of inertia fully fixed at A and C. Although classified as indeterminate to the third degree, the condition of symmetry reduces the necessary geometric equations to two. The physical conditions leading to the formation of two additional equations beyond those of static equilibrium are that no horizontal deflection occurs at C relative to B (or to A), and no rotation of the tangent occurs at C relative to the tangent to elastic curve at B. (By using B as a reference, only one quadrant of the semicircle enters the following integrations.) H and M_C are used as the redundants.

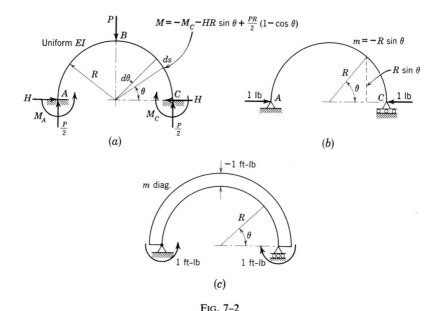

FIG. 7–2

To write the equations based on the above physical relations, two supplemental drawings are made of the structure and unit horizontal loads are applied in Fig. 2b and unit couples in Fig. 2c. The following equations may now be written.

Based on $1 \cdot \Delta = \int \dfrac{Mm\,ds}{EI} = 0$ we obtain

$$\int_0^{\pi/2} \left[-M_C - HR \sin\theta + \frac{P}{2} R(1 - \cos\theta) \right] [-R \sin\theta] R\, d\theta = 0$$

which upon integration and substitution of limits gives

$$M_C + \frac{HR\pi}{4} = \frac{PR}{4} \tag{7-1}$$

The second equation based on $1 \cdot \theta_c = \int \frac{Mm \, ds}{EI} = 0$ is

$$\int_0^{\pi/2} \left[-M_C - HR \sin \theta + \frac{P}{2} R(1 - \cos \theta) \right] [-1] R \, d\theta = 0$$

providing the following simplified equation after integration

$$M_C \frac{\pi}{2} + HR = -\frac{PR}{2} + \frac{PR\pi}{8} \tag{7-2}$$

Equations 7-1 and 7-2, when solved simultaneously, result in

$$H = \frac{P(\pi - 4)}{8 - \pi^2} = 0.458P$$

$$M_C = -0.11PR$$

where the negative sign indicates that M_C as well as M_A act in the opposite direction to that assumed.

Similar equations would result from the Castigliano approach. For example,

$$U = \frac{1}{2EI} \int_0^{\pi/2} \left[-M_C - HR \sin \theta + \frac{PR}{2} (1 - \cos \theta) \right]^2 R \, d\theta$$

Taking the derivative with respect to H, and setting it equal to zero, we obtain

$$\frac{\partial U}{\partial H} = \frac{1}{EI} \int_0^{\pi/2} \left[-M_C - HR \sin \theta + \frac{PR}{2} (1 - \cos \theta) \right] [-R \sin \theta] R \, d\theta = 0$$

and taking the derivative with respect to M_C and setting it equal to zero, we obtain

$$\frac{\partial U}{\partial M_C} = \frac{1}{EI} \int_0^{\pi/2} \left[-M_C - HR \sin \theta + \frac{PR}{2} (1 - \cos \theta) \right] [-1] R \, d\theta = 0$$

Since the integrals of these equations are identical to those found before, the results will be identical. Hence, it is a matter of personal preference as to the method used.

7-4. STRAIGHT AND CURVED MEMBERS COMBINED

Figure 3a illustrates a frame with vertical columns and a semicircular top member. For purposes of illustration it is assumed that EI is constant. The structure is indeterminate to the first degree and the horizontal reaction is taken as the redundant. The solution by $\int \frac{Mm\,ds}{EI}$ method is illustrated. Figure 3b indicates the unit loading required to involve the condition of zero horizontal deflection at the supports.

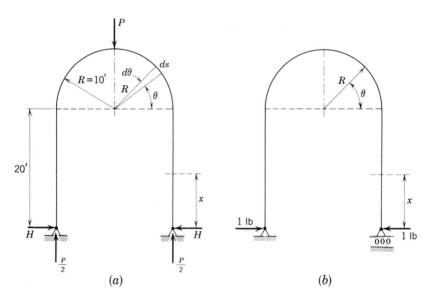

(a)   (b)

FIG. 7-3

With integration over one half the frame

$$EI\Delta = \int_0^{20} Hx \cdot x \, dx$$

$$+ \int_0^{\pi/2} \left[(20 + R \sin \theta)H - \frac{P}{2} R(1 - \cos \theta) \right] [20 + R \sin \theta] R \, d\theta = 0$$

Upon integration and substitution of $R = 10$ ft

$$H = 0.0597P$$

7-5. CLOSED RINGS

Figure 4a represents a closed ring of uniform EI loaded by two loads P along a diameter. It is further assumed that R is large relative to thickness of ring. The dotted outline indicates the elastic curve. The ring is indeterminate to the third degree but symmetry simplifies the problem to one redundant. The physical condition, wherein the horizontal deflection of the half ring above the horizontal diameter is equal to the horizontal deflection of the half ring below this diameter, precludes a horizontal shearing force in the ring at A and C in the present case. Owing to symmetry of loading and framing, the existence of internal forces and moments may be examined by using the principles of symmetry. The use of symmetry is shown in Fig. 4b, where the ring is cut apart on the horizontal diameter and the possibility of moment, shear, and normal force shown. Since forces on the lower half should be equal and oppositely directed to the forces on the upper half, the lower half should be super-imposable on the top half. If superposition is accomplished by rotating the lower half upward into congruency with the top half, it is found that moments and vertical forces superimpose but that the H forces do not. We may then conclude that H equal to zero is the only value compatible with symmetry and that H must vanish.

With only one redundant, the moment at A or C, one equation from geometry will suffice. To use the condition of zero rotation of the vertical tangent at A relative to the horizontal tangent at B, the m diagram of Fig. 4c is first drawn. Then

$$EI \cdot \theta = \int_0^{\pi/2} \left[-M_A + \frac{P}{2}(R - R\cos\theta) \right][-1]R\,d\theta = 0$$

from which we obtain

$$M_A = \left(\frac{\pi - 2}{2\pi} \right)PR = +0.182PR$$

The positive result indicates that M_A acts in the direction assumed in Fig. 4b. The final moment diagram is drawn in Fig. 4d.

The deflections of a ring are often needed, and to illustrate such a calculation the change in horizontal diameter during flexure, $2\Delta_x$, will be found. To proceed in the least involved manner the unit load method is adopted. This permits the ring to be simplified to the statically determin-ate form shown in Fig. 4e by the introduction of temporary hinges at A and C. Two unit loads are applied as shown, producing m moments equal to $\frac{1}{2}R\sin\theta$. After inducing the m moments, assume that the hinges are welded solid, thus restoring original continuity to the ring. With the

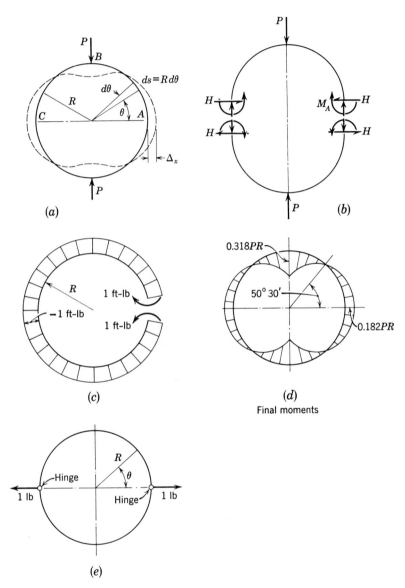

(a)

(b)

(c)

(d)

Final moments

(e)

Fig. 7–4

ring so restored, but with unit loads still acting, apply the real loads P, inducing the true flexural moments. The loads P produce horizontal deflection, and each of the unit loads does virtual work of $1 \cdot \Delta_x$. Hence

$$EI \cdot 1 \cdot \Delta_x = 2 \int_0^{\pi/2} \left[-M_A + \frac{P}{2}(R - R \cos \theta) \right] [+\tfrac{1}{2}R \sin \theta] R \, d\theta$$

where $M_A = \left(\dfrac{\pi - 2}{2\pi} \right) PR$, as determined priorly, will yield

$$\Delta_x = + \frac{0.068 PR^3}{EI}$$

By similar analysis but with the unit loads acting along a vertical diameter, one-half of the change in vertical diameter of ring is $0.075 PR^3/EI$. The checking of this is left as an exercise for the student.

7–6. CLOSED RING—ALTERNATE APPROACHES

As in most problems of structural analysis, alternate approaches also exist for the solution of the closed ring. The first given here is based on the Castigliano theorem.

Referring to Fig. 4a, $M = \left[-M_A + \dfrac{PR}{2}(1 - \cos \theta) \right]$. The strain energy for all quadrants of the ring is

$$U = \frac{4}{2EI} \int_0^{\pi/2} \left[-M_A + \frac{PR}{2}(1 - \cos \theta) \right]^2 R \, d\theta$$

and, owing to the fact that the tangent remains vertical at A, we may write

$$\frac{\partial U}{\partial M_A} = \int_0^{\pi/2} \left[-M_A + \frac{PR}{2}(1 - \cos \theta) \right] [-1] R \, d\theta = 0$$

which upon simplification gives

$$M_A = \frac{\pi - 2}{2\pi} PR \text{ as found before}$$

Since the ring is a closed loop, the elastic center method is also applicable. In Fig. 5a, the ring is reduced to a determinate base structure by severing at C. M_s, the statically determinate moment for right half, is $-PR \sin \theta$. Owing to symmetry, the elastic center is at the centroid of the ring, and hence in Fig. 5b, a pair of rigid arms extends from C to "O". (Arms are separated on drawing for clarity, whereas they actually coincide.) The physical conditions at C require that point "O" of the upper arm and point "O" of the lower arm coincide and that neither arm shall rotate, or, in other words, there can be no relative deflection or rotation of the two arms.

Since EI is constant let it equal unity in determining the following:

$$\mathbf{A} = \int \frac{ds}{EI} = \int ds = 2\pi R$$

$$\mathbf{I}_x = \int \frac{y^2 \, ds}{EI} = \int y^2 \, ds$$

$$\mathbf{I}_y = \int \frac{x^2 \, ds}{EI} = \int x^2 \, ds$$

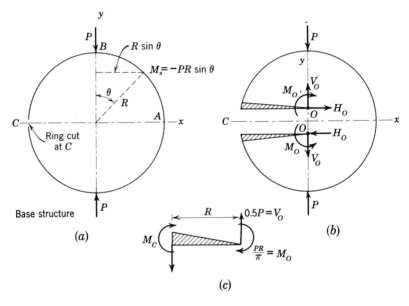

FIG. 7–5

Since $\mathbf{I}_x + \mathbf{I}_y = \mathbf{J}$, polar moment of inertia, and

$$\mathbf{J} = 2\pi R \cdot R^2 = 2\pi R^3$$

but with $\mathbf{I}_x = \mathbf{I}_y$, then

$$\mathbf{I}_x \quad \text{or} \quad \mathbf{I}_y = \pi R^3$$

With properties of the ring so determined, first evaluate

$$\int \frac{M_s \, ds}{EI} = -2 \int_0^{\pi/2} \frac{(PR \sin \theta)(R \, d\theta)}{EI}$$

$$= \frac{-2PR^2}{EI}$$

From the previous elastic center theory, equations 6-4, 6-5, and 6-6,

$$M_O = -\frac{\int \dfrac{M_s\,ds}{EI}}{\int \dfrac{ds}{EI}} = -\frac{(-2PR^2)}{2\pi R} = +\frac{PR}{\pi}$$

To determine V_O (although its value is evident in this instance), evaluate

$$\int \frac{M_s x\,ds}{EI} = 2\int_0^{\pi/2} [-PR\sin\theta][+R\sin\theta]R\,d\theta$$

$$= -\frac{\pi PR^3}{2}$$

Then

$$V_O = -\frac{\int \dfrac{M_s x\,ds}{EI}}{\int \dfrac{x^2\,ds}{EI}} = -\frac{-\dfrac{\pi PR^3}{2}}{\pi R^3} = +\frac{P}{2}$$

(where plus sign indicates that V_O acts in the assumed direction).

In addition, $H_O = 0$, owing to symmetry, or owing to the fact that $\int \dfrac{M_s \cdot y\,ds}{EI} = 0$ for the symmetrical loading.

To determine the bending moment at C in ring, a free body of the upper rigid arm is shown in Fig. 4c. By moments M_C equals $-0.182PR$ as determined before. Other bending moments can be determined by the same free body method.

The ring may also be solved by the column analogy method. The properties of area and moment of inertia were found in applying the elastic center method and will not be repeated. The ring severed at C is shown loaded with the M_s diagram in Fig. 6. The total upward force is shown as Q and is equal to $\int M_s\,ds$ or $2PR^2$.

The moment of Q about the y axis is $\int M_s x\,ds$ or $\dfrac{\pi PR^3}{2}$.

The moment of Q about the x axis is zero.

A routine analysis for stress on the cross-section of the analogous column at C follows as

$$M_{i(C)} = \frac{2PR^2}{2\pi R} - \frac{\pi PR^3 \cdot R}{2\pi R^3}$$

$$M_{i(C)} = \frac{PR}{\pi} - \frac{PR}{2} = -0.182PR$$

and since M_s was zero at C, the moment at C is $-0.182PR$, the minus sign representing a moment creating tension on the outside of the ring.

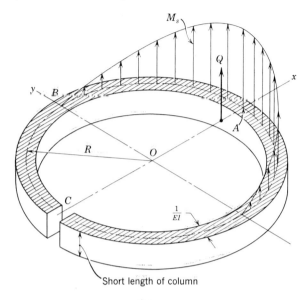

Short length of column

FIG. 7–6

7–7. CIRCULAR BEAMS

Curved frames and rings involved flexual distortions, whereas for circular beams, loaded in or out of the plane of beams axis, the flexural distortions are augmented by rotations or twists produced by torsion. A simple illustration is given in Fig. 7a. The beam AB lies in a horizontal plane and is supported and rigidly attached to its foundation at A. Static equilibrium requires a vertical reaction at A and a resisting couple T_A equal to $2RP$. This couple can be best represented by a moment vector using the right-hand rule to indicate its sense. To determine the bending moment and twisting moment at a general cross-section of the beam, such a cross-section is first located by angle θ and a free body is drawn in Fig. 7b as a plan view of the free body element. Considering equilibrium, the shear force at the cross-section must be P. The shear and the end load in a vertical plane produce a couple of $P[2R \sin (\theta/2)]$ which must be opposed by the equal couple "C" shown vectorally in the plan view. To effect a solution this couple must be resolved into its flexural and twisting components. Letting T represent the twisting couple

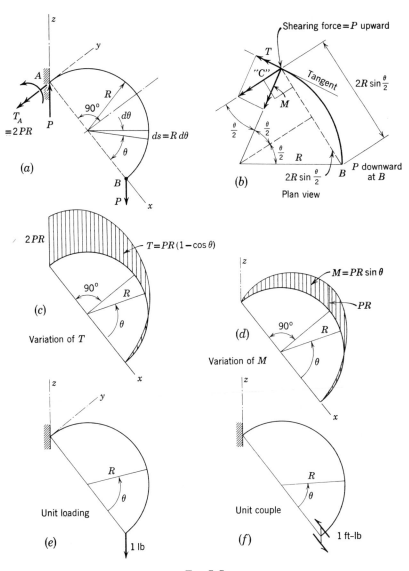

FIG. 7–7

component and M the flexural couple component or bending moment,

$$T = C \cdot \sin \frac{\theta}{2}$$

$$M = C \cdot \cos \frac{\theta}{2}$$

or with the value of C substituted

$$T = 2PR \sin^2 \frac{\theta}{2} = PR(1 - \cos \theta)$$

$$M = 2PR \sin \frac{\theta}{2} \cdot \cos \frac{\theta}{2} = PR \sin \theta$$

The variations of T and M are plotted in Figs. 7c and 7d.

The vertical deflection of point B will be considered in two parts, one due to the torsion effect and one due to the bending effect. The bending deflection, Δ_b, will follow from previous theory by the unit load method. The unit loading is shown in Fig. 7e.

$$EI\Delta_b = \int Mm \, ds$$

$$= \int_0^\pi (PR \sin \theta)(R \sin \theta)R \, d\theta$$

$$= PR^3 \int_0^\pi \sin^2 \theta \, d\theta$$

resulting in

$$EI\Delta_b = \frac{\pi PR^3}{2}$$

Before considering the deflection due to torsion, the unit load or virtual work concept must be extended to bars in torsion. When a round bar is twisted by a torque T the angle of twist is $d\phi = T \cdot ds/JG$ for an element ds long. J equals polar moment of inertia and G represents the shearing modulus of elasticity. If the vertical unit load is placed on the beam at B, as in Fig. 7e, a torque equal to t will be produced at each cross-section of $t = R(1 - \cos \theta)$. When the real load P is introduced, the torque T will twist the bar and t will do virtual work on an element ds long, of

$$t \cdot d\phi = \frac{Tt \, ds}{JG}$$

For the entire length of a bar, external work equals virtual internal strain energy stored, where Δ_t equals deflection of one-lb load due to P. Hence,

$$1 \cdot \Delta_t = \int \frac{Tt \, ds}{JG}$$

which is recognizable as an equation similar to $1 \cdot \Delta = \int \frac{Mm \, ds}{EI}$.

The vertical deflection at B due to the influence of torsion may now be found as follows: where t is produced by the unit loading of Fig. 7e

$$1 \cdot \Delta_t = \int_0^\pi \frac{[PR(1 - \cos \theta)][R(1 - \cos \theta)]R \, d\theta}{JG}$$

$$= \frac{PR^3}{JG} \int_0^\pi (1 - \cos \theta)^2 \, d\theta$$

resulting in

$$\Delta_t = \frac{3\pi PR^3}{2JG}$$

The total deflection at B is the sum of Δ_b and Δ_t, or

$$\Delta_B = \Delta_b + \Delta_t$$

The angle of twist at B, in the plane of the end cross-section, may be found by placing a unit couple at B, Fig. 7f, in the vertical plane of the end cross-section. Then for this unit couple, t', the torsion couple at any section located by angle θ, by resolution is

$$t' = 1 \cdot \cos \theta$$

and m, the bending moment couple at any section located by angle θ due to the unit couple, by resolution is

$$m' = -1 \cdot \sin \theta$$

determined by resolving the unit torsional vector at B into its torsion and bending moment components at the section. Thus, for a curved beam, the angle of twist at B is affected by both bending moment and torsion. The total angle of twist at B is

$$\phi_B = \int_0^\pi \frac{Mm' \, ds}{EI} + \int_0^\pi \frac{Tt' \, ds}{JG}$$

which may be expanded to

$$\phi_B = \frac{1}{EI} \int_0^\pi [PR \sin \theta][-\sin \theta]R \, d\theta + \frac{1}{JG} \int_0^\pi PR(1 - \cos \theta)(\cos \theta)R \, d\theta$$

$$\phi_B = -\frac{PR^2}{EI} \int_0^\pi \sin^2 \theta \, d\theta + \frac{PR^2}{JG} \int_0^\pi (\cos \theta - \cos^2 \theta) d\theta$$

After integration we obtain

$$\phi_B = -\frac{\pi PR^2}{2EI} - \frac{\pi PR^2}{2JG} = -\frac{\pi}{2} PR^2 \left[\frac{1}{JG} + \frac{1}{EI}\right]$$

The negative sign denotes that B rotates opposite to the direction of unit couple in Fig. 7f.

Figure 8a is a semicircular beam, round in cross-section, restrained at A and B and loaded with P at C. Bending moments and twisting moments, as well as the vertical reactions, will be developed at A and B. Moment

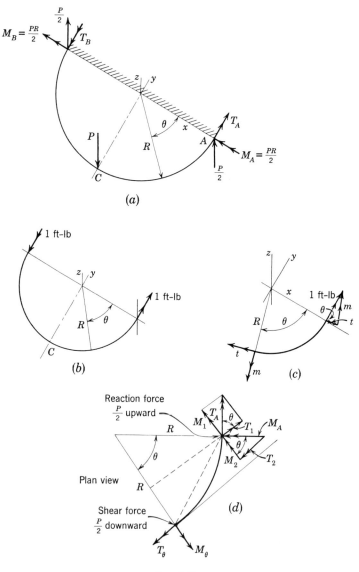

FIG. 7–8

equilibrium about the axis AB requires that M_A and $M_B = PR/2$. The twisting moments are the redundants. The physical conditions for determining the redundants are that cross-sections at A and B cannot twist.

The physical condition of zero twist angle at A will be utilized to formulate the one elastic equation required. First, place a unity twisting moment at A, Fig. 8b, and draw the free body of Fig. 8c. The unit couple is resolved into its components t and m, with t and m shown in the required sense at the section for static equilibrium. The directions at the section become the positive directions.

$$t = 1(\cos \theta); \qquad m = 1(\sin \theta)$$

With the unit twisting moments applied to the structure, lock the ends A and B against further twist in keeping with actual end conditions, and apply the load P at C. The loading develops the reactions of Fig. 8a, and the free body of Fig. 8d will assist in determining the effects at a general cross-section. The detailed resolution of the moment and twisting moment vectors at A and the resolution of the vertical shearing force couple follows and may be verified by referring to Fig. 8d.

The bending moment at a general section is

$$M_\theta = M_1 + M_2 - \frac{PR}{2} \sin \theta$$

$$M_\theta = T_A \sin \theta + M_A \cos \theta - \frac{P}{2} R \sin \theta$$

$$M_\theta = T_A \sin \theta + \frac{PR}{2} (\cos \theta - \sin \theta)$$

The twisting couple at a general section is

$$T_\theta = T_1 - T_2 + \frac{PR}{2} (1 - \cos \theta)$$

$$T_\theta = T_A \cos \theta - M_A \sin \theta + \frac{PR}{2} (1 - \cos \theta)$$

$$T_\theta = T_A \cos \theta - \frac{PR}{2} (\cos \theta + \sin \theta) + \frac{PR}{2}$$

Owing to symmetry, integrations may be performed from 0 to $\pi/2$ or for one-half of the beam. The elastic condition in general terms is one ft-lb times the angle of twist at A

$$1 \cdot \phi_A = \int_0^{\pi/2} \frac{Tt \, ds}{JG} + \int_0^{\pi/2} \frac{Mm \, ds}{EI} = 0$$

With appropriate substitutions, and with $ds = R\,d\theta$, this equation becomes

$$\frac{1}{JG}\left[T_A R \int_0^{\pi/2} \cos^2\theta\,d\theta - \frac{PR^2}{2}\int_0^{\pi/2}\cos^2\theta\,d\theta - \frac{PR^2}{2}\int_0^{\pi/2}\sin\theta\cos\theta\,d\theta \right.$$

$$+ \frac{PR^2}{2}\int_0^{\pi/2}\cos\theta\,d\theta \left.\right] + \frac{1}{EI}\left[T_A R \int_0^{\pi/2}\sin^2\theta\,d\theta \right.$$

$$+ \frac{PR^2}{2}\int_0^{\pi/2}\sin\theta\cos\theta\,d\theta - \frac{PR^2}{2}\int_0^{\pi/2}\sin^2\theta\,d\theta \left.\right] = 0$$

Integrations and simplifications are routine and are not repeated here. The above equation then yields the value of T_A

$$T_A = \frac{PR}{2} - \frac{PR}{\pi} = PR\!\left(\frac{\pi - 2}{2\pi}\right)$$

To find the bending moments and twisting moments at any cross-section, the general equations for M_θ and T_θ may be substituted into. For example, the bending moment at C when $\theta = \pi/2$ is

$$M_C = \left(\frac{PR}{2} - \frac{PR}{\pi}\right) \times 1 + \frac{PR}{2}\,(0 - 1) = -\frac{PR}{\pi}$$

The negative sign indicates that this moment is opposite to the positive direction assumed for M_θ in Fig. 8d.

For structures with noncircular cross-sections special formulas for the angle of twist must be used. Refer to standard texts on strength of materials.

7–8. TRUSSES—INDETERMINATE EXTERNALLY

A truss may be indeterminate with regard to outer reactions, internal forces, or any combination thereof. The calculation of selected redundants depends on properly relating the geometrical requirements as in other indeterminate problems. The truss has axially loaded members and the theoretical emphasis accordingly shifts from bending moment for flexural members to axial stress in elastic two-force members.

Figure 9a is a three-panel truss supported at A, C, and D. The support at C is removed, making V_C the redundant reaction, and Fig. 9b becomes the determinate base structure with axial forces in members denoted as S_s. Panel point C deflects Δ_C. If a unit load is placed at C, as in Fig. 9c, C deflects δ_C. The stresses produced in truss members by this unit load are denoted by "u". Then by virtual work or the unit load method,

$$\Delta_C = \sum \frac{S_s u L}{AE}$$

and

$$\delta_C = \sum \frac{u^2 L}{AE}$$

where summations are performed for all individual members of the truss, and A, E, and L represent area, modulus of elasticity, and length.

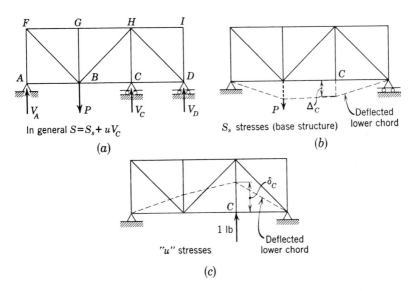

In general $S = S_s + u V_C$

(a)

S_s stresses (base structure)

(b)

"u" stresses

(c)

FIG. 7–9

Since V_C must be of such a value to return C to its position in Fig. 9a, we may write,

$$\Delta_C + V_C \cdot \delta_C = 0$$

or

$$V_C = -\frac{\sum \dfrac{S_s u L}{AE}}{\sum \dfrac{u^2 L}{AE}}$$

The same general equation may also be established by using total strain energy and Castigliano's theorem.

In Fig. 9a, the total strain energy with the summation performed for all members is

$$U = \sum \frac{(S_s + u V_C)^2 L}{2AE}$$

where $S = S_s + uV_C$ is the final stress in a member found by superimposing the statically determinate stress S_s and the stress produced by V_C represented by $u \cdot V_C$. Then, since the deflection at C must equal zero,

$$\frac{\partial U}{\partial V_C} = \sum \frac{(S_s + uV_C)uL}{AE} = 0$$

or

$$\sum \frac{S_s uL}{AE} + V_C \sum \frac{u^2L}{AE} = 0$$

This equation is general for determining one unknown reaction and is the same as the equation previously derived. The sign of V_C is automatic. If V_C is positive it acts in direction of the unit load, if negative it acts opposite to the unit load.

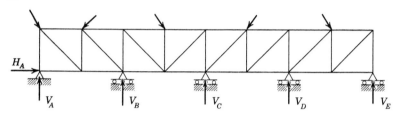

FIG. 7–10

The same principles may be applied to a truss with any number of redundant reactions. In Fig. 10, a truss with three redundant vertical reactions, assumed to be V_C, V_D, and V_E, may be solved by writing the total internal strain energy with three redundants as

$$U = \sum \frac{S^2L}{2AE}, \text{ where } S = S_s + u_cV_C + u_dV_D + u_eV_E$$

Then

$$U = \sum \frac{(S_s + u_cV_C + u_dV_D + u_eV_E)^2 \cdot L}{2AE}$$

The three partial derivatives of strain energy with respect to the redundants follow and each derivative is equal to a deflection in direction of the redundant, according to Castigliano's theorem. The supports are immovable, hence the deflections in direction of redundants are equal to zero. We then obtain

$$\frac{\partial U}{\partial V_C} = \sum (S_s + u_cV_C + u_dV_D + u_eV_E)\frac{u_cL}{AE} = 0$$

$$\frac{\partial U}{\partial V_D} = \sum (S_s + u_cV_C + u_dV_D + u_eV_E)\frac{u_dL}{AE} = 0$$

$$\frac{\partial U}{\partial V_E} = \sum (S_s + u_cV_C + u_dV_D + u_eV_E)\frac{u_eL}{AE} = 0$$

or, term by term, the equations reduce to

$$\sum \frac{S_s u_c L}{AE} + V_C \sum \frac{u_c^2 L}{AE} + V_D \sum \frac{u_c u_d L}{AE} + V_E \sum \frac{u_c u_e L}{AE} = 0$$

$$\sum \frac{S_s u_d L}{AE} + V_C \sum \frac{u_d u_c L}{AE} + V_D \sum \frac{u_d^2 L}{AE} + V_E \sum \frac{u_d u_e L}{AE} = 0$$

$$\sum \frac{S_s u_e L}{AE} + V_C \sum \frac{u_e u_c L}{AE} + V_D \sum \frac{u_e u_d L}{AE} + V_E \sum \frac{u_e^2 L}{AE} = 0$$

These equations may be rewritten, substituting a unit load deflection notation, δ_{ij}, for the last three summations of each equation, while recognizing that the first term of each of the above equations is the deflection of the fully loaded determinate base structure at C, D, and E which may be denoted as Δ_C, Δ_D, and Δ_E.

$$V_C \delta_{cc} + V_D \delta_{cd} + V_E \delta_{ce} = -\Delta_C$$

$$V_C \delta_{dc} + V_D \delta_{dd} + V_E \delta_{de} = -\Delta_D$$

$$V_C \delta_{ec} + V_D \delta_{ed} + V_E \delta_{ee} = -\Delta_E$$

In the above form the δ_{ij} terms are coefficients of the terms on the left. It should be observed that if a diagonal is drawn from upper left to lower right that a symmetrical arrangement of coefficients exists; for example, $\delta_{dc} = \delta_{cd}$, $\delta_{ec} = \delta_{ce}$, and $\delta_{ed} = \delta_{de}$, by direct inspection*. Thus, only six different coefficients representing deflection influence coefficients need be mathematically determined, in addition to the deflections of the base structure, before solving the three equations for the redundants.

The student should be aware that modern electronic computers exist to remove much of the burden of solving systems of equations. One of the keys to programming of these computers rests on the systematizing of the equations by the use of matrix algebra. This text can do no more than call attention to matrix notation, but the student will find matrix algebra a profitable area for independent study.

In matrix notation the form of the last three equations would be

$$\begin{bmatrix} \delta_{cc} & \delta_{cd} & \delta_{ce} \\ \delta_{dc} & \delta_{dd} & \delta_{de} \\ \delta_{ec} & \delta_{ed} & \delta_{ee} \end{bmatrix} \begin{bmatrix} V_C \\ V_D \\ V_E \end{bmatrix} = \begin{bmatrix} -\Delta_C \\ -\Delta_D \\ -\Delta_E \end{bmatrix}$$

The first matrix represents the δ_{ij} coefficients and is spoken of as the flexibility matrix. The second matrix is a column matrix and represents the redundant force matrix, and the matrix on the right-hand side is a

---

* The coefficients may also be shown to be equal by Maxwell's reciprocal deflection theorem given in Art. 8–3.

column matrix representing the deflection matrix of the statically determinate structure with algebraic sign taken into consideration. The student will note that the original equations may be easily reformulated by multiplying each row of the flexibility matrix into the left-hand column matrix. The elements of the right-hand column matrix are successively equated to the three products, formed from the three rows.

All problems of this nature may be generally stated in abbreviated form as

$$[\text{flexibility matrix}]\begin{bmatrix} \text{Redundant} \\ \text{matrix} \end{bmatrix} = -\begin{bmatrix} \text{deflection} \\ \text{matrix} \\ \text{determinate} \\ \text{structure} \end{bmatrix}$$

Example

As an example, the truss of Fig. 11a with the two reactions at B and C as redundants will be solved. First, remove the redundants as in Fig. 11b and calculate the S_s stresses. In Fig 11c the unit load is applied at B and the u_b stresses are determined. Values are given on the diagrams. A unit load at C will produce u_c stresses. Owing to symmetry the u_c stresses are equal to the u_b values if Fig. 11c is turned end for end.

The general equations for solution are written as

$$\sum \frac{S_s u_b L}{AE} + V_B \sum \frac{u_b^2 L}{AE} + V_C \sum \frac{u_b u_c L}{AE} = 0$$

$$\sum \frac{S_s u_c L}{AE} + V_B \sum \frac{u_c u_b L}{AE} + V_C \sum \frac{u_c^2 L}{AE} = 0$$

A tabular summation is shown in accompanying table, assuming for purposes of this problem only that $L/A = 1$ for all members. (This has been done to simplify solution for instruction purposes and is practically unrealistic.)

7-9. TRUSSES—INDETERMINATE INTERNALLY

A rigid truss having more than $2n - 3$ members, where n equals the number of joints, will be statically indeterminate internally. Figure 12a illustrates a simple case of this nature where two diagonals are employed, providing one more member than given by $2n - 3$. Either diagonal may be considered redundant, but for explanation purposes diagonal BD and its unknown stress X is taken as the redundant. Figure 12b indicates that

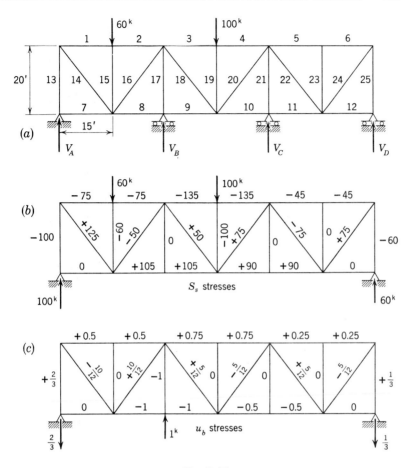

Fig. 7-11

member BD has been severed and also records the computed determinate stresses S_s. In general, the final total stress in a given member is

$$S = S_s + u_x X$$

where u_x is the stress produced by a tensile unit force in member BD. To determine the u_x values, Fig. 12c is drawn with a unit tensile stress in the member BD.

The establishment of the one necessary elastic equation is based on the geometric fact that if member BD of the loaded truss is severed, anywhere along its length, and its redundant force X retained as an active force applied to both sides of the cut section (see Fig. 12a), there will be no

Summary Table

Member	L/A	S_s, kips	u_b, kips	u_c, kips	$S_s u_b L/A$	$S_s u_c L/A$	$u_b^2 L/A$	$u_b u_c L/A$
1	1	−75	+0.5	+0.25	−37.5	−18.75	0.250	+0.125
2	1	−75	+0.5	+0.25	−37.5	−18.75	0.250	+0.125
3	1	−135	+0.75	+0.75	−101.25	−101.25	0.563	+0.563
4	1	−135	+0.75	+0.75	−101.25	−101.25	0.563	+0.563
5	1	−45	+0.25	+0.5	−11.25	−22.50	0.063	+0.125
6	1	−45	+0.25	+0.5	−11.25	−22.50	0.063	+0.125
7	1	0	0	0	0	0	0	0
8	1	+105	−1	−0.5	−105.00	−52.50	1.000	+0.500
9	1	+105	−1	−0.5	−105.00	−52.50	1.000	+0.500
10	1	+90	−0.5	−1.0	−45.00	−90.00	0.250	+0.500
11	1	+90	−0.5	−1.0	−45.00	−90.00	0.250	+0.500
12	1	0	0	0	0	0	0	0
13	1	−100	+0.667	+0.333	−66.70	−33.30	0.444	+0.222
14	1	+125	−0.833	−0.417	−104.13	−52.07	0.695	+0.348
15	1	−60	0	0	0	0	0	0
16	1	−50	+0.833	+0.417	−41.65	−20.85	0.695	+0.348
17	1	0	−1	0	0	0	1.000	0
18	1	+50	+0.417	−0.417	+20.85	−20.85	0.174	−0.174
19	1	−100	0	0	0	0	0	0
20	1	+75	−0.417	+0.417	−31.08	+31.08	0.174	−0.174
21	1	0	0	−1.0	0	0	0	0
22	1	−75	+0.417	+0.833	−31.08	−62.16	0.174	+0.348
23	1	0	0	0	0	0	0	0
24	1	+75	−0.417	−0.833	−31.08	−62.16	0.174	+0.348
25	1	−60	+0.333	+0.667	−20.00	−40.00	0.111	+0.222
Summations					−904.87	−829.81	+7.893	+5.124

Note: $\sum \dfrac{u_c^2 L}{A}$ also equals 7.893, due to symmetry. The two equations to be evaluated are: $7.893 V_B + 5.124 V_C = 904.87$ and $5.124 V_B + 7.893 V_C = 829.81$, from which $V_B = 79.9$ kips and $V_C = 53.4$ kips.

relative movement between the severed ends.   This enables the following elastic equation to be written with the summation for all members of the truss.

$$\sum \frac{(S_s + u_x X)u_x L}{AE} = 0$$

which may be expanded to

$$\sum \frac{S_s u_x L}{AE} + X \sum \frac{u_x^2 L}{AE} = 0$$

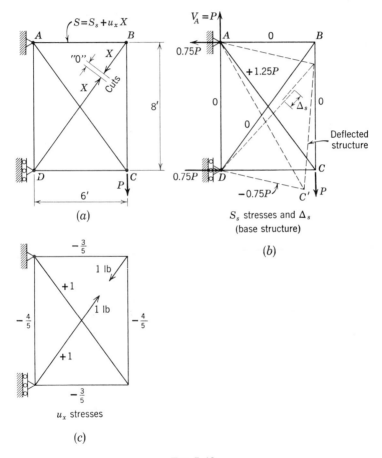

(a)

S_s stresses and Δ_s
(base structure)

(b)

u_x stresses

(c)

FIG. 7–12

The first term of the expanded equation is physically interpreted as equal to Δ_s, Fig. 12b, the closing of the cut section which equals the relative movement of joint B with respect to joint D with the truss in its statically determinate condition. The second term represents the combined effect of the change in length of BD under its stress X and the relative movement of B with respect to D due to the changes in length accounted for by forces $u_x X$ working in all other members of the truss.

A detailed solution is not provided, but a tabulation similar to the previous problem would suffice to evaluate the terms of the equation. A structure which is redundant internally requires as many elastic equations as redundants. The equations would be formulated in a manner similar to the procedures outlined.

REFERENCES

1. Archer, J. S., "Digital Computation for Stiffness Matrix Analysis," *J. Structural Division, Am. Soc. Civil Engineers*, **84**, No. ST 6 (October 1958).
2. Sutherland, H., and H. L. Bowman, *Structural Theory*, fourth edition, Chapter 10, New York: John Wiley, 1950.
3. Hoff, N. J., *The Analysis of Structures*, Part 4, pp. 332–380, New York: John Wiley, 1956.
4. Seely, F. B., and J. O. Smith, *Advanced Mechanics of Materials*, Chapter 16, New York: John Wiley, 1952.
5. Spofford, C. M., *The Theory of Continuous Structures and Arches*, Chapter 1, New York: McGraw-Hill, 1937.
6. Van den Broek, J. A., *Elastic Energy Theory*, second edition, New York: John Wiley, 1942.

PROBLEMS

7–1. Calculate the vertical and horizontal displacement of point B due to P. Constant I.

$$Ans. \; \Delta_V = \frac{PR^3}{4EI}(3\pi - 8); \quad \Delta_H = \frac{PR^3}{2EI}.$$

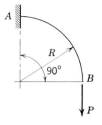

Prob. 7–1   Prob. 7–2

7–2. Calculate the vertical and horizontal displacement of point B due to P. Constant I.

$$Ans. \; \Delta_V = \frac{\pi PR^3}{4EI}; \quad \Delta_H = \frac{PR^3}{2EI}.$$

7–3. Calculate the horizontal displacement of C due to P.   Constant I.

$$\text{Ans. } \Delta = \frac{3PR^3}{2EI}.$$

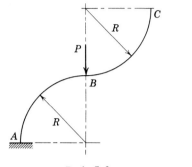

Prob. 7–3

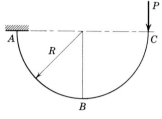

Prob. 7–4

7–4. Calculate the vertical displacement of C due to P.   Constant I.

7–5. Calculate the rotation and the resultant displacement of B.   Constant I.

$$\text{Ans. Displacement } = \frac{M_B R^2}{EI} (\pi^2 + 4)^{1/2}; \ \theta_B = \frac{\pi M_B R}{EI}.$$

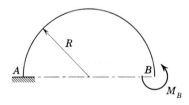

Prob. 7–5

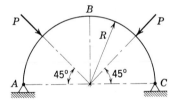

Prob. 7–6

7–6. Determine the horizontal reactions and calculate the bending moment at B.   Constant I. $\text{Ans. } H = 0.095P$.

7–7. Determine the horizontal reactions and plot the final bending moment curve.   Constant I.

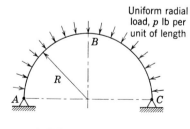

Prob. 7–7

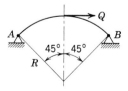

Prob. 7–8

7–8. Determine the reactions at A and B. Constant I. *Ans. $H_B = 0.5Q$.*

7–9. Determine the reactions at A and B. The load p is uniform per unit of vertical height. Constant I.

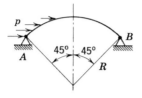

Prob. 7–9

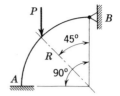

Prob. 7–10

7–10. Determine the reactions at A and B. Constant I.

Ans. $H_B = 0.471P$; $V_B = -.118P$.

7–11. Determine the reactions at A and C. Constant I.

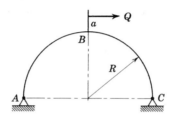

Prob. 7–11

7–12. Fix the ends A and B of the structure of problem 7–6 and solve for the fixed-end moments.

7–13. Fix the end B of the structure of problem 7–10 and solve for the fixed-end moments.

7–14. A couple T is applied in the plane of the structure. Determine the moments at A and C. Constant I.

Ans. $V_A = 2T/\pi R \cdot M_A = T(4 - \pi)/2\pi$.

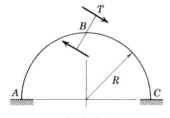

Prob. 7–14

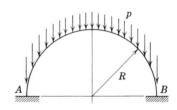

Prob. 7–15

7–15. Determine the moments at A and B.   The loading is p lb per unit of horizontal projection.   Constant I.

7–16. Determine the horizontal reactions.   Constant I.   *Ans. H = 0.38P.*

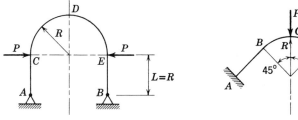

Prob. 7–16   Prob. 7–17

7–17. Determine the reactions and the moment at C.   Constant I.

7–18. For the closed ring, what is the moment at A?   Plot the moment diagram for the quadrant ABC.   Constant I.   *Ans. $M_A = 0.196QR$.*

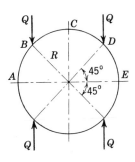

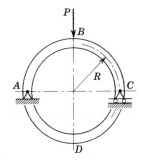

Prob. 7–18   Prob. 7–19

7–19. For the closed ring supported as shown, determine the redundants at D. What is the change in length of the horizontal diameter?

Ans. $M_D = PR(4 - \pi)/4\pi$; $H_D = P/2\pi$; $\Delta = PR^3(4 - \pi)/4\pi EI$.

7–20. For the closed ring, determine the redundants at E.   What is the vertical deflection at B?   Constant I.   (See figure next page.)

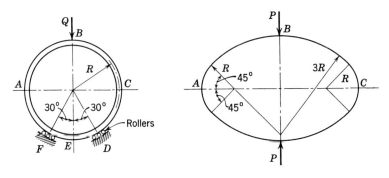

Prob. 7–20   Prob. 7–21

7–21. Determine the moment at *C*. Constant *I*.

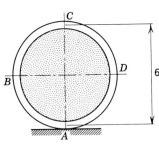

Prob. 7–22

7–22. A round pipe, just full of water at no additional head, is supported at point *A*. Compute the moments and internal forces in pipe at *B* and *D* for a section of pipe 1 ft long. Use *R* as 3 ft and let *w* designate the weight of water per cubic foot.

Ans. $M_D = wR^3(2 - \pi)/4$; $V_D = wR^2(4 - \pi)/4$;
$$H_D = 7wR^2/16.$$

7–23. For problem 7–19 cut the ring at point *D* to provide a determinate base structure and solve for the redundants at *D* by the column analogy method.

7–24. Determine the redundants at *C* by the column analogy method. The ring of constant *I* is supported at *A*.

Ans. $M = \dfrac{QR}{\pi}(4 - \pi)$; $H = 3Q/\pi$.

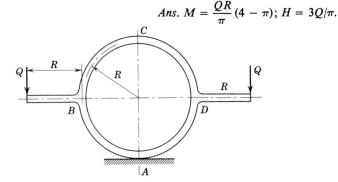

Prob. 7–24

7-25. A full semicircular beam (in a horizontal plane) is cantilevered from A. Determine the deflection of B in the z direction and also the rotation of the cross-section at B. The beam is round and of constant diameter "d".

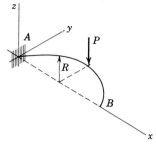

$$Ans. \; \Delta = \frac{PR^3}{2}\left[\frac{1}{EI} + \frac{\pi - 1}{JG}\right].$$

Prob. 7–25

7-26. A full semicircular beam with a constant round cross-section is loaded as shown. Assuming full restraint at A and B, determine the redundants at C.

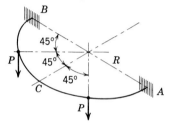

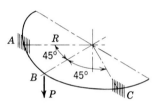

Prob. 7–26   Prob. 7–27

7-27. A circular beam represented by one quadrant of a full circle is fixed at A and C. Determine the moments and reactions at A and C. Constant I.

7-28. A truss, wherein for purposes of this problem, all members have an L/A value equal to 5, where L and A are in inches and inch^2 units respectively, is supported as shown. Determine the horizontal reaction at B.   *Ans.* $H_B = 2.5$ kips.

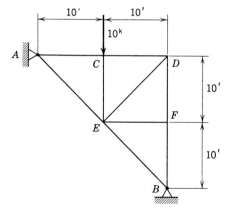

Prob. 7–28

7–29. With all members considered to have a cross-sectional area of 10 in.^{2}, calculate the horizontal reactions. *Ans. H_A = 8.13 kips.*

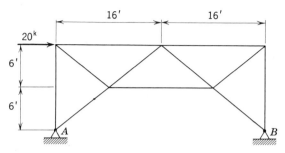

Prob. 7–29

7–30. With all L/A values taken as 10, where L is in inches and area is in square inches, calculate the reaction at B. *Ans. V_B = 74.10 kips.*

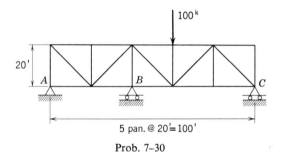

Prob. 7–30

7–31. Assume that the L/A values are as follows where L is in inches and A in square inches. All chord members 20; all web members 15. Calculate the value of the reactions.

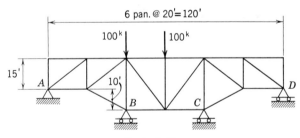

Prob. 7–31

7-32. Assume that the area of ADC is 10 in.², and the area of all other members is equal to 8 in.².   Calculate the reactions.

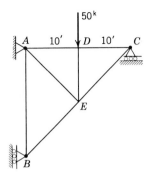

Prob. 7-32

7-33. With CG as the redundant diagonal, compute its stress.   For purposes of this problem assume L/A for all members as 2, where L is in feet and A in square inches.   *Ans.* −4.42 kips.

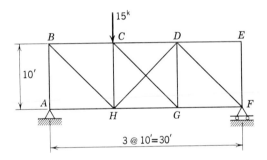

Prob. 7-33

7-34. With CE and BF as the redundant members, determine their stress and the stress in all other members.   For purposes of this problem use the following areas:

$$ABC, DEF = 10 \text{ in.}^2 \qquad BD \text{ and } AE = 5 \text{ in.}^2$$
$$CD \text{ and } BE = 5 \text{ in.}^2 \qquad AF = 10 \text{ in.}^2$$
$$CE \text{ and } BF = 10 \text{ in.}^2$$

(See figure next page.)

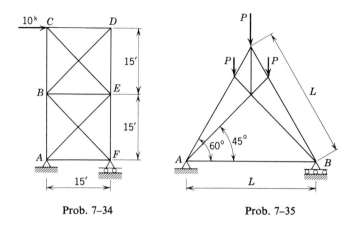

Prob. 7–34   Prob. 7–35

7–35. Calculate the tension in the tie bar *AB*.   Assume all L/A values as unity for purposes of this problem.

7–36. Compute the tension in cable *EH*.   Assume that both cables have a cross-sectional area of 0.5 in.² and that all other members have an area of 10 in.². The effective modulus of elasticity of the cable is 20×10^6 psi, and the modulus for all other members is 30×10^6 psi.

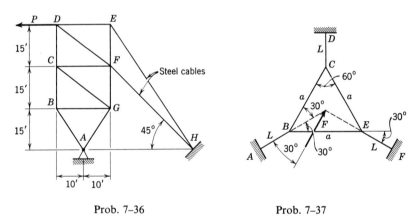

Prob. 7–36   Prob. 7–37

7–37. A *rigid* triangular structure is supported by three equal length and equal area members capable of resisting either tension or compression.   A force *F* is applied through the centroid of the triangle.   Compute the stress in each of the supporting members.

Influence Lines

8

8-1. INTRODUCTION

The determination of the maximum and minimum structural effects of loads and forces applied to structures is basic to analysis and design. In a bridge, the moving loads imposed by traffic may assume many positions and invariably produce cyclic maximum and minimum effects on the trusses, the stringers, the floor beams, or other structural elements. In a building, relatively large but movable static loads representing floor live load may be superimposed on the dead loads in varying patterns or combinations to produce maximum effects in beams, girders, and columns.

As an aid to determining the effects of moving or movable loads the influence line concept is introduced. The student has probably been previously introduced to the general nature of the influence line for statically determinate structures, thus, for the most part, this chapter will deal with the problems associated with indeterminate structures.

8-2. NATURE OF INFLUENCE LINES

For purposes of review Fig. 1a indicates a simple span AB, and it is required to plot a curve representing the influence of a moving unit vertical load on the left-hand reaction V_A. Locating the unit or 1-lb load at a variable distance x from B, then by moments $V_A = 1 \cdot x/L = x/L$. The quantity x/L is the influence of the unit load on V_A and it is termed an ordinate of the influence line to be plotted under the unity load. The influence line is shown in Fig. 1b. As the unity load moves to the left, x increases and the influence ordinate increases linearly with x. Thus, the influence line for the left-hand reaction is a single straight line with the

maximum influence line ordinate at A. Simple and obvious interpretations result; namely, a load at any point contributes to V_A; if the live loads
are concentrated and in a connected system of loads, the heaviest loads
shall be placed at or near A to produce the maximum reaction; and
finally, if the live loading is a uniform load unlimited in length, the entire
span must be loaded to find the maximum reaction. It should be apparent
that the influence line ordinate may either be used as a ratio or as pounds of
force per pound of load.

Figures 1c and 1d extend the influence line principle to vertical shear at
section C of the beam. Again x may vary from O to L as the unity load
moves from B towards A. Two free bodies, one to the left and one to the
right of a section, cut through the beam at C, and static equilibrium
considerations will disclose that the influence line for shear has a vertical

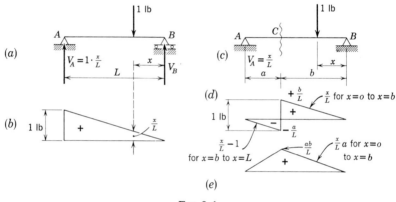

FIG. 8–1

break at C of unity or 1 lb, dividing the beam span into segments of
positive and negative influence ordinates. The equations representing the
influence values for the two segments are shown on Fig. 1d. It is now
apparent that fractional span loadings are required for determining maximum positive and negative shear effects.

Figure 1e is the influence line for the bending moment at section C,
determined by bending moment principles. The influence line ordinate
is maximum at C and occurs with the unity load at C. Consequently, an
influence line is simply a graphical representation of the effect of a moving
unit load on various quantities relating to the structure. In its simplest
application, the unit load is considered to move over the entire length of the
structure and the relevant quantities calculated by principles of statics.
The student will be aware that, based upon this viewpoint, various criteria

for the placement of complicated concentrated load patterns were derived.

For statically determinate structures the influence diagrams for forces and moments are always composed of straight lines. For statically indeterminate structures the influence diagram will always be formed by curved lines or an approximation thereto in the case of trussed structures.

8-3. MAXWELL'S RECIPROCAL THEOREMS

Prior to a discussion of influence lines for indeterminate structures it is advisable to introduce the student to the Maxwell reciprocal theorems, which will relate the elastic movements of two points of a structure.

Although a general development of the reciprocal relations may be made, it is best for initial purposes to restrict the development to specific cases of loading. In Fig. 2a, place a load P_1 vertically at some point (1)

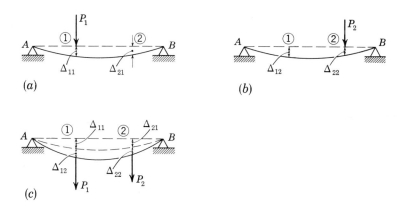

Fig. 8-2

on the axis of beam AB. Point (2) is any other point on axis of the beam. Since the beam is elastic, deflections will occur and the specific deflections at points (1) and (2) are indicated. A Δ_{ij} notation is followed for denoting deflections to define location of the deflection and the position of the load that produced it. The first subscript i is reserved for defining the point at which the deflection is measured and the j subscript reserved for defining the position of the load. In Fig. 2b, a load P_2 is placed at point (2) and the deflections at both points (1) and (2) indicated according to the above notation.

Consideration may now be given to the external work done on the beam by slowly applying the force P_1 alone in Fig. 2a. The external work is

$$W_1 = \frac{P_1\Delta_{11}}{2}$$

Then, with P_1 already on the beam, P_2 is slowly applied at point (2) with additional external work being done on the beam of

$$W_2 = P_1\Delta_{12} + \frac{P_2\Delta_{22}}{2}$$

In the configuration shown in Fig. 2c, with both P_1 and P_2 applied, the total external work done on the beam is

$$W_3 = W_1 + W_2$$

$$W_3 = \frac{P_1\Delta_{11}}{2} + P_1\Delta_{12} + \frac{P_2\Delta_{22}}{2}$$

It will be recalled that the elastic configuration of Fig. 2c was the result of superposing the deflection due to P_1 and P_2, assuming that superposition applies. If superposition applies then this is the same configuration that would be obtained by placing loads P_1 and P_2 on the beam simultaneously. For a simultaneous slow application of these loads the external work done on the beam would equal

$$W_4 = \frac{P_1(\Delta_{11} + \Delta_{12})}{2} + \frac{P_2(\Delta_{21} + \Delta_{22})}{2}$$

Since the final elastic configuration is the same as in Fig. 2c, indicating an equality of internal strain energy, and hence an equality of external work done, we may write

$$W_3 = W_4$$

we may now substitute the expanded values of W_3 and W_4 and obtain

$$\frac{P_1\Delta_{11}}{2} + P_1\Delta_{12} + \frac{P_2\Delta_{22}}{2} = \frac{P_1(\Delta_{11} + \Delta_{12})}{2} + \frac{P_2(\Delta_{21} + \Delta_{22})}{2}$$

And when simplified,

$$P_1\Delta_{12} = P_2\Delta_{21} \tag{8-1}$$

which, in terms of a generalized principle, states that the external work done by the forces of the first system moving through the displacements produced by the second system is equal to the external work done by the forces of the second system moving through the displacements produced by the first system.

Equation 8-1, for purposes of future discussion of influence lines, can be simplified by assuming that $P_1 = P_2$, resulting in

$$\Delta_{12} = \Delta_{21} \qquad (8\text{-}2)$$

which is a useful reciprocal relationship between deflections at two specific points on the beam when $P_1 = P_2$. P_1 and P_2 can be applied in any direction as long as Δ_{12} is measured in the direction of P_1 and Δ_{21} is measured in the direction of P_2.

A reciprocal relationship between deflection and rotation may also be developed. Refer to Figs. 3a and 3b. In Fig. 3a a vertical load is placed at point (1), producing Δ_{11} at (1) and a rotation of the tangent to the elastic curve of θ_{21} at point (2). In Fig. 3b, a moment at (2) produces Δ_{12} at (1) and a rotation of the tangent of θ_{22} at (2). It is desired to show that

$$P \cdot \Delta_{12} = M \cdot \theta_{21}$$

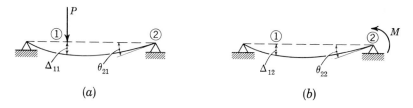

(a) $\qquad\qquad\qquad\qquad$ (b)

FIG. 8–3

If we follow the procedures previously used, the external work done on the beam by P in Fig. 3a is

$$W_1 = \frac{P}{2} \cdot \Delta_{11}$$

If the moment or couple M is now applied at (2), with P still on the beam at (1), the additional external work done is

$$W_2 = P \cdot \Delta_{12} + \frac{M}{2} \cdot \theta_{22}$$

The total external work done is

$$W_3 = W_1 + W_2$$

Next determine the total external work done on the beam if P and M are slowly but simultaneously applied as

$$W_4 = \frac{P}{2}(\Delta_{11} + \Delta_{12}) + \frac{M}{2}(\theta_{21} + \theta_{22})$$

Recognizing that the beam assumes the same deflected configuration regardless of the sequence of loading, we may write the equality

$$W_3 = W_4$$

Then by substitution

$$\frac{P}{2}\Delta_{11} + P\Delta_{12} + \frac{M}{2}\theta_{22} = \frac{P}{2}(\Delta_{11} + \Delta_{12}) + \frac{M}{2}(\theta_{21} + \theta_{22})$$

And when simplified, we obtain,

$$P \cdot \Delta_{12} = M \cdot \theta_{21} \tag{8-3}$$

In equation 8-3 the force is multiplied by a displacement and the moment by a rotation where the deflection is produced by the moment and the rotation is produced by the force. On first inspection there appears to be a conflict of units in the above equality; however, the angle of rotation is measured in radians, a dimensionless ratio, and if P is in pounds and Δ in inches, M would be expressed in inch-pounds, and necessary dimensional relations preserved.

For purposes of influence line construction, the reciprocal relationship of equation 8-3 may be simplified by letting P be equal numerically to M and obtain

$$\Delta_{12} = \theta_{21} \tag{8-4}$$

One further reciprocal relation is useful. It is stated here with the proof left to the student. If points (1) and (2) are any two points on an elastic structure, then M_1 applied at (1) will produce a rotation at (2) equal to θ_{21}, and M_2 applied at (2) will produce θ_{12} at (1). Show that

$$M_1 \cdot \theta_{12} = M_2 \cdot \theta_{21} \tag{8-5}$$

8–4. INFLUENCE LINES—INDETERMINATE STRUCTURES

Although the basic concept of an influence diagram is no different for an indeterminate structure than for a determinate structure, the effect of the unit load moving on the structure in the indeterminate situation can best be ascertained by dealing with the elastic displacements of the structure. It will be found that the elastic curve of the structure under certain prescribed geometric conditions will be the influence line or diagram. These influence lines will be found to be curved lines. Professor Müller-Breslau is generally credited with the following concepts stemming from Maxwell's reciprocal theorems.

The most basic of influence lines is the influence line for vertical reaction. For the two-span beam of Fig. 4a, let it be required to construct an influence line for the vertical reaction V_A.   This could be done by placing a load P, equal to unity or one lb, on the beam and computing V_A for a series of definite positions of P and plotting the values of V_A as ordinates in the

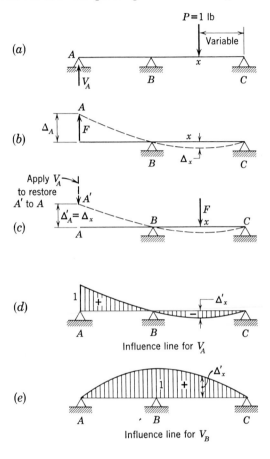

FIG. 8–4

usual manner.   However, a new view of this problem can be obtained if the reaction at A of the unloaded beam is first removed as in Fig. 4b, and A is deflected upward by an arbitrary force F by an amount equal to Δ_A. The dotted line indicates the elastic curve produced by F.   At any given point x along the beam the deflection is denoted as equal to Δ_x and this deflection is directly proportional to F.

In Figure 4c, place a load equal to F at x, with reaction at A removed and A free to deflect to A'. A will deflect upward by Δ'_A.   By the Maxwell reciprocal relationship of equation 8-2, $\Delta'_A = \Delta_x$.   The final question may now be asked, what is the value of V_A to return A' to its original position of A in Fig. 4c, since this value of V_A will be the reaction produced by a force F at x, if A had been held in place as in Fig. 4a.   In Fig. 4b, a force of F pushed A up Δ_A, hence accepting deflection as proportional to force we may write

$$\frac{V_A}{F} = \frac{\Delta'_A}{\Delta_A} = \frac{\Delta_x}{\Delta_A}$$

where the ratio of V_A to F is the equivalent of the influence value that would be determined by computation if a load P equal to unity had been placed at x in Fig. 4a.   The ratio of Δ_x/Δ_A remains constant regardless of the magnitude of F.   If, in Fig. 4b, Δ_A had been made an arbitrary unity value by applying an appropriate force F', then with the new deflection at x equal to Δ'_x, we may express the influence line value as the ratio

$$\frac{V_A}{F'} = \frac{\Delta'_x}{1} = \Delta'_x$$

A valuable interpretation of this relationship is shown in Fig. 4d, where, if $\Delta_A = 1$, then Δ'_x, the deflection of x, equals the influence value or ratio for V_A with a unity load at x.   That is, the elastic curve of Fig. 4d, representing the deflection of all points, is the influence line required since x is a general notation applying to any given point along the beam.   The signs of the influence line ordinates are given as positive if an upward reaction is produced by the vertical unity load, and negative if a downward reaction is developed at A.

The influence line for V_B is shown in Fig. 4e.   To develop this curve, deflect point B upward a unity distance.   The elastic curve so produced becomes the influence line for V_B in accordance with the above principles. Note that the unity displacement is made in the positive direction of the reaction.

The influence lines as drawn or sketched in Fig. 4 are qualitative only; however, any deflection method would suffice to determine quantitative values.   If deflection calculations should become too complex, then a model could be made and the deflections measured directly, after the introduction of a known displacement in the direction of the reaction.

Figure 5a illustrates a two-span beam with end A fixed.   To develop the influence line for V_A, A must be displaced a distance of unity in the vertical direction to A' while rotation of the tangent to elastic curve is prevented. It is essential that this rotational restraint remain in effect during the vertical translation and the proof of this important requirement follows.

In Fig. 5b, let A be displaced Δ_A, by a force F, while a moment or couple M sufficient to restrain the tangent at A from rotating is also applied. Then the external work done on the beam due to F alone, since the couple M does not rotate during the translation, is

$$W_1 = \frac{F \cdot \Delta_A}{2}$$

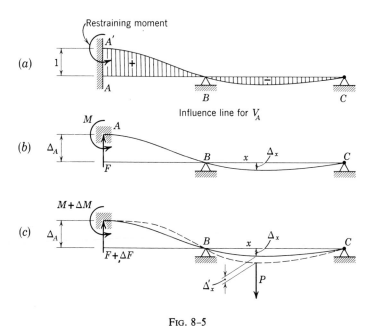

FIG. 8-5

In Fig. 5c, a load P is superimposed on the configuration of Fig. 5b, modifying the configuration as shown. If we note that A is still fully restrained at its initial displacement of Δ_A, then the only additional external force doing work is P, although a ΔM and a ΔF have been produced at A, and this work is

$$W_2 = \frac{P}{2} \cdot \Delta'_x$$

The total external work cumulatively done on the beam at this stage is

$$W_1 + W_2$$

We may now resort to the technique of applying all moments and forces to the beam simultaneously to achieve the configuration of Fig. 5c, and the external work done is

$$W_3 = \frac{(F + \Delta F)}{2} (\Delta_A) + \frac{P}{2} (\Delta_x + \Delta'_x)$$

Since $W_3 = W_1 + W_2$, we find, by substitution and simplification, that

$$\frac{\Delta F}{P} = \frac{\Delta_x}{\Delta_A}$$

The ratio of ΔF to P is the influence value for the vertical reaction since ΔF is due to P at point x. The procedure indicated in Fig. 5a is thus verified since, if Δ_A is made unity, the ordinates to the elastic curve numerically representing the ratio of Δ_x/Δ_A are influence line ordinates. The important geometric consideration in this instance is the preventing of M and ΔM from doing work, which is accomplished by maintaining the fixed end at A, thus letting only vertical displacement at A and the vertical redundant enter the external work equations as A is translated. This requirement demonstrates the necessity of permitting one and only one redundant to do external work on the structure during displacement. Zero displacements, or rotations, are necessary in the direction of all of the redundants except the redundant for which the influence line is being generated. This is a definite requirement when determining quantitative influence lines by model wherein the displacements are physically obtained by the application of arbitrary forces in the direction of the redundants.

Influence lines for fixed-ended moments may be explained by referral to Fig. 6. Figure 6a is a beam fixed at A and B. To explain the analysis of an influence line for M_A, first modify the supporting condition at A to a hinge, and with a couple M rotate the tangent to elastic curve at A through a clockwise angle of θ_A. This rotation produces the elastic curve with a deflection of Δ_x at point x. In Fig. 6c, place a load of P at x while end A is hinged. The tangent at A rotates through an angle θ'_A. To find the fixed-ended moment at A produced by P, determine the counterclockwise or negative moment M_A required to rotate the tangent of Fig. 6c back to the horizontal. By proportionality of moments and rotations we obtain

$$M_A = -\frac{\theta'_A}{\theta_A} (M)$$

To simplify this expression it will be recalled from Maxwell's reciprocal relationships, where x is one point on the elastic curve and A is the second point, and if P is equal numerically to M, that

$$\Delta_x = \theta'_A$$

thus, by substitution and recognizing that Δ_x is negative,

$$M_A = \frac{\Delta_x}{\theta_A}(P)$$

or

$$\frac{M_A}{P} = \frac{\Delta_x}{\theta_A}$$

where the quotient M_A/P equals the influence line ordinate for moment at A with P at x and the sign of the ordinate assumes the same sign as Δ_x.

(a)

(b)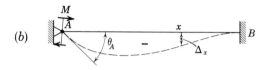

(c)

(d)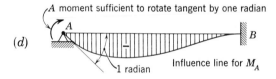

FIG. 8–6

The construction of the influence line is facilitated by imposing a value of θ_A equal to one radian.   Then

$$\frac{M_A}{P} = \frac{\Delta_x}{1} = \Delta_x$$

where the units are understood to be in terms of moment per unit of applied load.   If the deflection is in inch units the moment is in inch pounds.   Thus, the influence line for moment is easily constructed by rotating the end tangent through a clockwise angle of one radian as shown in Fig. 6d.   The end rotation was produced by a clockwise moment at A in order that the sign of Δ_x would represent the correct sign for M_A by the beam convention. Hence, the practice followed requires downward deflection to be termed negative.

To this point the influence line concept has been limited to external reactions, end shears, and end moments. The principles will now be expanded to influence lines for shear and moment at sections within the spans. Figure 7a represents a two-span beam and Figs 7b and 7c represent

(a)

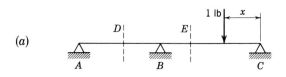

(b)

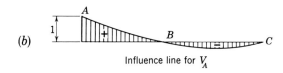

Influence line for V_A

(c)

Influence line for V_B

(d)

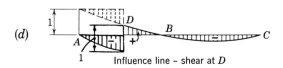

Influence line – shear at D

(e)

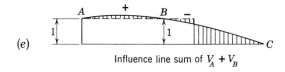

Influence line sum of $V_A + V_B$

(f)

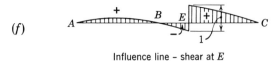

Influence line – shear at E

FIG. 8–7

respectively the influence lines for outer reactions V_A and V_B, drawn or sketched in a qualitative manner by the methods outlined previously. To sketch the influence line for the beam shear at section D in span AB consider the vertical static equilibrium of a free body to the left of D. As long as the 1-lb load on the beam is to the right of D the shear at C is equal to the left-hand reaction V_A, thus the portion of the influence line for V_A to the right of D also represents the influence line for the shear at C. This is shown in Fig. 7d. When the 1-lb load is to the left of D then the shear at D is $V_A - 1$. This graphical subtraction is shown in Fig. 7d, with the resultant values shown as a negative part of the influence curve to the left of D.

A similar reasoning will establish the influence line for vertical shear at section E. The shear at E, when the load is to the right of E, must be equal to the algebraic sum of V_A and V_B. If the two influence lines for V_A and V_B are combined an influence line for their algebraic sum results. To sketch the effect of this superposition, each support, A and B, may be simultaneously displaced a unit distance upward. The elastic curve so determined would be as in Fig. 7e. The graphical subtraction of the unity load from the sum of V_A and V_B, when the load is to the left of E, is shown. The conventional type of resultant influence line for shear at E is then drawn in Fig. 7f. It should also be observed that the vertical discontinuity at the sections D and E is equal to unity and that the tangents to the positive and negative branches at the section are parallel.

Based on the foregoing observations, a more general procedure may be developed to permit qualitative influence lines to be rapidly sketched. Referring to Fig. 8a, again assume that it is desired to sketch the influence line for shear at E. To provide distortions compatible with the physical requirement of parallel tangents to the elastic curve at E as noted previously, first cut the beam at E and install a two-bar linkage as shown in auxilliary sketch. If a relative displacement of Δ is given the adjacent cross-sections of the beam so linked together by two oppositely directed forces F, then the linkage system distorts into a parallelogram form, automatically establishing the requirement of equal rotation of the left and right tangents at section E. This is important, as moments imposed on adjacent sections by forces in the bars of the linkages, although equal and oppositely directed, rotate in the same directions and produce a total combined work of zero on the beam, thus eliminating themselves from the external work summations. The beam sections at E, cut and reconnected as described by the linkage, are displaced a relative amount Δ by a pair of arbitrary shear forces F impressing the elastic curve shown and producing at any given point x a deflection Δ_x. To permit the use of Maxwell's reciprocal relations, a force equal to F is placed at x in Fig. 8b, while the beam remains in its cut and linked condition at E. The linkage system will respond with a relative vertical displacement equal to Δ_x according to the reciprocity theorem. To complete the analysis it must be realized that section E must actually be restored to continuity. To pull or push the ends at E back into coincidence a pair of oppositely directed internal shearing forces, as shown dotted in Fig. 8b, are applied to the two vertical sections at E. The magnitude of these forces will be the magnitude of shear at E caused by a force F at point x. Letting V equal the internal shear, then

$$\frac{V}{F} = \frac{\Delta_x}{\Delta}$$

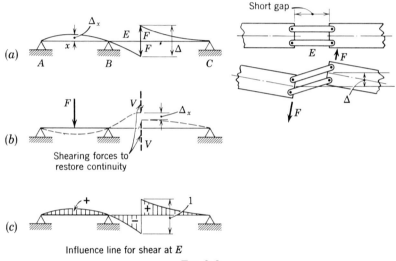

FIG. 8–8

Since the shearing forces of F initially displaced the ends of the section at E an arbitrary distance of Δ, one final simplification is then apparent. If Δ is made unity the ratio of V to F, which is the influence value, is equal to Δ_x, hence influence line ordinates for shear are the ordinates to the elastic curve produced in Fig. 8c with Δ equal to unity. Figure 8c is the influence diagram for shear at E developed by this simplified procedure. With the unity displacement introduced as shown, upward or positive deflection corresponds to positive shear, and downward or negative deflection corresponds to negative shear.

The influence line for moment at a section of a beam may also be represented by an elastic curve. In Fig. 9a, let E represent a point for which the moment influence line is to be determined. First, sever the beam at E and install a hinged joint. Two equal but oppositely directed couples M are applied, one to the right and the other to the left-hand beam segments at the hinge. These arbitrary valued couples cause E to rise and the tangents to elastic curve at E to rotate relatively by θ_E. Deflections are produced throughout the beam and at a point x the deflection is Δ_x. In Fig. 9b, with the hinge still in action at E, place a load

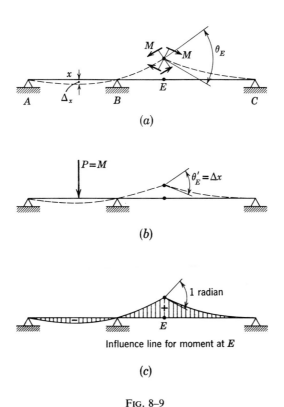

(a)

(b)

Influence line for moment at E

(c)

FIG. 8–9

P equal numerically to M at point x. The resultant effect is to cause the tangents to the elastic curve at point E to rotate through the relative angle θ'_E. θ'_E equals Δ_x by Maxwell's reciprocal theorem if P equals M. Since the beam at E is not actually hinged, but continuous, the continuity in Fig. 9b may be restored at E by applying two equal and oppositely directed moments, which would be equal to M_E, the actual bending moment at E with P at x, and of such magnitude to reduce θ'_E to zero. Since

M in Fig. 9a rotated the tangents at E through a relative angle of θ_E, then

$$\frac{M_E}{M} = \frac{\theta'_E}{\theta_E}$$

or

$$\frac{M_E}{P} = \frac{\theta'_E}{\theta_E} = \frac{\Delta_x}{\theta_E}$$

This quotient is interpreted as an influence line value and it may be further simplified by making θ_E equal to one radian. The influence line for moment at E then appears as shown in Fig. 9c, where the Δ_x ordinates of the deflection curve become influence line values. The sign of moment produced at E is designated positive or negative by beam convention. If the unit angle is induced as shown, upward or positive deflection agrees with positive moment and downward or negative deflection is conversely interpreted as negative moment.

8–5. BEAM INFLUENCE LINES AND UNIFORM LIVE LOADS

Whenever the live loading of a structure is capable of moving, or inducing partial loadings, it is essential to determine the placement of the live loads for maximum or minimum effects. If the live load may be represented by a uniform load and it is also postulated that this loading may exist in the patterns required, then influence lines will determine the load patterns to produce the maximum and minimum effects.

Let Fig. 10a represent a four-span continuous beam having a uniform dead load of w lb per ft. The live loading is a uniform load of w' lb per ft. Whereas the dead load is a fixed load, the live load may be placed as required. For purposes of illustration, determine the patterns for placing the live load for maximum reaction at B and for maximum bending moments at B, at C, and for a section near the center line of span BC. As long as only full span loadings are required, the construction of the influence lines may be qualitative.

Figure 10b represents the influence line for V_B. The pattern of loading for maximum positive V_B is shown above the curve. Figure 10c is the influence line for M_B with the loading pattern for maximum negative moment at B. For the maximum possible live load positive moment at B, only span CD would be loaded. Figure 10d is the influence line for M_C with the load pattern for maximum negative moment at C.

Figure 10e is the influence line for moment at the centerline of span BC, indicating that spans BC and DE should be loaded for maximum positive moment (pattern shown) and only spans AB and CD for maximum negative moment. These two separate loading patterns will determine the maximum positive or maximum negative moment for any cross-section of the beam between the two points F and G, due to the similarity of the qualitative influence lines for moment at any section between these two points. The points F and G are termed the left and right fixed points for span BC.

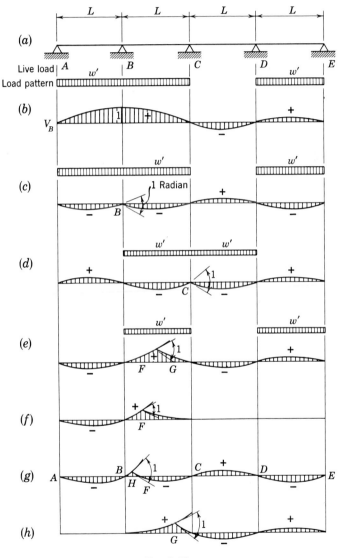

FIG. 8–10

To illustrate the true nature of the fixed points, first sketch the influence line for moment at F, in Fig. 10f, on the assumption that no deflection will occur in spans CD and DE. Contrast the elastic curve in these two spans with the elastic curve for the same two spans in Fig 10c and 10e. Note that the sign of deflection in these spans is reversed, indicating that the influence line drawn for moment at some section between the center line and the support B should have a zero deflection value in spans CD and DE, since in changing from a positive value to a negative value, the deflections of the elastic curve must pass through zero. This observation further implies that a load anywhere on spans CD and DE will produce zero moment at F. The student may verify that zero moment will exist at F by placing a load anywhere on either span CD or DE and noting that a point of inflection or zero moment will always occur at point F. This

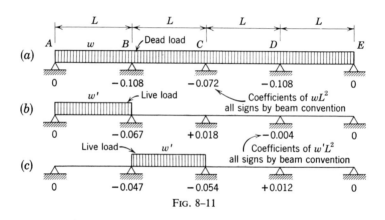

FIG. 8–11

verification may be performed by a moment distribution analysis. An influence line for moment for a section H between B and F will appear approximately as shown in Fig. 10g. Whereas full span live loadings may be used on spans AB, CD, and DE, a split span live loading is indicated for span BC. For short spans, such as those occurring in buildings, the split span loading is generally not used, owing to time required for an analysis to locate the load divide point, but for longer bridge spans the use of the split span loading may be essential. Figure 10h, illustrates the influence line for the right-hand fixed point G. Particularly note that a load in span AB produces zero moment at G. The term *fixed* applied to points F and G implies that they have a definite location in keeping with the structural restraint of adjacent spans. They are approximately 0.2L from the supports in span BC if the beam is of uniform section.

To illustrate the procedure, take the four-span beam of Fig. 10 and develop the moment curves for span *BC*. Let the uniform dead load be 1 kip per ft and the live load be 2 kips per ft. Figure 11 tabulates the results of moment distribution for the various loading patterns shown. The moment values at the supports are coefficients of wL^2 or $w'L^2$ with signs according to the beam convention. The data is sufficient since results of Figs. 11*b* and 11*c* may be interpreted for loads on symmetrically placed spans.

The solution starts by determining the maximum positive moment

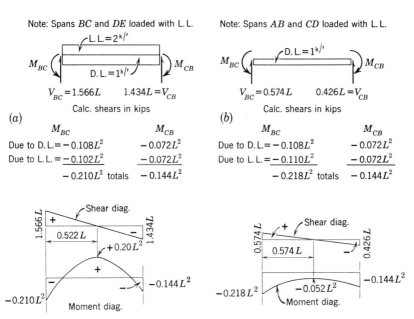

FIG. 8–12

within span *BC*. Figure 12*a* shows the summarizing of moment co-efficients times load intensity for full dead load and with live load on spans *BC* and *DE*. The moment diagram for span *BC* is drawn. Figure 12*b* is a similar set of calculations for determining the value of a maximum negative moment at or near the center line of span *BC*. The shear and moment diagrams are constructed following conventional procedures and are shown in Figs. 12*a* and 12*b*.

The maximum combined negative moment at *B*, using the live load pattern of Fig. 10*c*, is $-0.344L^2$. This result should be checked by the student.

8–6. BEAM INFLUENCE LINES AND CONCENTRATED LOADS

In Art. 8-5 the live load was a uniform load, but, as is often the case, the live loads may be concentrated loads. The influence line can no longer be qualitative and the ordinates must be quantitatively determined to sum up the effects of a series of concentrated loads. Influence line principles remain the same, but a deflection analysis must be used to obtain the ordinates.

The beam of Fig. 13 is taken as a first example where it is desired to determine the ordinates of the influence line for the fixed-ended moment at A. After hinging A, rotate the tangent at A through an angle equal to one radian. Then $M_A = 4EI/L$ from previous relations of equation 4-2, and since B is fixed, $M_B = 2EI/L$, the carry-over moment.

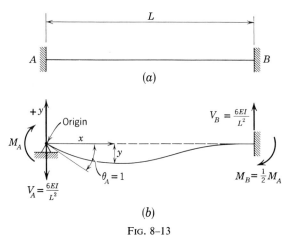

$$(a)$$

$$(b)$$

Fig. 8–13

Using A as an origin with y positive upward, we start the deflection analysis by writing

$$EI\frac{d^2y}{dx^2} = +M_A - V_A \cdot x$$

and by substitution obtain

$$EI\frac{d^2y}{dx^2} = +\frac{4EI}{L} - \frac{6EI}{L^2} \cdot x$$

Dividing both sides of the equation by EI we have

$$\frac{d^2y}{dx^2} = +\frac{4}{L} - \frac{6}{L^2} \cdot x$$

The first integration gives

$$\frac{dy}{dx} = +\frac{4}{L}x - \frac{6x^2}{2L^2} + [C_1 = -1]$$

since $\qquad \frac{dy}{dx} = -1$ radian when $x = 0$

From the second integration we obtain

$$y = +\frac{4x^2}{2L} - \frac{6x^3}{6L^2} - x + [C_2 = 0]$$

since $\qquad y = 0$ when $x = 0$

Then $\qquad y = +\frac{2x^2}{L} - \frac{x^3}{L^2} - x$

where the final equation is the equation for the required influence line for M_A by which all ordinates of the curve may be determined.

If one concentrated load P is to be placed for maximum moment at A, then the location of this load may be found by finding the location of the maximum deflection ordinate at a point of zero slope. The solution follows by setting the slope equal to zero and solving for x.

$$\frac{dy}{dx} = \frac{4x}{L} - \frac{3x^2}{L^2} - 1 = 0$$

Then $\qquad x = \tfrac{1}{3}L$

The determination of the influence line ordinates for a continuous beam involves considerable calculation but follows the principles discussed. Let the influence line for M_B be chosen as an example for the beam of uniform section of Fig. 14. Introduce a pin in the beam at B for analysis purposes and apply the equal but oppositely directed couples M to the adjacent sections. The tangents at B will rotate so that the angle between them is $\theta_B = \theta_B{}^L + \theta_B{}^R$. M can be chosen large enough for θ_B to equal unity or one radian, and if angle θ_B is unity the elastic curve represents the influence line for moment at B.

To determine M in terms of EI/L, first perform the moment distribution of Fig. 14b, where moment signs are given in terms of direction on joints. With the final moments known throughout, then $\theta_B{}^L$ and $\theta_B{}^R$ may be determined by moment-area or conjugate-beam methods.

For span AB

$$\theta_B{}^L = \frac{ML}{3EI}$$

and for span BC

$$\theta_B{}^R = \frac{0.865ML}{3EI}$$

and since θ_B must equal unity, we stipulate that

$$\theta_B = \frac{1.865ML}{3EI} = 1$$

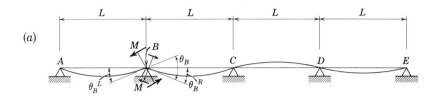

(a)

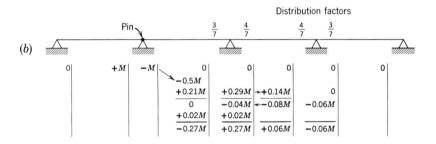

(b)

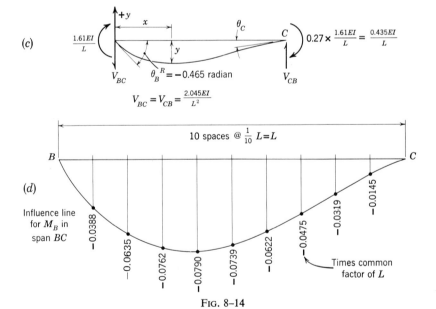

(c)

(d)

Influence line
for M_B in
span BC

FIG. 8–14

Solving, $$M = \frac{1.61EI}{L}$$

to produce an angle θ_B equal to one radian, or by substitution

$$\theta_B{}^L = 0.535 \quad \text{and} \quad \theta_B{}^R = 0.465 \text{ radian}$$

The ordinates to the elastic curve for span BC may now be analytically determined since the analysis has provided the slope at B. Using span BC as an example and referring to Fig. 14c,

$$EI\frac{d^2y}{dx^2} = \frac{1.61EI}{L} - \frac{2.045EI}{L^2} x$$

which may be simplified by dividing through by EI to

$$\frac{d^2y}{dx^2} = \frac{1.61}{L} - \frac{2.045}{L^2} x$$

From the first integration we obtain

$$\frac{dy}{dx} = \frac{1.61}{L} x - \frac{2.045}{2L^2} x^2 + [C_1 = -0.465]$$

since $$\frac{dy}{dx} = -0.465 \text{ when } x = 0$$

By a second integration we obtain the deflection as well as the influence ordinate

$$y = \frac{1.61x^2}{2L} - \frac{2.045x^3}{6L^2} - 0.465x + [C_2 = 0]$$

since $y = 0$ when $x = 0$.

The method is general and may be repeated for each span using the moment values found in Fig. 14b related to

$$M = \frac{1.61EI}{L} \text{ at } B$$

Furthermore, the point of zero slope or point of maximum ordinate may be located in each span by setting the $\frac{dy}{dx}$ equation equal to zero. The final influence line results calculated from the elastic curve equation are given in Fig. 14d, for M_B. The influence line for M_C may be developed by the same procedure used for M_B.

Although there is merit attached to introducing unit displacements or rotations by means of the internal shears, reactions, or moments, since deflection ordinates become direct influence line ordinates, such a procedure is not essential. The influence line ordinate, as has been demonstrated, is the ratio of the deflection ordinate to the impressed displacement, wherein the impressed displacement may be a deflection due to a force or an angle due to a moment of any arbitrary value.

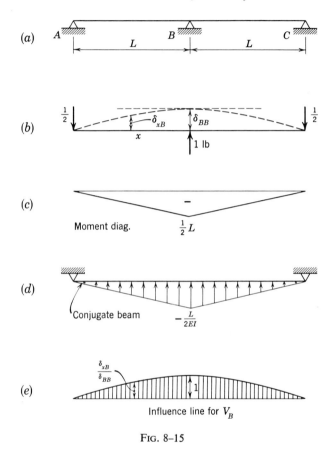

Fig. 8–15

This new approach is illustrated first by determining the influence line for vertical reaction at B of the two-span beam shown in Fig. 15a.

In Fig. 15b a 1-lb force is introduced in the direction of V_B, producing the elastic curve and the displacement δ_{BB} in the direction of the 1-lb force. The deflection at any other point x is denoted as δ_{xB} by use of the δ_{ij} notation. The moment curve is constructed in Fig. 15c, and in Fig. 15d, the conjugate beam is shown loaded with the M/EI diagram. As the student will recall, the bending moment in the conjugate beam is equal to deflection, and δ_{xB} may be found at as many points as may be required. The influence line ordinates are then computed and plotted in Fig. 15e as the ratio δ_{xB}/δ_{BB}.

The influence line for moment at the centerline of span AB of Fig. 16a may be developed by introducing a hinge at D and applying two oppositely directed unit couples. The influence line for M_D is determined by the quotient δ_{xD}/θ_{DD}. The conjugate-beam method is illustrated in Fig. 16b, where the bending moment diagram is placed on the conjugate beam

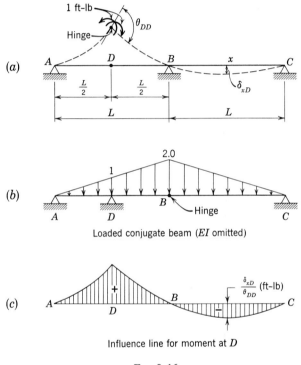

FIG. 8–16

which extends continuously over a support at D, with a hinge at B, in accordance with conjugate-beam principles. Routine conjugate-beam procedures suffice to determine the reaction at D, which will be equal to θ_{DD}, and the deflections at selected points along the span. Figure 16c is the influence line determined from the ratios δ_{xD}/θ_{DD}.

The influence line for shear at the midspan of AB, Fig. 17a, may be established by first cutting at D and inserting the parallelogram linkage of Fig. 8, thus spreading the two adjacent sections apart by two oppositely

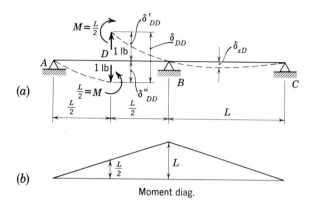

(a)

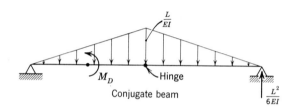

(b)

Moment diag.

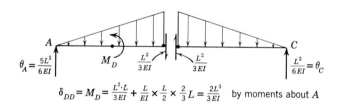

Conjugate beam

(c)

$$\delta_{DD} = M_D = \frac{L^2 \cdot L}{3EI} + \frac{L}{EI} \times \frac{L}{2} \times \frac{2}{3} L = \frac{2L^3}{3EI} \quad \text{by moments about } A$$

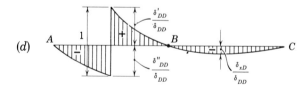

(d)

FIG. 8–17

directed unit shearing forces. The mechanical device maintains parallel tangents to the elastic curve at D and the inserted linkage develops moments of $1 \times L/2$ ft-lb as may be shown by analysis of AD as a free body. The complete moment diagram follows, Fig. 17b.

To proceed by conjugate-beam techniques the conjugate beam of Fig. 17c is constructed with a hinge at B. Owing to the relative deflection of δ_{DD} at D, a couple must be installed at D to provide for a break in the moment curve of the conjugate beam equal to this relative deflection. This moment is the moment required by equilibrium principles and detail of computation is shown below Fig. 17c. The bending moments in conjugate beam, equal to δ_{xD}, may then be computed as required to determine the influence line. The influence line for V_D is given in Fig. 17d.

8–7. INFLUENCE LINES USING MOMENT DISTRIBUTION

Müller-Breslau procedures for quantitative influence lines may often be laborious, especially if an indeterminate structure remains after establishing the initial displacement conditions in the direction of the given redundant. Moment distribution procedures will easily establish influence line ordinates for moment at the supports. With such influence line ordinates determined it is a simple matter to calculate all other influence lines for reactions, moments, and internal shearing forces by using the equations of static equilibrium.

Let it be required to determine the influence line for moment over support B of the four-span beam shown in Fig. 18a. Since a load in span BC would generate fixed-ended moments at B and C, determine two separate moment distribution systems for unit fixed-ended moments applied separately to the right of B and to the left of C.

Figure 18b establishes distributed moments due to a unit fixed-ended moment acting on B, and Fig. 18c establishes the same for a unit fixed-ended moment acting on C. Then these two separate solutions may be combined as follows:

$$M_{BC} = 0.46M_{BC}^F + 0.11M_{CB}^F$$

To develop the influence line for M_B let a unit vertical load be located a distance of x from B in span BC, as in Fig. 18d, and compute fixed-ended moments (see equations 3-1)

$$M_{BC}^F = \frac{x(L-x)^2}{L^2}; \qquad M_{CB}^F = \frac{x^2(L-x)}{L^2}$$

With $x = 0.10L$

$$M_{BC}^F = \frac{0.10L(L - 0.10L)^2}{L^2} = 0.081L$$

$$M_{CB}^F = \frac{(0.10L)^2(L - 0.10L)}{L^2} = 0.009L$$

Then $M_{BC} = 0.46 \times 0.081L + 0.11 \times 0.009L = +0.0372L$

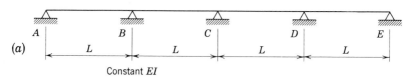

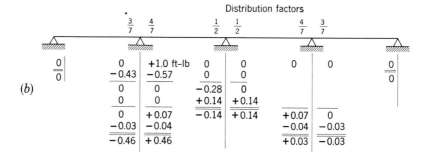

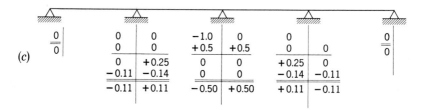

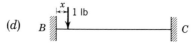

FIG. 8–18

The positive sign for the moment must be interpreted with reference to the sign convention used in the moment distribution method and indicates a clockwise moment acting on joint B.   By the beam convention the moment over the support B is then negative.

With $x = 0.5L$

$$M_{BC}^F = M_{CB}^F = \frac{0.5L \times (0.5L)^2}{L^2} = 0.125L$$

Then $\qquad M_{BC} = (0.46 + 0.11)(0.125L) = +0.0714L$

The computed values of M_B or M_{BC} compare favorably with values obtained by deflection procedures in Fig. 14d, and would compare exactly if high arithmetic precision is carried out in both solutions. Normally, moment distribution is faster and precise enough. The above calculations may be extended to as many points as is required, and to loads in all spans, by extending the unit moment procedure to all spans at all supports. It should again be noted that moment-distribution signs have been observed in the above example; the positive moment M_B is a negative moment by beam convention. The final influence line is as given previously in Fig. 14d for M_B at $\frac{1}{10}L$ intervals for span BC. Influence lines for other spans are left unsolved for student drill. As previously stated, all other influence lines may be determined from statics after establishing the influence lines for the moments at the supports.

8-8. INFLUENCE LINES FOR RIGID FRAMES

The general principles of influence line construction holds for frames as well as for beams. The frame of Fig. 19a is redundant to the first degree, and with H at A designated as the redundant a qualitative influence line may be developed by translating A a unity distance, as in Fig. 19b, in the direction of H by applying a force F. Maxwell's reciprocal deflection theorem or Müller-Breslau principles permits the designation of the deflections as values of the influence line for H. For example, at point y the vertical deflection Δ_y gives the effect of a unit vertical load acting down at y on H, since in Fig. 19c H must be of such a value as to return A' to A. Similarly, Δ_x is the effect of a unit horizontal load placed at x on H. The student will realize that a sign convention is important and the external load placed and directed as shown at x in Fig. 19d will require H to act to the left or, in the negative sense, to restore A' to A. If H must act to the right the effect is positive, since this was the assumed direction of F in making the initial unit displacement at A. The final influence diagram for H with signs is shown in Fig. 19e, with the assumption that vertical loads act downward and that horizontal loads on structure act to the right.

Figure 20a indicates a rigid frame with fixed column bases. In this instance choose the vertical reaction, horizontal reaction, and the moment

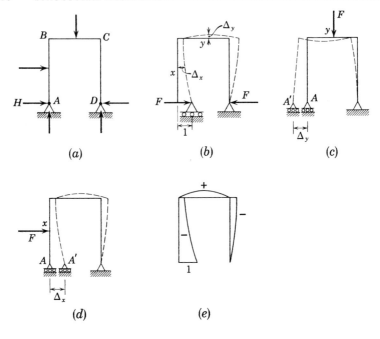

F_{IG.} 8–19

at A as the redundant quantities for which influence lines are required. The influence lines may be qualitatively sketched for each of these redundants by introducing the basic unit displacements as shown, recalling that a displacement in the direction of only one of the redundant quantities can be involved at a time. This is essential to prevent the other redundants, two redundants in this case, from doing work. Thus, while A is displaced vertically one unit A is also fixed or held against rotation with horizontal

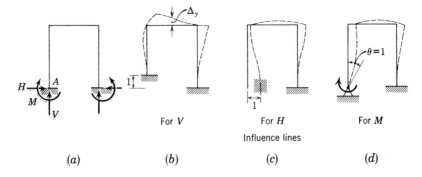

F_{IG.} 8–20

displacement also prevented. This is shown in Fig. 20*b*. A similar analysis will indicate that the geometric conditions during displacement for the other two influence lines of Figs. 20*c* and *d* are also theoretically correct.

The general principles of influence lines are illustrated in Fig. 21 for interior sections of a frame. Figure 21*a* is the influence line for moment at *B*, where the tangents to elastic curves meeting at a temporary hinge at *B* are given a relative rotation of one radian. Sidesway was prevented. Figure 21*b* is the influence line for moment at section *F* at midlength of *BC*. Figure 21*c* is the influence line for beam shear at *F* with sidesway prevented.

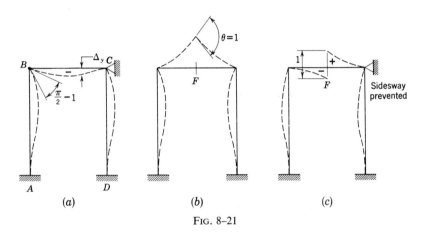

FIG. 8–21

In a building of many floors and columns the same problem exists in determining maximum and minimum live load effects in girders and columns. Figure 22*a* is a three story bent rigidly framed. The influence line for moment at Section *F* is shown. The tracing of deformations through the bent may be accomplished by thinking in terms of moment distribution analysis. For example, the effect of the moments at *F* to induce $\theta = 1$ radian at *F* would be to produce moments at *A* and *D* which would rotate the tangents to the elastic curve at *A* and *D* as shown. With such rotation taking place against the restraint of the attached columns and adjacent girders, balancing moments must be developed in these attached members. A moment distribution sign convention may be adopted and these initial moments filtered out into the remainder of the structural system by mentally or actually carrying over and distributing with due regard to signs. As the distribution proceeds the related elastic curve may be sketched. The sign of the influence line ordinate for moment at F is shown.   An interesting result appears in describing the loading pattern for maximum positive live load moment at F.   If dealing with uniform loading the pattern is as shown in Fig. 22*b*.   For negative moment the loading must be on the spans designated as minus.   Although the pattern shown is the complete theoretical requirement, the contribution of loads on floors remote to the one containing the section F is normally

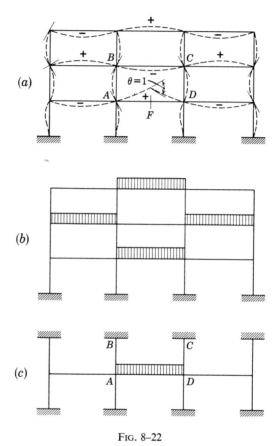

FIG. 8–22

negligible.   For that reason the loading can usually be simplified to that shown in Fig. 22*c*, for maximum positive moment at F.

The simplified framing of Fig. 23*a* assumes that the far ends of the columns are fixed and that sidesway is prevented.   The influence line of Fig. 23*a* for M_{AD} is generated by introducing a temporary hinge just to the right of the column intersection.   The influence line indicates that adjacent spans must be loaded for the maximum negative moment M_{AD}.   Figure

$23b$ is the influence line for shear V_{AD}.   Thus, if the live load is uniform
in character the maximum negative moment and maximum positive shear
is caused by identical loading.   One more influence line of interest may be
drawn for moment in the column just below the intersection A in Fig. $23c$.
Note that spans adjacent to the column when loaded produce moment of
opposite sign.   One or the other of these spans fully loaded will produce
maximum live load moment.

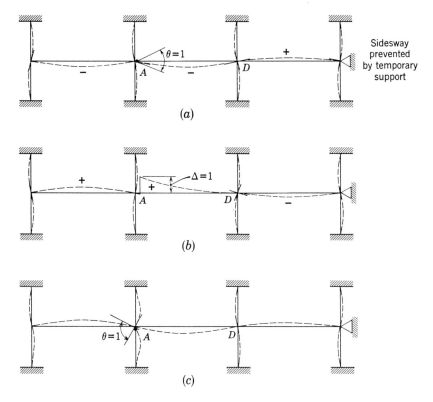

Fig. 8–23

The quantitative influence lines are required when concentrated loads
are involved.   They may best be developed for a frame by successively
placing a unit load at intervals along the spans involved and the required
influence line ordinates calculated by the moment distribution method.
In this manner full cognizance may be taken of sidesway influence.   The
qualitative considerations previously discussed will, however, guide the
detailed solution.

8–9. INFLUENCE LINES—CONTINUOUS TRUSSES

The Maxwell reciprocal deflection theorems and Müller-Breslau principles are valid for any elastic structure if superposition is permissible, hence the general principles of influence line construction are applicable to trussed structures. Whereas in the beam and frame, flexural considerations determined the elastic curve, in the truss, axial deformations of the members establish the basis for deflection calculation.

In the discussion to follow, a floor system delivering loads only to the lower panel points is assumed. If the live loading is a uniform load wherein full span loadings may be employed in required loading patterns, then a qualitative approach to influence lines will suffice. For a concentrated load system quantitative influence lines are generally needed.

Figure 24 is a two-span continuous truss. The influence line for the vertical reactions V_B and V_A may be qualitatively visualized by displacing B or A a unit vertical distance as shown in Figs. 24b and 24c. If the quantitative influence line is to be constructed, the deflections of Fig. 24b may be computed by the unit load approach. First remove V_B and introduce an upward force F_B whose magnitude is sufficient to deflect B unity upward. The deflection at B is

$$\Delta_{44} = \sum \frac{F_B u_B u_B L}{AE}$$

where u_B is the axial stress in a given truss member owing to a unit load of 1 lb upward at B. Then for

$$\Delta_{44} = 1$$

$$F_B = \frac{1}{\sum \dfrac{u_B{}^2 L}{AE}}$$

The deflection produced at any other panel point by F_B would be determined by a second application of the unit load method. For example, to find Δ_{24} (the deflection produced at 2 with the load at 4) first place a unit load at 2 and then introduce F_B. The unit load may be placed acting up or acting down, but here the unit load is assumed to act up as the movement of 2 is up and u_2 is the axial stress produced in a given member. Then

$$\Delta_{24} = F_B \sum \frac{u_B \cdot u_2 \cdot L}{AE} = \frac{\sum u_B \cdot u_2 \dfrac{L}{AE}}{\sum \dfrac{u_B{}^2 L}{AE}}$$

For constant E

$$\Delta_{24} = \frac{\sum \dfrac{u_B u_2 L}{A}}{\sum \dfrac{u_B{}^2 L}{A}}$$

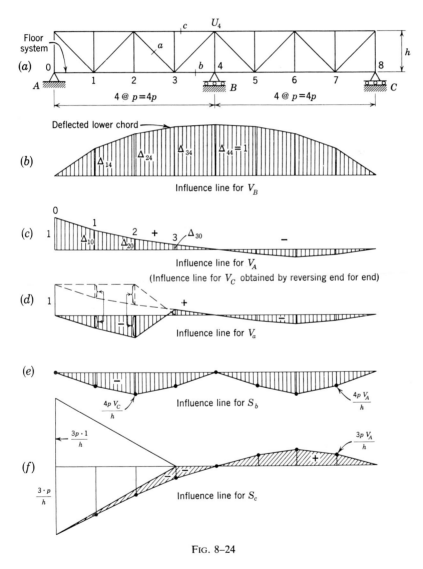

Fig. 8–24

An orderly tabular solution generally will be employed to determine these deflections, or it is possible to use a Williot-Mohr construction by using the changes in member length due to the stresses produced by F_B.

Since design must precede analysis to permit an evaluation of L/A an approximation in the design stage would be to assume L/A equal for all members. Then

$$\Delta_{24} \approx \frac{\sum u_B u_2}{\sum u_B{}^2}$$

To avoid tedious summations the continuous truss could be rationalized to a continuous girder of uniform section. Neither this approximation nor the previous one would or could be expected to give accurate absolute deflections. Since influence line construction deals with the ratio of deflections, these ratios will usually be quite satisfactory for initial design purposes.

In Fig. 24d, the influence line for the vertical component of stress in diagonal a is evolved by treating the vertical static equilibrium of a free body to the left of a section through panel 2-3. With a 1-lb load on the floor system to the right of panel point 3, the vertical component V_a is equal to V_A. As the load moves to the left into the panel, its influence is felt on the free body. This effect is depicted in Fig. 24d by graphical subtraction. The resultant influence line is shown.

The influence line for S_b, the stress in a lower chord in panel 3-4, may also be devised from the influence line for V_A as shown in Fig. 24e. A section is passed through panel 3-4 and moments are taken about the upper panel point 4. By using a free body to the left when the unit load is in the span, BC enables the influence values for S_b to be written in terms of the influence line values for the left-hand reaction V_A. In the symmetrical situation depicted in Fig. 24a the influence diagram for the reaction V_C is the reverse of the diagram for V_A; hence, to complete the influence diagram for S_b when the unity load is on the span AB, the free body to the right of panel 3-4 may be employed with S_b written as a function of V_C. The influence line for S_b is given in Fig. 24e.

The influence line for S_c, an upper chord bar, is shown in Fig. 24f. Moments about the lower panel point 3 with a section through panel 3-4 involves, with a free body to the left, the effect of V_A and the effect of the 1-lb load on the floor system when that load is to the left of panel point 3. The two parts of the moment calculation with their respective effects on S_c are shown along with the resultant influence line. Generally then, for continuous trusses, the influence lines for the reactions may be computed or qualitatively drawn and the influence lines for all other members derived by superimposing the proper terms after an analysis indicates the proper constituent terms.

REFERENCES

1. Beggs, G. E., "Design of Elastic Structures from Paper Models," *Proc. Am. Conc. Inst.*, **19**, 53–66 (1923).
2. Beggs, G. E., "The Use of Models in the Solution of Indeterminate Structures," *J. Franklin Inst.* (March 1927).
3. Cross, H., and N. D. Morgan, *Continuous Frames of Reinforced Concrete*, Chapters 6 and 8, New York: John Wiley, 1932.
4. Eney, W. J., "Determining the Deflection of Structures with Models," *Civil Engineering* (March 1942).
5. Eney, W. J., "Model Analysis of Continuous Girders," *Civil Engineering*, (September 1941).

PROBLEMS

8–1. For this determinate beam, draw the influence lines for V_A, V_B, and V_C.

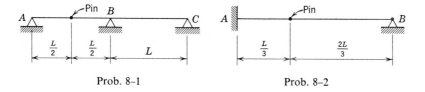

Prob. 8–1   Prob. 8–2

8–2. For this determinate beam, draw the influence lines for V_B and M_A.

8–3. Using one stationary deflection measuring gage, how could you determine the deflections at points B, C, D, and E?

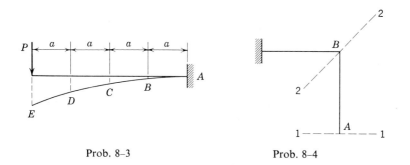

Prob. 8–3   Prob. 8–4

8–4. If a force P at point A in the 1–1 direction displaces B in the 2–2 direction a distance of Δ, how much would A move in the 1–1 direction if P is applied at B in the 2–2 direction?

8–5. If a couple T (in the plane of the beam) at A causes the tangent at B to rotate θ_{BA}, prove that if the couple is applied at B the rotation θ_{AB} of the tangent at A equals θ_{BA}.

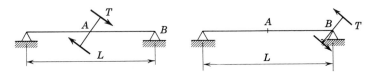

Prob. 8–5

8–6. Sketch the qualitative influence lines for V_A, V_B, and V_C, by using Müller-Breslau principles.

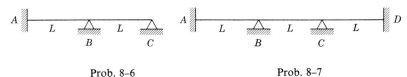

Prob. 8–6   Prob. 8–7

8–7. Sketch the qualitative influence lines for V_A and V_B by Müller-Breslau principles.

8–8. Sketch the qualitative influence lines for V_A and V_C by Müller-Breslau principles.

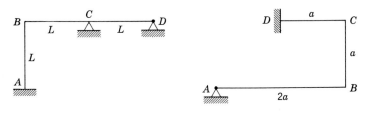

Prob. 8–8   Prob. 8–9

8–9. Sketch the qualitative influence line for V_A by Müller-Breslau principles.

8–10. Sketch the qualitative influence lines for M_A, M_B, M_D, and for shear at section D, and for a section just to the right of B.   If the live loading is a uniform load, develop the loading patterns for maximum and minimum values.

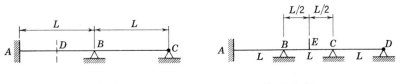

Prob. 8–10   Prob. 8–11

8–11. Sketch the qualitative influence lines for M_A and V_C and for shear and moment at section E.   If the live loading is a uniform load, develop the loading patterns for maximum and minimum values for all quantities for which the influence lines were sketched.

8–12. Sketch the qualitative influence lines for M_A, M_B, V_A, V_{AB}, and for shear at the center line of span AB.   Assuming that the uniform dead load of beam is 1.5 kips per ft, and that the uniform live load is 4 kips per ft, develop and draw (a) the bending moment curve for span AB with all maximum and minimum values, and (b) the shear curve for span AB with all maximum values.

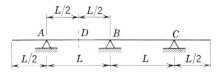

Prob. 8–12

8–13. Sketch the qualitative influence lines for M_B, V_B, and V_{BC}.   Assuming a uniform dead load of 1.0 kips per ft and a uniform live load of 3 kips per ft, determine (a) the maximum and minimum reaction at B, (b) the maximum shear V_{BC}, and (c) the controlling bending moments for span BC and draw the moment diagrams.   *Ans.* Max $V_B = 7.265L$ (in kip units).

Prob. 8–13

8–14. Assuming a uniform dead load of 1.2 kips per ft, and a live load of 3.6 kips per ft, develop the controlling bending moment curve for span CD.   Also develop the curve for maximum shear for span CD.
             Ans. Max negative $M_C = -0.495L^2$; max shear $V_{CD} = 2.724L$.

Prob. 8–14   Prob. 8–15

8–15. Develop a quantitative influence line for M_B by the procedure followed in Fig. 8–14 of the text.   Determine the numerical influence line quantities for span AB only.
                    Ans. Influence line equation $\dfrac{3x}{7L}\left(\dfrac{x^2}{L} - x\right)$.

8–16. Develop a quantitative influence line for M_B by the procedure followed in Fig. 8–14 of the text. Determine the numerical influence line quantities for span BC only.

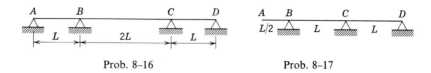

Prob. 8–16   Prob. 8–17

8–17. Develop the quantitative influence lines for M_B and M_C by the procedure followed in Fig. 8-14 of the text. Determine the numerical influence line quantities for spans AB and BC only.

8–18. Develop the quantitative influence line for V_B of the beam of problem 8–15 by the conjugate-beam method.

8–19. Develop the quantitative influence line for M_B of the beam of problem 8–15 by the conjugate-beam method. Assume a hinge at B and introduce two equal and opposite unit couples at B.

8–20. Develop the quantitative influence line for shear at the center line of span AB of the beam of problem 8–15, by the conjugate-beam method.

8–21. Develop the quantitative influence line for moment at the center line of span BC of the beam of problem 8–16, by the conjugate-beam method.

8–22. Develop a quantitative influence line for M_B of the beam of problem 8–15 by the moment distribution procedure outlined in Fig. 8–18 of text. Determine numerical values for span AB only.

8–23. Develop a quantitative influence line for M_B of the beam of problem 8–16 by the moment distribution procedure outlined in Fig. 8-18 of text. Determine numerical values for span BC only.

8–24. Develop a quantitative influence line for M_B by the moment distribution procedure outlined in Fig. 8-18 of text.

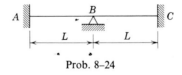

Prob. 8–24

8–25. Develop a quantitative influence line for M_C of the beam of problem 8–17 by the moment distribution procedure. Determine numerical values for span BC only.

8–26. Sketch the qualitative influence lines for M_A, H_A, and V_A.

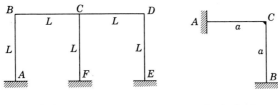

Prob. 8–26   Prob. 8–27

8–27. Sketch the qualitative influence lines for M_B, H_B, and V_B.

8–28. Sketch the qualitative influence lines for moment and shear at section G.

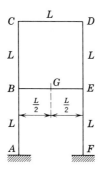

Prob. 8–28

8–29. Sketch the qualitative influence lines for moment and shear at section E.

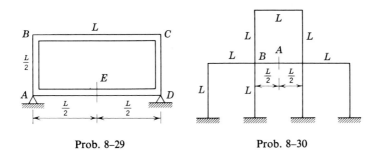

Prob. 8–29   Prob. 8–30

8–30. Sketch the qualitative influence line for moment at A and for moment at B just below the joint.

8–31. Develop the quantitative influence lines for M_A and H_A by a procedure of your choice. The frame is free to sway. Constant I. Determine numerical values along BC only at $\frac{1}{10}$ span intervals.

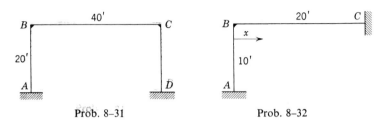

Prob. 8–31   Prob. 8–32

8–32. Develop the quantitative influence lines for M_B by a procedure of your choice. Determine numerical values along BC only. Constant I.

$$Ans. \ \tfrac{1}{600}(-x^3 + 40x^2 - 400x).$$

8–33. Determine the quantitative influence line for moment M_{BC}. Isolate a story above and below the floor level $ABCD$, and assume the ends of columns fixed.

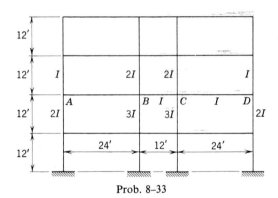

Prob. 8–33

8–34. Sketch the qualitative influence lines for V_A and V_B. Also sketch the qualitative influence lines for the vertical component of diagonal a, and for the stress in lower chord member b. The vertical live loading may be considered as being applied at the lower chord level.

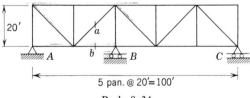

5 pan. @ 20′= 100′

Prob. 8–34

8–35. The swing bridge is in closed position with the floor system at the lower chord level.   Sketch the qualitative influence lines for the vertical component of diagonal *a*, and for total stress in the upper chord member *b*.

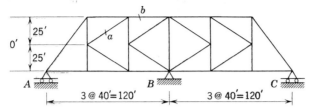

Prob. 8–35

8–36. Sketch the qualitative influence lines for stress in chord member *a*, and for the vertical component of diagonal *b*.   Sketch the qualitative influence line for stress in chord member *c*.   Live loads at lower chord level.

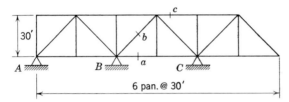

Prob. 8–36

Variable Moment of Inertia

9

9–1. INTRODUCTION

The previous chapters have purposely avoided introducing members of variable moment of inertia, owing to the desire on the part of the author to put first principles first without complicating these fundamentals by time consuming exercises.   However, the member of variable moment of inertia is the rule in many structures rather than the exception.   In the sections to follow the deflection methods and methods of analysis, previously presented, are modified as necessary to accommodate the variation in moment of inertia.

9–2. DEFLECTIONS

Let Fig. 1 represent a cover-plated beam, with the center section having a moment of inertia of I_1 and the outer segments a moment of inertia of I_2. An analysis follows for the slopes and deflections by the double integration method.   Due cognizance must be given to the discontinuity in moment of inertia.

For $x = 0$ to $x = a$,

$$EI_2 \frac{d^2y}{dx^2} = M = \frac{P}{2}x$$

$$EI_2 \frac{dy}{dx} = \frac{Px^2}{4} + C_1$$

$$EI_2 y = \frac{Px^3}{12} + C_1x + [C_2 = 0]$$

since $y = 0$ when $x = a$ where C_1 must remain unevaluated for the present.

For $x = a$ to $2a$,

$$EI_1 \frac{d^2y}{dx^2} = \frac{Px}{2}$$

$$EI_1 \frac{dy}{dx} = \frac{Px^2}{4} + [C_3 = -Pa^2]$$

since $\frac{dy}{dx} = 0$ when $x = 2a$.

Then

$$EI_1 y = \frac{Px^3}{12} - Pa^2x + C_4$$

where C_4 must remain unevaluated for the present.

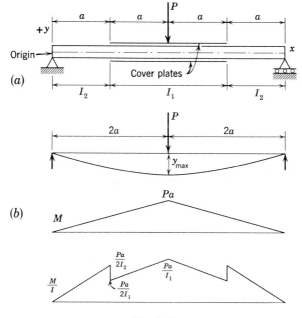

FIG. 9–1

At $x = a$ the slopes and deflections have a common value, providing the two conditions needed for evaluation of C_1 and C_4. Hence, for the slope condition

$$\frac{1}{I_2}\left(\frac{Pa^2}{4} + C_1\right) = \frac{1}{I_1}\left(\frac{Pa^2}{4} - Pa^2\right)$$

$$C_1 = -\frac{Pa^2}{4}\left(3\frac{I_2}{I_1} + 1\right)$$

and for the deflection condition and substituting C_1 we determine

$$\frac{Pa^3}{12I_2} - \frac{Pa^3}{4I_2}\left(3\frac{I_2}{I_1} + 1\right) = \frac{Pa^3}{12I_1} - \frac{Pa^3}{I_1} + \frac{C_4}{I_1}$$

$$C_4 = \frac{Pa^3}{6}\left[-\frac{I_1}{I_2} + 1\right]$$

With C_4 known the center or maximum deflection may be obtained from the second equation for y when $x = 2a$ as

$$y\max = \frac{Pa^3}{12EI_1I_2}[-14I_2 - 2I_1]$$

The same problem will now be solved by the moment-area procedure. Figure 1b indicates the bending moment curve and the M/I diagram taking into account the variable moment of inertia. Then

$$E \cdot y\max = \frac{Pa \cdot a}{4I_2}\left(\frac{2}{3}a\right) + \frac{Pa}{2I_1}(a)\left(\frac{3}{2}a\right) + \left(\frac{Pa}{I_1} - \frac{Pa}{2I_1}\right)\left(\frac{a}{2}\right)\left(a + \frac{2}{3}a\right)$$

$$y\max = \frac{Pa^3}{12EI_1I_2}[14I_2 + 2I_1]$$

This is the same as obtained by the double integration method with the exception of signs. The same result of course will be obtained by the conjugate-beam method or any other deflection method. The choice of method is often an individual preference, but, in general, if the equation of the elastic curve is needed an integration method will be used.

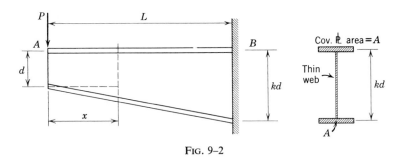

FIG. 9–2

When moment of inertia varies in a way to complicate the integration, or the determination of areas and centroids of the M/EI diagram, the mathematical effort is increased. As an example of this, assume the cantilever beam of Fig. 2a to have its depth varying linearly from a minimum depth at the free end to the maximum depth at the support.

It is further assumed for this problem that the beam cross-section is composed of a thin web and two cover plates, and for ease in formulating the moment of inertia variation the small effect of the thin web plate has been neglected.

The moment of inertia at A is

$$I_A = \frac{Ad^2}{2}$$

and at a section x distance from the free end

$$I_x = 2A \left[\frac{d + \frac{xd}{L}(k - 1)}{2} \right]^2$$

which simplifies to

$$I_x = I_A \left[1 + \frac{x}{L}(k - 1) \right]^2$$

The vertical deflection at A may be determined by the unit load method, $\int \frac{Mm\, dx}{EI}$, where $M = -Px$ and $m = -1 \cdot x$, as

$$\Delta_A = \frac{P}{EI_A} \int_0^L \frac{x^2\, dx}{\left[1 + \frac{x}{L}(k - 1) \right]^2}$$

Fortunately this is a standard integral for which a ready solution exists and is given in published tables of integrals. Integration provides

$$\Delta_A = \frac{P}{EI_A} \left[\int_0^L \frac{1}{\left(\frac{k-1}{L}\right)^3} \left\{ 1 + \left(\frac{k-1}{L}\right)x - 2\log_e\left(1 + \left(\frac{k-1}{L}\right)x\right) \right. \right.$$

$$\left. \left. - \frac{1}{1 + \frac{x}{L}(k-1)} \right\} \right]$$

and with limits substituted the general equation is

$$\Delta_A = \frac{PL^3}{(k-1)^3 EI_A} \left[k - \frac{1}{k} - 2\log_e k \right]$$

and for a beam where $k = 2$

$$\Delta_A = \frac{PL^3}{EI_A} \left[2 - \frac{1}{2} - 2 \times 0.693 \right]$$

$$\Delta_A = \frac{0.114 PL^3}{EI_A}$$

As a point of mathematical interest, as well as an illustration of checking the final form of a complex equation, let $k = 1$, thus reducing the cantilever beam to a beam of uniform moment of inertia. If the general solution is correct the result $\Delta_A = PL^3/3EI_A$ should be obtained. However, if $k = 1$ is substituted in the general equation, an indeterminate or meaningless value of $\Delta_A = 0/0$ results. This signifies that the evaluation of the equation for $k = 1$ must be determined by applying the rule known as L'Hospital's rule.* The rule is stated as: To find the value of a fraction which takes the form $0/0$ when $k = 1$, replace the numerator and the denominator each by its derivative and substitute $k = 1$. If the new fraction is also $0/0$, repeat the process.

With respect to the given situation, we want the limit when $k \to 1$, thus,

$$\lim_{k \to 1} \frac{k - 1/k - 2 \log_e k}{(k - 1)^3} = \lim_{k \to 1} \frac{1 + 1/k^2 - 2/k}{3(k - 1)^2}$$

$$= \lim_{k \to 1} \frac{-2/k^3 + 2/k^2}{6(k - 1)} = \left[\frac{+6/k^4 - 4/k^3}{6} \right]_{k=1} = \frac{1}{3}$$

where three successive differentiations are necessary, since the first two differentiations still lead to $0/0$ for $k = 1$, but after the third differentiation a finite limit of one-third is obtained. Thus, for

$$k = 1$$

$$\Delta_A = \frac{PL^3}{3EI_A}$$

and since this checks the known solution for a beam of uniform section with an end load P, it can be said that the general solution is correct.

A general solution may usually be obtained by the integration approach. The more involved the integration the greater is the effort required for solution. Since many cases of nonuniform moment of inertia may have variations in moment of inertia not falling into easily expressed analytical functions, a numerical approach is often mandatory, as discussed in the following section.

9–3. DEFLECTIONS BY NUMERICAL INTEGRATION

To illustrate numerical integration refer to Fig. 3, wherein the beam of Fig. 2 is again used as the example with $k = 2$. The successive steps

---

* See standard texts in advanced calculus.

leading to the final solution are graphically shown in Fig. 3. Since it is the intent to solve for $\varDelta$ at A by moment-area procedures the M/I diagram is needed. The M diagram is drawn and a decision is made as to the

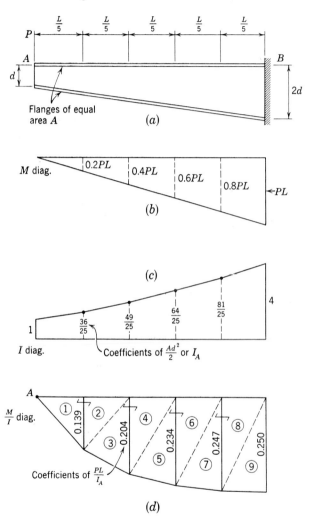

FIG. 9–3

number of finite intervals to be used in the numerical integration. Five equal intervals have been chosen to minimize the arithmetic in this instance, but in many problems 10, 20, or more intervals should be chosen to secure a close approximation. With intervals as described the ordinates to the M and I curve have been computed and recorded on the diagrams for the common points of adjacent intervals. The M/I diagram has ordinates, found by dividing M by I, stated as coefficients of PL/I_A for convenience in later calculations. The plotted values of ordinates are connected with straight lines providing an assumed linear variation of M/I within each interval. This assumption is the source of error which is minimized by taking more intervals.

The moment-area principle may now be applied by taking the moments of the areas of M/I diagram about A. Diagonal construction lines on this diagram identify definite triangular areas where the area of each triangle is equal to its base times half of the altitude. A simplification appears if two adjacent triangles are linked together, since with a common base (an ordinate) and equal altitude equal to the length of the interval, the combined area of the two triangles equals the common base times the equal altitude. The centroid of these common areas is located at the common base. All triangles in Fig. 3d can be so combined in pairs except triangle number 9.

The moments of the areas about A to obtain the deflection at A are tabulated in the following table. With equal finite intervals the arithmetic is simplified by tabulating areas and moment arms as if the interval were equal to unity. The final summation for moment is corrected for the true interval length of $L/5$ by multiplying by $L/5 \times L/5$.

Areas Designated	Area ÷ PL/I_A	Arm	Area × Arm
1–2	0.139	1	0.139
3–4	0.204	2	0.408
5–6	0.234	3	0.702
7–8	0.247	4	0.988
9	0.125	4.67	0.585
			2.822

$$E \cdot \Delta_A = 2.822 \times \frac{PL}{I_A} \times \frac{L}{5} \times \frac{L}{5}$$

$$\Delta_A = \frac{0.113PL^3}{EI_A}$$

The answer for Δ_A may be compared with $0.114PL^3/EI_A$, previously determined by integration, showing that arithmetical solution is in error by about 1 per cent in this case.

9-4. MOMENT DISTRIBUTION MODIFICATIONS

Moment distribution techniques required fixed-ended moments, absolute or relative stiffness, distribution factors, and carry-over factors. These values were easily formulated for uniform moment of inertia sections but, although basic definitions remain valid, the variable moment of inertia section involves considerable labor. The problem illustrated in Fig. 4, will serve as the first example. Since it is the same beam illustrated in Fig. 1, much of the prior solution will be incorporated in this example.

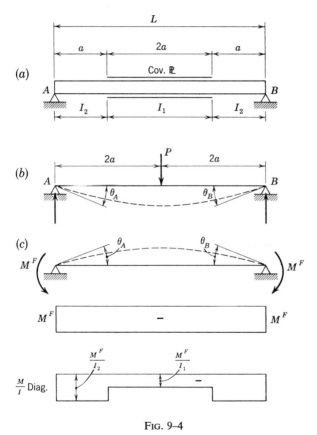

FIG. 9-4

Place a load P at the center line of the span and compute the fixed-ended moments. In Fig. 4b, the slopes or rotation of the end tangents of a simply supported beam are found from previous solution as

$$\theta_A = \theta_B = \frac{Pa^2}{4EI_2}\left(3\frac{I_2}{I_1} + 1\right)$$

Since the fixed-ended moments will restore the tangents to the horizontal position, these moments, termed M^F, are placed on both ends of the beam as in Fig. 4c to produce rotations equal but in opposite direction to the simple span rotations. Following moment-area procedures and working with the M/I diagram,

$$E \cdot \theta_A = \frac{M^F}{I_2} \cdot a + \frac{M^F}{I_1} a$$

$$\theta_A = \frac{M^F \cdot a}{E} \left(\frac{1}{I_2} + \frac{1}{I_1} \right)$$

equating the two separate values of θ_A we determine the fixed-ended moment as

$$M^F = \frac{Pa}{4} \frac{(3I_2 + I_1)}{(I_2 + I_1)}$$

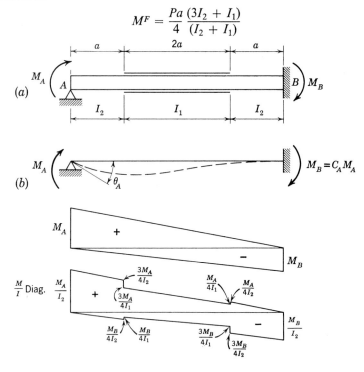

FIG. 9–5

Figure 5a depicts the same beam simply supported at A and fixed at B. If the moment M_A is applied at A the moment generated at B is M_B, which is in some ratio to M_A. That ratio has previously been defined as the carry-over factor, hence $M_B = C_A \cdot M_A$ where C_A is the carry-over factor from A to B. To interrelate M_A and M_B, first sketch the moment diagram in parts and develop the M/I diagram shown. Then moments of these areas about A set equal to zero (since deflection of A from tangent to B is zero) will provide in the final simplified form

$$C_A = \frac{M_B}{M_A} = \frac{5/I_2 + 11/I_1}{19/I_2 + 13/I_1}$$

This expression may be tested by making $I_1 = I_2$. Then $C_A = \frac{1}{2}$ as it should be for a beam of uniform moment of inertia.

If $I_1 = 2I_2$

$$C_A = 0.412$$

If cover plates were over the two outer segments of the span instead of over the center one-half, with $I_2 = 2I_1$, then

$$C_A = 0.600$$

Thus, it is observed that the carry-over factor for the variable moment of inertia members deviates from the value of one-half determined for uniform moment of inertia members.

From Fig. 5, the value of θ_A may now be found by moment areas to be

$$\theta_A = \frac{M_A \cdot a}{E} \left[\frac{1}{I_2} + \frac{1}{I_1} - C_A \left(\frac{1}{I_2} + \frac{1}{I_1} \right) \right]$$

Then substituting $a = L/4$

$$M_A = \frac{4E \cdot \theta_A}{L \left[\frac{1}{I_2} + \frac{1}{I_1} - C_A \left(\frac{1}{I_2} + \frac{1}{I_1} \right) \right]}$$

This may be tested for the uniform moment of inertia case for $I_1 = I_2 = I$, where $C_A = \frac{1}{2}$, giving $M_A = 4EI \cdot \theta_A/L$, as established previously by equation 4-2.

If $I_1 = 2I_2$, and $C_A = 0.412$ from the previous calculation,

$$M_A = \frac{4.54EI_2}{L} \cdot \theta_A$$

For purposes of moment distribution the basic *absolute stiffness* K_A is defined as the value of M_A to make θ_A equal to unity while B is restrained, hence

$$K_A = \frac{4.54EI_2}{L}$$

It can be generalized that K_A will always be of the form

$$K_A = \frac{k_A EI_0}{L}$$

where k_A is a coefficient, dependent on the member's moment of inertia variation, and I_0 represents the minimum moment of inertia in the span. For uniform moment of inertia $k_A = 4$.

9–5. MOMENT DISTRIBUTION MODIFICATIONS—GENERAL

The general relationships of stiffness and carry-over factors may be formalized as follows. In Fig. 6, a haunched beam is shown, and for an unsymmetrical variation of moment of inertia the stiffness as well as

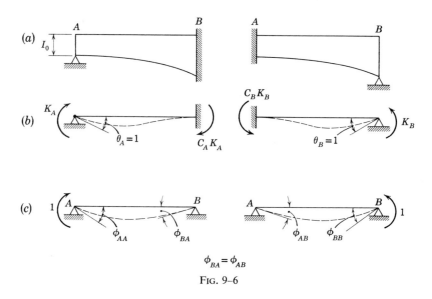

$$\phi_{BA} = \phi_{AB}$$

Fig. 9–6

carry-over factors are unequal for ends A and B. If a moment K_A is applied at A while B is fixed, the tangent will be rotated through an angle θ_A equal to one radian and a moment induced at B equal to $C_A K_A$. In the same manner, a moment K_B, applied at B with A fixed, rotates the tangent at B one radian and induces a moment $C_B K_B$ at A. These relationships are shown in the two figures of Fig. 6b.

Several general relationships may be derived through geometric reasoning. First, to assist in the discussion to follow, separately apply a unit moment at A and at B as shown in the two diagrams of Fig. 6c, denoting the angular rotations by the nomenclature shown. (The subscripts follow the usual ij convention.)

We may note in the left-hand diagram Fig. 6b that the rotation of the tangent at B is zero when B is fixed. With zero rotation as a condition at B, we may write an equation which combines or superimposes the separate rotations due to K_A and $C_A K_A$. The rotation at B due to K_A is $K_A \phi_{BA}$ and the rotation at B due to $C_A K_A$ is $C_A K_A \phi_{BB}$. The resulting equation is

$$K_A \cdot \phi_{BA} - C_A K_A \cdot \phi_{BB} = 0$$

or
$$C_A = \frac{\phi_{BA}}{\phi_{BB}}$$

Furthermore, by Maxwell's reciprocal relation $\phi_{AB} = \phi_{BA}$, hence

$$C_A = \frac{\phi_{AB}}{\phi_{BB}} \tag{9-1}$$

Similarly, from the condition of zero rotation at A when K_B is applied at B, and by Maxwell's reciprocal relation, it may be proven that

$$C_B = \frac{\phi_{BA}}{\phi_{AA}} \tag{9-2}$$

Thus, the carry-over factors for a beam AB are found to be the ratio of the rotations produced by a unit moment at B for C_A and by a unit moment at A for C_B. Since a ratio is involved, any moment value would provide the same result, or, for a model of the beam, the ratio of the angles for any arbitrary applied but unmeasured moment would suffice.

A further development of fundamental relations is based on requiring angle θ_A at A to be equal to unity when B is fixed. In terms of the quantities of Figs. 6b and 6c we obtain

$$K_A \cdot \phi_{AA} - C_A K_A \cdot \phi_{AB} = 1$$

Since the absolute stiffness is required we solve for K_A in the form

$$K_A = \frac{1}{\phi_{AA} - C_A \cdot \phi_{AB}} = \frac{1}{\phi_{AA} - (\phi_{AB}/\phi_{BB}) \cdot \phi_{AB}}$$

which finally simplifies to

$$K_A = \frac{\phi_{BB}}{\phi_{AA}\phi_{BB} - \phi_{BA}^2} \tag{9-3}$$

Similarly, at B with A fixed, we obtain

$$K_B \cdot \phi_{BB} - C_B K_B \cdot \phi_{BA} = 1$$

and
$$K_B = \frac{\phi_{AA}}{\phi_{AA}\phi_{BB} - \phi_{BA}^2} \tag{9-4} .$$

and by making use of the reciprocal theorem requiring $\phi_{AB} = \phi_{BA}$, we obtain

$$K_B = \frac{\phi_{AA}}{\phi_{AA}\phi_{BB} - \phi_{AB}^2}$$

Thus the following ratio may be found by dividing 9-3 by 9-4.

$$\frac{K_A}{K_B} = \frac{\phi_{BB}}{\phi_{AA}}$$

which is still true in the following form since $\phi_{AB} = \phi_{BA}$

$$\frac{K_A}{K_B} = \frac{\phi_{BB}}{\phi_{AA}} \cdot \frac{\phi_{BA}}{\phi_{AB}} = \frac{\phi_{BA}}{\phi_{AA}} \cdot \frac{\phi_{BB}}{\phi_{AB}} = \frac{\phi_{BA}}{\phi_{AA}} \cdot \frac{1}{\phi_{AB}/\phi_{BB}}$$

and recognizing the separate angle ratios as carry-over factors,

$$\frac{K_A}{K_B} = \frac{C_B}{C_A}$$

and finally

$$K_A C_A = K_B C_B \tag{9-5}$$

The latter relationship is of great value for providing a check on separately computed values of the stiffness and carry-over factors.

Thus it may be seen that the factors involved in moment distribution may be obtained by properly relating the angular rotations produced by unit moments as denoted in Fig. 6c. Although many excellent charts and tables* have catalogued calculated values for these factors, the following example will illustrate their determination by numerical integration.

Example

Figure 7 is a frame supporting one concentrated load P. For purposes of this example the thickness of concrete perpendicular to the view shown is assumed to be 1 ft. Before a moment distribution may be performed the constants of the member AB must be determined. Following previous theory, AB is shown in Fig. 8, with a unit moment at A. The M, I, and M/I diagrams follow. I is calculated in ft⁴ units. Note that the length of beam is taken to the center line of the column and that depth at A is measured as shown. By using the M/I diagram with ten divisions, each 2 ft in length, the following table is a summation of moments of areas about B to determine Δ_{BA}, the deflection of B from a tangent to elastic

---

* Suggested references: R. A. Coughey and R. S. Cebula, *Constants for Design of Continuous Girders with Abrupt Changes in Moments of Inertia*, Bulletin 176, Iowa Engineering Experiment Station, Ames, Iowa; see also ref. 2 of this chapter.

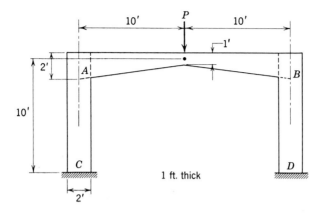

FIG. 9–7

curve at A, and thence ϕ_{AA}. The process is repeated by a summation of moments about A to determine ϕ_{BA}. To prepare for future calculations the total area of the M/I diagram is also found. Note that the interval of 2 ft is introduced only in final calculation steps and that the ordinates used in formulating area are equal to $M/12I$ instead of M/I.

Areas Designated	Area	Moments about B		Moments about A	
		Arm	Moment	Arm	Moment
1	0.063	9.67	0.610	0.33	0.021
2–3	0.155	9	1.395	1	0.155
4–5	0.191	8	1.528	2	0.382
6–7	0.255	7	1.785	3	0.765
8–9	0.347	6	2.082	4	1.388
10–11	0.500	5	2.500	5	2.500
12–13	0.231	4	0.924	6	1.386
14–15	0.110	3	0.330	7	0.770
16–17	0.048	2	0.096	8	0.384
18–19	0.017	1	0.017	9	0.153
	1.917		11.267		7.904

Adjusting summations for the 2 ft interval and by a factor of 12 since this was omitted in calculations of I, we find

$$\text{Area of } \frac{M}{I} \text{ diagram} = 1.917 \times 2 \times 12 = 46.01$$

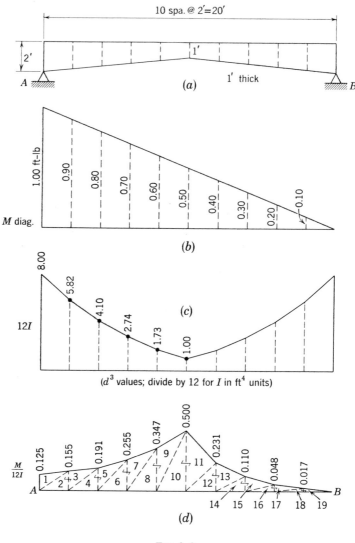

Fig. 9–8

$$E \cdot \phi_{AA} = \frac{11.267 \times 2 \times 2}{20} \times 12 = 27.04$$

$$E \cdot \phi_{BA} = \frac{7.904 \times 2 \times 2}{20} \times 12 = 18.97$$

The carry-over factor from B to A, or from A to B due to symmetry, by equation 9-2, is

$$C_A = C_B = \frac{\phi_{BA}}{\phi_{AA}} = \frac{18.97}{27.04} = 0.70$$

Owing to symmetry, a unit couple at B would produce a rotation ϕ_{BB} equal to ϕ_{AA}, and $\phi_{AB} = \phi_{BA}$, hence, by equation 9-4, the *absolute stiffness* at B is

$$K_B = \frac{\phi_{AA}}{\phi_{AA} \cdot \phi_{BB} - \phi^2_{AB}} = \frac{27.04/E}{27.04/E \times 27.04/E - (18.97/E)^2}$$

$$K_B = 0.0718E$$

also $K_A = 0.0718E$

If a numerical value of *absolute stiffness* is required E would be substituted in pounds per square foot, since all lengths and the moment of inertia have been used in units of feet. This is seldom necessary due to the fact that comparative stiffnesses are sufficient to determine distribution factors. For the frame of this example the absolute stiffness K_A for the vertical column AD would be

$$K_A = \frac{4EI}{L} = \frac{4 \times E \times (1 \times 2^3)/12}{10} = 0.265E$$

Distribution factors at A

to $AB = \dfrac{0.0718}{0.0718 + 0.265} = 0.213$

to $AD = \dfrac{0.265}{0.3368} = 0.787$

If the student has a copy of *Handbook of Frame Constants* the above constants for AB may be checked. From p. 18 of that publication the carry-over factor equals 0.692 in comparison with 0.70 calculated. For determining *absolute stiffness*, k in the following general equation for stiffness is found in the tables to be 17.34.

$$K_A = \frac{kEI \text{ min}}{L}$$

$$= \frac{17.34 \cdot E \cdot \dfrac{1 \cdot 1^3}{12}}{20}$$

$$K_A = 0.0722E$$

to compare with $0.0718E$ calculated.

All quantities for moment distribution are now available except for the fixed-ended moments. Figure 9 is drawn and the M/I curve determined for simply supported ends.   One-half of the area of the latter diagram divided by E will equal the end slope since the member is symmetrical and symmetrically loaded.   Note that ordinates of Fig. 9d equal $M/12I$ instead of M/I.

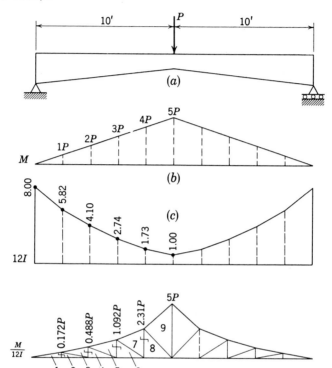

FIG. 9–9

The one-half area is found by the following summation:

Area Designated	Area
1–2	0.172P
3–4	0.488P
5–6	1.092P
7–8	2.310P
9	2.500P
	6.562P

One-half area $= 6.562P \times 2 \text{ ft} \times 12 = 157.49P$

Then the fixed-ended moment may be found by making the end slopes equal to zero.   Previously calculated values of ϕ_{AA} and ϕ_{BA} are used.

$$M_{AB}^F\left(\frac{27.04}{E}\right) + M_{AB}^F\left(\frac{18.97}{E}\right) = \frac{157.49P}{E}$$

or
$$M_{AB}^F = 3.42P$$

This moment may also be determined from published tables, as $3.33P$.

Figure 10 demonstrates the moment distribution for the frame of Fig. 7 with $P = 10$ kips.   No further explanation of the method is necessary except to note that the carry-over factor for the beam is 0.7 in contrast to 0.5 for the columns.

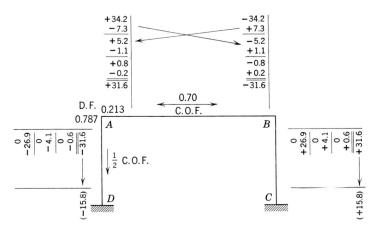

FIG. 9–10

9–6. TRANSLATION—VARIABLE MOMENT OF INERTIA

A single member AB of uniform moment of inertia with restrained ends develops equal end moments of $M = (6EI\Delta/L^2)$, when B is translated a distance Δ with respect to A.   Figure 11a illustrates the physical situation where R is an angle.   Figure 11b is a member of nonuniform moment of inertia under similar circumstances where M_A and M_B will be unequal. The fixed-ended moments may be thought of as turning the tangents to the elastic curve through the angle R at each end.   At A, from the previous geometric notation of Fig. 6c the angle R is expressed as

$$R = M_A \cdot \phi_{AA} - M_B \cdot \phi_{AB}$$

and at B
$$R = M_B \cdot \phi_{BB} - M_A \cdot \phi_{BA}$$

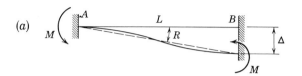

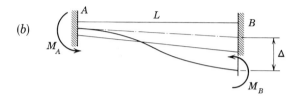

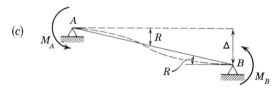

FIG. 9–11

Equating the two expressions for R

$$M_A \cdot \phi_{AA} - M_B \cdot \phi_{AB} = M_B \cdot \phi_{BB} - M_A \cdot \phi_{BA}$$

and simplifying to a ratio, we obtain

$$\frac{M_A}{M_B} = \frac{\phi_{BB} + \phi_{AB}}{\phi_{AA} + \phi_{BA}} \tag{9-6}$$

or modifying the right-hand side by dividing both the numerator and denominator by ϕ_{BB}, so that

$$\frac{M_A}{M_B} = \frac{(\phi_{BB} + \phi_{AB})/\phi_{BB}}{(\phi_{AA} + \phi_{BA})/\phi_{BB}} = \frac{1 + C_A}{\phi_{AA}/\phi_{BB} + C_A}$$

and recognizing that $\phi_{AA}/\phi_{BB} = K_B/K_A$ and that $C_B K_B = C_A K_A$, we find by substitution and simplification that the ratio becomes

$$\frac{M_A}{M_B} = \frac{1 + C_A}{K_B/K_A + C_A} = \frac{K_A(1 + C_A)}{K_B + C_A K_A} = \frac{K_A(1 + C_A)}{K_B(1 + C_B)} \tag{9-7}$$

and an alternate form of the ratio may also be expressed as

$$\frac{M_A}{M_B} = \frac{C_B(1 + C_A)}{C_A(1 + C_B)} \tag{9-8}$$

Thus the fixed-ended moments at A and B, during any arbitrary translation of $\varDelta$, bear a fixed ratio to one another, with the ratio determinable by either equation 9-7 or 9-8.

If only the ratio of M_A to M_B is required rather than absolute values of moment, then the above expression in terms of carry-over factors will suffice. However, if absolute values of M_A and M_B are required for a given translation $\varDelta$, then it may be shown that

$$M_A = \frac{K_A \cdot \varDelta}{L} (1 + C_A)$$

and

$$M_B = \frac{K_B \cdot \varDelta}{L} (1 + C_B)$$

since moments are directly proportional to the angle R which equals $\varDelta/L$. It will also be recalled that the values of *absolute stiffness* at A and B are

$$K_A = \frac{k_A EI \min}{L}; \qquad K_B = \frac{k_B EI \min}{L}$$

where k_A and k_B are obtained from tables or evaluated by methods previously outlined.

An example of the treatment of sidesway for a nonuniform moment situation follows for the frame shown in Fig. 12a. The beam is the same as that used in Figs. 7, 8, and 9, hence constants previously calculated for AB may be used. However, the columns are tapered and in lieu of working out the constants they have been taken from the *Handbook of Frame Constants* and shown in a separate sketch. The ratio of end moments for column AD for an arbitrary displacement of $\varDelta$ will be

$$\frac{M_A}{M_D} = \frac{C_D(1 + C_A)}{C_A(1 + C_D)} = \frac{0.834(1 + 0.294)}{0.294(1 + 0.834)} = 2$$

In Fig. 12b, trial moments in this ratio have been chosen as a basis for sidesway analysis.

Distribution factors around joints A and B must be determined. By previous calculations, the *absolute stiffness* at A of member AB was found to be $K_A = 0.0718E$. For the column AD, using k_A found from the handbook, the *absolute stiffness* is

$$K_A = \frac{19.46 \times E \times (1 \cdot 1^3)/12}{10} = 0.162E$$

Distribution factors based on these *absolute stiffnesses* are shown directly on Fig. 12b. The distribution routinely follows previous procedures with use of proper carry-over factors. Only three cycles of distribution are shown. A free body of column AD gives

$$H_D = \frac{85.3 + 68.2}{10} = 15.35 \text{ kips}$$

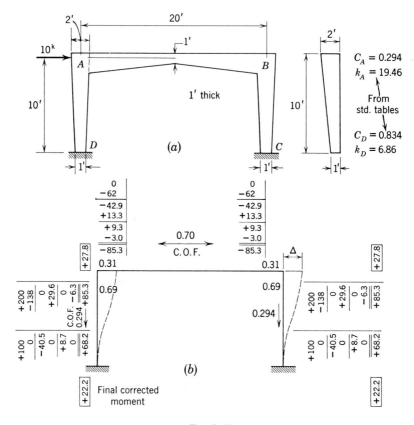

FIG. 9–12

With H_C also equal to 15.45 kips, the correction factor is

$$\frac{10 \text{ kips}}{2 \times 15.35 \text{ kips}} = 0.326$$

Corrected moments as would be induced by a 10-kip lateral load at A are shown in the rectangles adjacent to the preliminary moments based on the arbitrary displacement.

9-7. SLOPE DEFLECTION EQUATIONS—VARIABLE MOMENT OF INERTIA

Although the slope deflection method has had its chief field of application in instances of uniform moment of inertia, the standard equations may be modified so as to apply to cases of variable moment of inertia.

The standard equations for members of uniform section from Chapter 5 are

$$M_{AB} = \frac{2EI}{L}(-2\theta_A - \theta_B + 3R) \pm M_{AB}^F$$

and $$M_{BA} = \frac{2EI}{L}(-2\theta_B - \theta_A + 3R) \pm M_{BA}^F$$

(9-9)

which are restated in the following form:

$$M_{AB} = -\left(\frac{4EI}{L}\theta_A + \frac{2EI}{L}\theta_B - \frac{6EI\Delta}{L^2}\right) \pm M_{AB}^F$$

$$M_{BA} = -\left(\frac{4EI}{L}\theta_B + \frac{2EI}{L}\theta_A - \frac{6EI\Delta}{L^2}\right) \pm M_{BA}^F$$

(9-10)

Then these equations may be reviewed for necessary modifications so as to be applicable to the variable moment of inertia members. For the uniform moment of inertia, $K_A = 4EI/L$ since k was equal to 4. Hence, in general, substitute $k_A EI \min/L$ for this term in order to adjust the equation for variable moment of inertia. $2EI/L$ may be likened to $K_B \cdot C_B$, since $K_B = 4EI/L$ for the member of uniform moment of inertia and C_B is equal to one-half under similar circumstances. Thus, for $2EI/L$ substitute $C_B \cdot (k_B EI \min/L)$. The term $6EI\Delta/L^2$, a uniform moment of inertia, has its counterpart in $(k_A EI \min/L)(\Delta/L)(1 + C_A)$ for variable moment of inertia. Hence, M_{AB} by substitution modifies to

$$M_{AB} = -\left[\frac{k_A EI \min}{L}\theta_A + C_B k_B \frac{EI \min}{L}\theta_B\right.$$

$$\left. - \frac{k_A EI \min}{L} \cdot \frac{\Delta}{L}(1 + C_A)\right] \pm M_{AB}^F$$

$$M_{AB} = \frac{EI \min}{L}[-k_A\theta_A - C_B \cdot k_B\theta_B + k_A \cdot R(1 + C_A)] \pm M_{AB}^F$$

(9-11)

or the preferred form is

$$M_{AB} = [-K_A\theta_A - C_B K_B\theta_B + K_A R(1 + C_A)] \pm M_{AB}^F \qquad (9\text{-}12)$$

and it may also be shown that

$$M_{BA} = [-K_B\theta_B - C_A K_A\theta_A + K_B R(1 + C_B)] \pm M_{BA}^F \qquad (9\text{-}13)$$

An inspection of the last two equations indicates that their derivation could have been arrived at directly by the superposition of moments as was the case in the original derivation for uniform moment of inertia. For example, the moment at A required to turn the tangent through an angle of θ_A is $K_A \cdot \theta_A$, when it is recalled that K_A is the *absolute stiffness* or the moment required to rotate the tangent at A by one radian while B is restrained. The term $C_B K_B \cdot \theta_B$ is the moment produced at A, or carried over to A, with A restrained against further rotation, by the moment $K_B \cdot \theta_B$ at B. The moment $K_B \cdot \theta_B$ is required to turn the tangent at B through an angle of θ_B while A is restrained. The expression for effect of relative displacement is the term $K_A \cdot R(1 + C_A)$, a fixed-end moment induced in the member by translation.

The slope deflection equations may be formulated by reference to tables of constants for variable moment of inertia sections. Their application will not be demonstrated, owing to the complete similarity of procedure as explained in Chapter 5.

9–8. ALTERNATE APPROACH FOR STEPPED MEMBERS

Many structural members are stepped or composed of uniform moment of inertia segments, such as those shown in Fig. 13. Prior discussion has viewed these members as one continuous member with discontinuities in moment of inertia; however, it is often expeditious to analyze these members by the moment distribution method as follows:

In Fig. 13b, point C is displaced through an arbitrary deflection Δ while restraint is applied to prevent rotation of C and to maintain Δ. Fixed-ended moments are developed in the two segments. In span L_1

$$M_1 = \frac{6EI_1 \Delta}{L_1{}^2}$$

and in span L_2

$$M_2 = \frac{6EI_2 \Delta}{L_2{}^2}$$

therefore

$$\frac{M_1}{M_2} = \frac{L_2{}^2}{L_1{}^2} \times \frac{I_1}{I_2}$$

Letting $I_1 = 2I_2$, the ratio of the fixed-ended moments is

$$\frac{M_1}{M_2} = \frac{10^2}{20^2} \times \frac{2I_2}{I_2} = \frac{1}{2}$$

Trial moments are assumed in this ratio and moment distribution performed. The distribution factors are one-half to each segment at C.

The moments obtained are trial moments and free bodies of each span enable trial shears to be calculated. The shears in span L_1 and L_2 are 6.875 kips and 16.25 kips. The vertical downward force to be furnished by the temporary support at C is 23.125 kips. The correction factor to reduce moments, to that which would have been produced by a 10-kip load at C, is $10/23.125 = 0.432$. Corrected final moments are recorded in the rectangles.

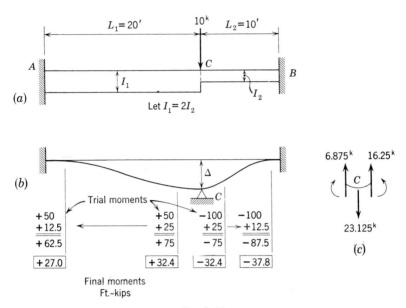

FIG. 9–13

Member AB of Fig. 14a is a stepped member of a rigid frame or a single span of a continuous beam. The *absolute stiffness* and carry-over factors are needed. Figure 14b illustrates the essential quantities for end A. These quantities will be determined by treating AB like a two-span beam, recognizing that no forces may exist at C and that Δ_C is an undetermined quantity. The solution must be approached by superposition. First, temporarily hold point C from translation, but fixed against rotation, and apply a convenient arbitrary moment at A. Moment distribution is performed as shown in Fig. 14c, and the force on temporary support at C may be computed to be 15 kips up for the moments shown. Since moments are wanted without this temporary support, a corrective set of moments must be determined, which is produced by a force of 15 kips in the down direction. This set of corrective moments must be determined with A free to rotate. Figure 14d is a solution to obtain these corrective

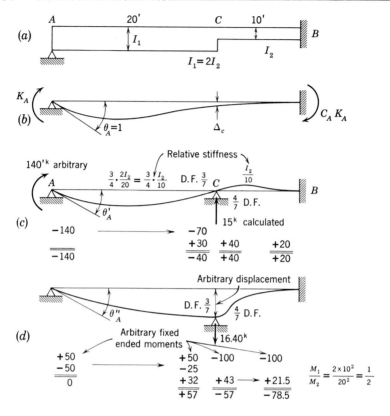

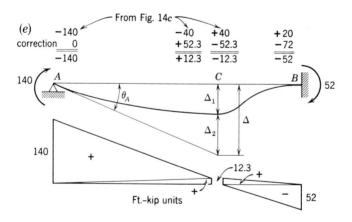

FIG. 9–14

moments, where C was first translated down an arbitrary amount with all joints fixed against rotation.   The moment distribution should be clear. The force on beam at C may be determined to be 16.40 kips down in Fig. 14d.   Final correction moments may be obtained by using a factor of $15/16.40 = 0.915$ and final moments are summarized in Fig. 14e for the beam free of support at C.   This figure also portrays the elastic curve with C free from support.   The carry-over factor from A to B or C_A may now be determined.

$$C_A = \frac{52}{140} = 0.37$$

Before K_A the stiffness at A can be found, the value of θ_A must be determined.   Figure 14e suggests the solution, wherein Δ is found by finding Δ_1 and Δ_2 for the separate segments by moment-area procedures. Moment diagrams using the beam sign convention are shown.   The deflection and θ_A calculations are summarized as

$$2EI_2 \cdot \Delta_1 = 3050$$
$$2EI_2 \cdot \Delta_2 = 19520$$
$$\overline{2EI_2 \cdot \Delta = 22570}$$

$$\theta_A = \frac{22570}{2EI_2 \times 20} = \frac{562}{EI_2}$$

Since the *absolute stiffness* K_A is the moment required to produce a value of θ_A equal to one radian when B is fixed, we may write, based on the direct proportionality of rotation to moment, that

$$\frac{K_A}{140} = \frac{1}{562/EI_2}: \qquad K_A = 0.249EI_2$$

Similar repetitive procedures wherein A is kept fixed will enable C_B and K_B to be found.

Although the foregoing is a valid and physical means of obtaining these constants it would have been more expeditious in the present instance to have used a basic moment-area approach as illustrated previously.

REFERENCES

1 Cross, H., and N. D. Morgan, *Continuous Frames of Reinforced Concrete*, Chapter 3, pp. 65–70, and Chapter 5, New York: John Wiley, 1932.
2. *Handbook of Frame Constants*, Portland Cement Association, Chicago, Ill.
3. Hanson, W. E., and W. F. Wiley, "Constant-Segment Method for Analysis of Non-uniform Members," *Trans. Am. Soc. Civil Engineers*, **121**, 1317–1336 (1956).

PROBLEMS

9–1. Derive an expression for the slope and vertical deflection at B.   Check your final result by letting I_1 equal I_2.

$$Ans. \ \theta_B = \frac{PL^2}{8EI_1I_2} (3I_2 + I_1); \ \Delta_B = \frac{PL^3}{24EI_1I_2} (7I_2 + I_1).$$

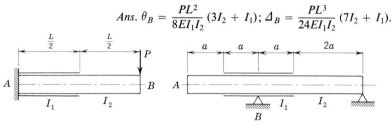

Prob. 9–1   Prob. 9–2

9–2. The cover-plated beam shown is subjected to a uniform load of w lb per ft.   Determine the slope and deflection at A.

9–3. For the cover-plated beam shown, determine the slope at A and the maximum deflection.

$$Ans. \ \text{Deflection} = \frac{Pa^3}{12EI_1I_2} (55I_2 + 4I_1).$$

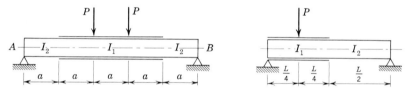

Prob. 9–3   Prob. 9–4

9–4. For the cover-plated beam shown, locate the point of maximum deflection if $I_1 = 2I_2$.

9–5. A solid cantilever beam has constant depth, but the width varies as shown. Derive an expression for the vertical deflection at A by a mathematical integration approach. $Ans. \ 6PL^3/bD^3E$

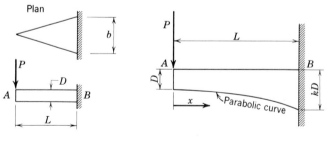

Prob. 9–5   Prob. 9–6

9–6. A cantilever beam has constant width, but the depth varies with x. Derive an expression for the vertical deflection at A by a mathematical integration approach.

9–7. Assume that the top and bottom flange plates each have an area of A sq. in., and that the moment of inertia of the web plate can be neglected. Let I_0 represent the moment of inertia at center line. Derive an expression for the vertical deflection at center line by a mathematical integration approach. Check your final expression by taking the limit as $k \to 1$.

$$Ans. \frac{PL^3}{16EI_0}\left[\frac{1 - k^2 + 2k \log_e k}{(1 - k)^3}\right].$$

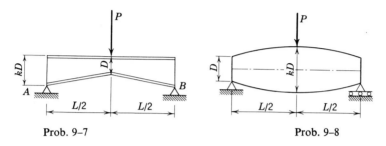

Prob. 9–7   Prob. 9–8

9–8. The beam has a constant width and the depth varies. Assume that the top and bottom surfaces are defined by second degree parabolas. Derive an expression for deflection at the center line.

9–9. Solve problem 9–5 by numerical integration. Use six equal intervals.

9–10. Solve problem 9–6 by numerical integration. Use four equal intervals.

9–11. Solve problem 9–7 by numerical integration. Use four equal intervals in each half span.

9–12. Determine the slope at A and the deflection at the center line. The width of the beam is 12 in. and $D = 24$ in., and $L = 32$ ft. Use E as 3×10^6 psi and evaluate the slope at A and the deflection in terms of P. Use the numerical integration method.

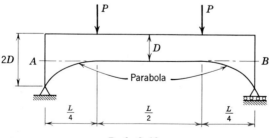

Prob. 9–12

9–13. Assuming that A and B are fixed, calculate the fixed-end moments for the beam of problem 9–12. Use the numerical integration method.

9–14. Assuming that A and B are fully restrained, calculate the fixed-end moments for the beam of problem 9–7 if $k = 2$. Use the numerical integration or calculus methods. *Ans. 0.153PL.*

9–15. Assuming that A and B are fully restrained, calculate the fixed-end moments for the beam of problem 9–3. Use the numerical integration method. Use $I_1 = 2I_2$. *Ans. Pa.*

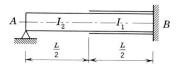

9–16. Calculate the carry-over factors C_A and C_B and also the stiffness factors K_A and K_B if $I_1 = 2I_2$. Check your results, using $C_A K_A = C_B K_B$.

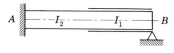

Ans. $C_A = \frac{2}{3}$; $C_B = 0.4$; $K_A = 48EI_2/11L$; $K_B = 80EI_2/11L$.

Prob. 9–16

9–17. For the unloaded beam of problem 9–7, calculate the carry-over and stiffness factors if $k = 2$.

9–18. For the prismatic tapered member of constant width, calculate C_A, C_B, K_A, and K_B.

Ans. $C_A = 0.834$; $C_B = 0.294$; $K_A = \dfrac{6.86EI \text{ min}}{L}$; $K_B = \dfrac{19.46EI \text{ min}}{L}$.

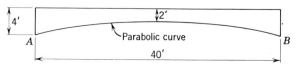

Prob. 9–18

9–19. Using relevant calculations made for problems 9–12 and 9–13, calculate the carry-over factor and stiffness factor for the beam of problem 9–12.

9–20. Assuming that the symmetrical beam is of constant width, calculate the carry-over and stiffness factor. *Ans. $C = 0.694$; $K = \dfrac{12.03EI \text{ min}}{L}$.*

Prob. 9–20

9–21. Assuming that this structural member has constant width, calculate C_A, C_B, K_A, and K_B. Check your results using $C_A K_A = C_B K_B$.

$$Ans.\ C_A = 0.462;\ C_B = 0.742;\ K_A = \frac{7.43EI\ \text{min}}{L}.$$

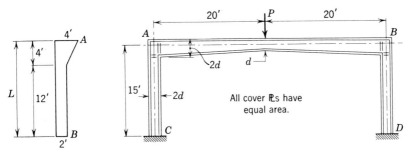

Prob. 9–21   Prob. 9–22

9–22. Determine the moments for this frame by moment distribution, using all relevant calculations from problems 9–14 and 9–17, if these problems were assigned. If these problems were not assigned, then calculate the fixed-end moments, carry-over factor, and stiffness for member AB.

9–23. The same as problem 9–22, except consider C and D as hinged instead of fixed.

9–24. Determine the moments for this frame by moment distribution. Use all relevant calculations from problems 9–18 and 9–20 if these problems were assigned. If they were not assigned, then calculate the essential factors, or, at the option of the instructor, select these factors from standard tables. Fixed-ended moments at A and B equal $0.1025wL^2$.

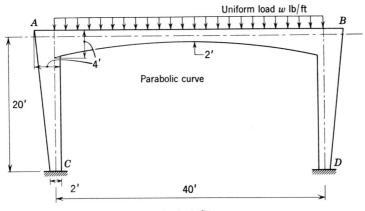

Prob. 9–24

9–25. Using calculated factors obtained in problem 9–18 for the tapered member, determine the moments in this frame by moment distribution. The frame has constant width.

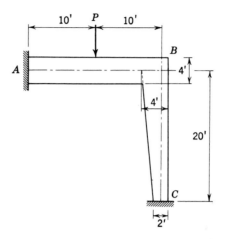

Prob. 9–25

9–26. In this problem the side span members are the same as the member used in problem 9–21, and the center span member is the same as the member of problem 9–20. Use the data from these two problems as well as fixed-end moments computed in problem 9–24, or, at the option of the instructor, select needed factors from standard tables. Compute the final moments at the supports by moment distribution.

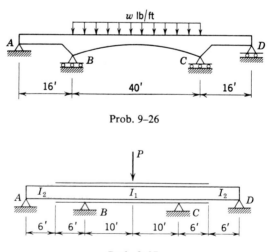

Prob. 9–26

Prob. 9–27

9–27. A cover-plated beam as shown, with $I_1 = 2I_2$, is to be solved by moment distribution for moments at B and C. Use factors calculated in problem 9–16 for the side span member.   *Ans.* $M_B = M_C = 1.72P$.

9–28. Remove the vertical load from the top member of problem 9–22 and reintroduce the load P as a horizontal force acting to the right at A. Calculate the bending moments by moment distribution.

9–29. The frame is free to sway. Calculate the bending moments by moment distribution. For dimensions of structure, see problem 9–24.

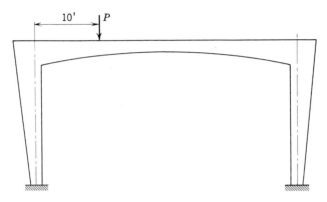

Prob. 9–29

9–30. Using data of problem 9–21, for AB, compute the moments in this frame. Constant width. Compute other factors needed or select from tables.

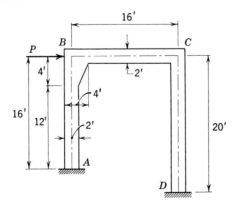

Prob. 9–30

9–31. Compute the fixed-end moments at *A* and *B* by displacing *C* an arbitrary distance downward and making the necessary corrections. See your solution for problem 9–18 for basic factors or, at the option of the instructor, take the factors from a standard table.   *Ans. 6.68P.*

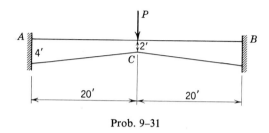

Prob. 9–31

9–32. Compute the fixed-end moments at *A* and *B* by displacing *C* and *D* by an equal arbitrary distance downward and making the necessary corrections. How much will *C* and *D* actually deflect in terms of P, E, I_2, and L?

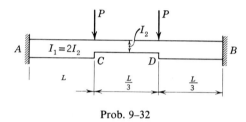

Prob. 9–32

9–33. Introduce a temporary support at *C* and calculate the necessary force to prevent *C* from deflecting. Then correct this solution by letting *C* move freely. Calculate final moments at *A* and *B*. $I_1 = 1.5I_2$.

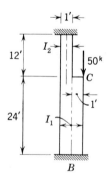

Prob. 9–33

Arch Theory

10

10–1. INTRODUCTION

The arch provides resistance to external loading by internal axial compression as well as by shear and moment. It is further characterized by the requirement that it must be supported so as to induce both vertical and horizontal reaction components. Strictly speaking, if horizontal reaction components cannot be developed, the structure is simply a curved beam.

The arch may also be a trussed structure composed of two-force members. However, as discussed in this chapter, the arch will be considered as a rib of steel or concrete. Furthermore, the following elementary treatment will ignore the effect of roadway and roadway columns.

10–2. GENERAL

An arch span may be three-hinged, two-hinged, or fixed. Figure 1 illustrates these conditions. The three-hinged arch is statically determinate and causes no new problems in analysis. The two-hinged arch is indeterminate to the first degree and the fixed arch is indeterminate to the third degree.

The ideal arch would be one in which the geometric centerline of the rib exactly fits the thrust line. Figure 2a shows three equal vertical forces, P_1, P_2, and P_3, symmetrically spaced. A funicular polygon, Fig. 2b, is constructed by the graphical construction shown. This should be familiar to the student as the process wherein the components of the various forces are substituted for the forces. An arbitrary pole p was selected so as to maintain symmetry, with the rays drawn in the force diagram providing an

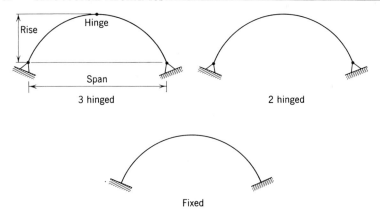

FIG. 10–1

arbitrary set of components of P_1, P_2, and P_3.   In general, these components may be substituted for P_1, P_2, and P_3 in the space diagram, but to bring out the nature of the arch thrust, reverse the direction of these components, as shown in Fig. 2a, and designate them as N_1, N_2, N_3, and N_4.   This construction develops the thrust line, as these are the directions for which these particular arbitrary valued and directed components provide separate equilibrants equal and opposite to P_1, P_2, and P_3, and, thus, provide static equilibrium.   Note that N_1 and N_2 hold P_1 in equilibrium.   If a

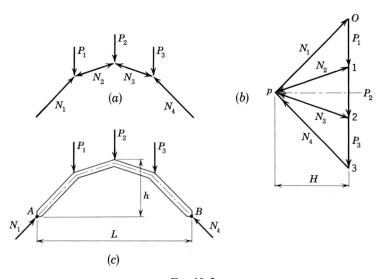

FIG. 10–2

polygonal structure where the axes of the members coincide with N_1, N_2, N_3, and N_4 could be built, as in Fig. 2c, a polygonal arch would result, and N_1, N_2, N_3, and N_4 would represent the true arch thrusts furnished by the axial load resistance of the polygonal members. Points A and B represent end supports and h indicates the rise, where the rise is dependent in this case on the arbitrary location of the pole. It must be realized, of course, that if any one of the external forces were changed in magnitude, or removed, the polygonal form of arch just drawn would no longer fit the new theoretical thrust line. In this event, the thrust line would fall outside of the center line of the polygonal arch, and bending moments would be created in the arch rib.

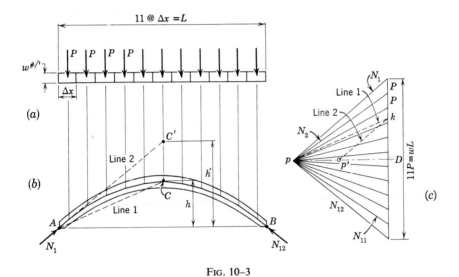

FIG. 10–3

If the number of external forces is increased, as in Fig. 3a, the polygon representing the thrust line closely simulates a curve, owing to the shortness of individual chords. If the loading is uniform per unit of horizontal length, a parabolic curve would be an exact fit of the thrust line. Figure 3b illustrates the construction of the funicular polygon assuming 11 equal spaces and loads. To construct a symmetrical funicular polygon the pole p in the force diagram of Fig. 3c must be located on a horizontal line through D, where D is at the midlength of the vector representing the eleven vertical loads. The position of p on the line through D is arbitrarily taken. The rays are drawn and a funicular polygon constructed passing through points A, C, and B of Fig. 3b. The arch structure whose axis closing fits the polygon is shown in Fig. 3b. Since the pole p was arbitrarily taken, the rise of the arch h is also arbitrary. For a given situation h may be given, then the pole may be found by trial-and-error procedure to provide a funicular polygon satisfying this condition. To avoid trial-and-error procedure to fit a new rise h', draw line 1, in Fig. 3b, and make line 1 in the force diagram parallel to it, locating a point k. Also draw line 2 in Fig. 3b, and then line 2 in force diagram parallel to it from point k. If new rays were drawn (not shown) using pole p', a new funicular polygon could be drawn passing through AC'B. The theoretical justification of this procedure may be found in many standard texts. It may be stated, however, that the theory is based on the principles pertaining to equilibrium polygons.

Although the above procedure has merit in establishing a shape for the arch axis, the shape would be the ideal one only in the event that the loads remained fixed in value. In early arch construction the dead weight of the arch and the roadway, often supported on an earth fill over the arch, comprised the major part of the load. Since live loads were only a small part of the total load, their variation produced only small deviations of the thrust line from the ideal arch axis. Owing to these minor effects, early arches were successfully made of wedge-shaped stone voussoirs set dry face to face, and the thrust line could be held within the middle one-third of the cross-section, with compressive stress over the entire cross-section. Modern arches, on the other hand, are generally loaded through columns supporting the roadway and live loading is a greater proportion of the total load. For these reasons it is rare for the arch axis to fit the theoretical thrust line and the thrust line will often fall well outside the cross-section, requiring an arch cross-section that will resist bending as well as thrust.

10–3. ELASTIC TWO-HINGED PARABOLIC ARCH

In the discussion to follow, the radius of the arch axis is considered to be large relative to the depth of the arch rib, in order to assume that basic formulas for the bending of a straight beam apply. This assumption is essentially true in all practical arches.

In Fig. 4a, the arch axis conforms to a second degree parabolic curve, and the moment of inertia is assumed to vary from a minimum at the crown, I_c, to a maximum at the reactions. To permit a simple mathematical solution, it is assumed that I at any section equals $I_c \sec \theta$, where θ is the inclination of the tangent to the arch axis with respect to the horizontal. A general loading is assumed.

The two-hinged arch is redundant to the first degree, and H, the horizontal components of the reactions, is taken as the redundant.   Figure 4b indicates the statically determinate *base structure* on rollers at B. H must be sufficient to return B' to the original position B.

Letting M_s equal the bending moment in the rib for the determinate base structure and calculating Δ_x by the unit load approach where $m = -y$ from Fig. 4c,

$$\Delta_s = \Delta_x = \int_A^B \frac{M_s \cdot (-y)ds}{EI}$$

The equation of the parabola with origin at A is

$$y = 4h\left(\frac{x}{L} - \frac{x^2}{L^2}\right)$$

and ds is expressible in terms of dx as

$$ds = dx \sec \theta$$

Hence, the Δ_x equation may be transformed to

$$\Delta_x = \int_A^B -\frac{M_s \cdot y \cdot dx \sec \theta}{EI_c \sec \theta} = -\frac{1}{EI_c}\int_A^B M_s y\, dx$$

Letting Δ'_x be the horizontal deflection of B due to H alone where "δ" $= \frac{1}{EI_c}\int_A^B y^2\, dx$ for H equal to 1 lb, we obtain this deflection as

$$\Delta'_x = \frac{H}{EI_c}\int_A^B y^2\, dx$$

Geometrically, the deflection condition at B is expressed as

$$\Delta_x + \Delta'_x = 0 \tag{10-1}$$

owing to the fixed span distance between A and B.   It is to be noted that Δ_x is negative since it represents a movement opposite to the direction of the unit load in Fig. 4c.   By similar reasoning Δ'_x is positive.

Equation 10-1 may be restated by substitution and eliminating EI_c as

$$-\int_A^B M_s y\, dx + H\int_A^B y^2\, dx = 0$$

Then
$$H = \frac{\int_A^B M_s y\, dx}{\int_A^B y^2\, dx} \tag{10-2}$$

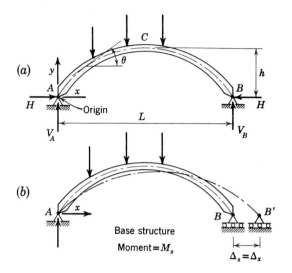

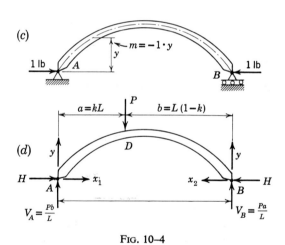

FIG. 10-4

To evaluate H for a single load P, Fig. 4d, first write M_s, taking advantage of two origins for x.

From A to D, for $x_1 \lesseqgtr a$; $M_s = \dfrac{Pb}{L}(x)$

From B to D, for $x_2 \lesseqgtr b$; $M_s = \dfrac{Pa}{L}(x)$

The numerator of equation 10-2 is determined as follows.

$$\int_A^B M_s y \, dx = \int_A^D M_s y \, dx + \int_B^D M_s y \, dx$$

$$= \frac{4Ph}{L} \left[\int_0^a bx \left(\frac{x}{L} - \frac{x^2}{L^2} \right) dx + \int_0^b ax \left(\frac{x}{L} - \frac{x^2}{L^2} \right) dx \right]$$

$$\int_A^B M_s y \, dx = \frac{Ph}{3} ab \left[\left(\frac{a}{L} \right)^3 + \left(\frac{b}{L} \right)^3 + \frac{4ab}{L^2} \right]$$

The denominator of equation 10-2 is evaluated as

$$\int_A^B y^2 \, dx = 16h^2 \int_0^L \left(\frac{x}{L} - \frac{x^2}{L^2} \right)^2 dx = \frac{16h^2 L}{30}$$

Then for a single load P equation 10-2 reduces to

$$H = \frac{5Pab}{8hL} \left[\left(\frac{a}{L} \right)^3 + \left(\frac{b}{L} \right)^3 + \frac{4ab}{L^2} \right] \tag{10-3}$$

which is a useful general equation as, if more than one concentrated load is applied to the arch, the total H may be found by superposition. If $a = kL$ and $b = (1 - k)L$, the general equation for H may be written in simplified form as

$$H = \frac{5PLk(1 - k)(1 + k - k^2)}{8h} \tag{10-4}$$

If the load is uniform at w per unit of horizontal length over the entire span, M_s of equation 10-2 equals the moment as found for a simple span of length L. We may then evaluate the numerator of equation 10-2 as

$$2 \int_0^{L/2} M_s y \, dx = 8wh \int_0^{L/2} \left(\frac{Lx}{2} - \frac{x^2}{2} \right) \left(\frac{x}{L} - \frac{x^2}{L^2} \right) dx$$

After integrating, substituting, and simplifying, the numerator of equation 10-2 is

$$\int_0^L M_s y \, dx = \frac{whL^3}{15}$$

Then equation 10-2 will become

$$H = \frac{whL^3/15}{16h^2L/30} = \frac{1}{8} \frac{wL^2}{h} \tag{10-5}$$

for a uniformly loaded arch of parabolic outline.

The result, $H = \frac{1}{8}wL^2/h$, may also be inferred directly from statics since it was previously demonstrated that the axis of a parabolic arch was an exact fit of the thrust line under uniform load conditions and that no bending moment existed in rib at any section. A free body of the arch to the left of the arch center line will be in moment equilibrium under action of $V_A = wL/2$, the uniform load over the half-span, and H. Taking moments at the crown of the arch,

$$H \cdot h + \frac{wL}{2}\left(\frac{L}{4}\right) - \frac{wL}{2}\left(\frac{L}{2}\right) = 0$$

from which

$$H = \frac{wL^2}{8h}$$

10–4. EFFECT OF AXIAL CHANGE IN LENGTH

Precise arch analysis must take into account the elastic change in length of the arch rib due to axial compression. In Fig. 5a, a segment of an arch

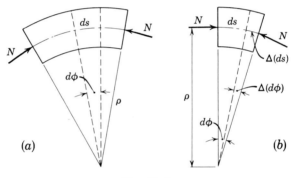

Fig. 10–5

rib having a cross-sectional area of A is shown. A differential length along the axis, defined as ds, is isolated from the rib in Fig. 5b, with only the normal forces shown. (Note: the effect of bending alone was considered in Art. 10-3.) The shortening of this element due to the thrust N is

$$\Delta(ds) = \frac{N\,ds}{AE}$$

Then the differential change in the central angle $\Delta(d\phi)$ will be the angular rotation of the cross-section caused by this shortening, or,

$$\Delta(d\phi) = \frac{\Delta(ds)}{\rho} = \frac{N\,ds}{AE\rho}$$

From this fundamental view of rib shortening and rib rotation, the effect on arch deflection and rotation may be accounted for in a number of ways.   For a geometrical analysis refer to Fig. 6a, where only the effect of rotation at one cross-section on the rotation and deflection of B is considered.   The tangent to arch axis at B rotates by $\Delta(d\phi)$, causing B to translate to B'.   The components of this translation are

$$d(\Delta x) = -y'\Delta(d\phi) \quad \text{and} \quad d(\Delta y) = x'\,\Delta(d\phi)$$

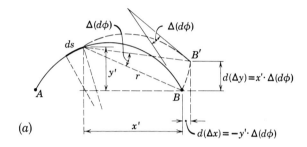

(a)

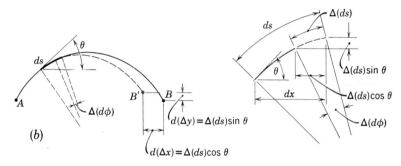

(b)

FIG. 10–6

The reader should note that the sign of $d(\Delta x)$ must be termed negative since this movement is opposite in direction to the unit load at B in Fig. 4.

In Fig. 6b, the translation of B caused by the shortening of one differential element is shown to be

$$d(\Delta x) = +\Delta(ds)\cos\theta \quad \text{and} \quad d(\Delta y) = (ds)\sin\theta$$

The movements of Figs. 6a and 6b, must be combined to account for total movements at B or

$$d(\Delta x) = -y'\,\Delta(d\phi) + \Delta(ds)\cos\theta$$
$$d(\Delta y) = x'\,\Delta(d\phi) + \Delta(ds)\sin\theta$$

$$(10\text{-}6)$$

where θ is the angle the tangent to the axis has with respect to the horizontal direction, and x' and y' are positively measured coordinates of the ds section with respect to B.

Now, since all ds elements along axis deform, then total movement of B

$$\Delta_x = -\sum y' \, \Delta(d\phi) + \sum \Delta(ds) \cos \theta$$
$$\Delta_y = \sum x' \, \Delta(d\phi) + \sum \Delta(ds) \sin \theta \tag{10-7}$$

but since $\sin \theta = \dfrac{dy}{ds}$ and $\cos \theta = \dfrac{dx}{ds}$,

and
$$\Delta(ds) = \frac{N \, ds}{AE}$$

$$\Delta(d\phi) = \frac{N \, ds}{AE\rho}$$

then equations 10-7 simplify to

$$\Delta_x = -\int_A^B y' \frac{N \, ds}{AE\rho} + \int_A^B \frac{N \, dx}{AE}$$

and
$$\Delta_y = \int_A^B x' \frac{N \, ds}{AE\rho} + \int_A^B \frac{N \, dy}{AE} \tag{10-8}$$

The rotation of the tangent at B is accounted for from Fig. 6a by the angle

$$\Delta(d\phi) = \frac{N \, ds}{AE\rho}$$

and the total rotation accumulated from A to B is

$$\Delta(\phi) = \int_A^B \frac{N \, ds}{AE\rho} \tag{10-9}$$

The evaluation of the above equations entails a number of mathematical complications if a full integration method is adopted. In general, a trial analysis would be made on the basis of flexural distortions alone and then an arithmetic integration applied to the above equations to determine secondary effects. In a parabolic arch the error may be in the order of 2 per cent if axial distortion is neglected.

10–5. EFFECT OF RISE AND FALL OF TEMPERATURE

Expansion and contraction of the material forming the arch acts to change the length along the axis. This is similar to rib lengthening and rib shortening and creates additional horizontal reaction effects for the two-hinged arch.

Figure 7a illustrates a rise in temperature with B on rollers moving to B'. Figure 7b is an enlarged view of one differential ds length along the axis, undergoing expansion owing to a rise in temperature of ΔT degrees. If C equals the coefficient of thermal expansion, then the ds length expands $C(\Delta T)ds$. The horizontal component of this expansion is $C \Delta T\, ds\,(\cos \theta)$. Over the entire axis length the summation of these effects gives

$$\Delta_x = \int_A^B C \Delta T\, ds \cos \theta = C \Delta T \int_A^B \cos \theta\, ds$$

However, a geometric interpretation of the integral shows that it equals L, the span length.   Hence

$$\Delta_x = C \Delta T L$$

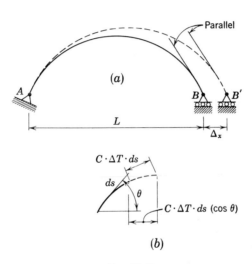

(a)

(b)

FIG. 10-7

This is true no matter what the configuration of the axis is between A and B.   It may also be noted that the tangent at B moves parallel to itself during the displacement to B', since dotted outline is expanded in length only with all length elements still parallel to their original position.

The horizontal force due to this expansion for a parabolic arch by equation 10-2 is

$$H = \frac{C \Delta T L}{\displaystyle\int_0^L \frac{y^2\, dx}{EI_c}} \qquad (10\text{-}10)$$

10–6. INFLUENCE LINES—TWO-HINGED PARABOLIC ARCH

The influence line concept is as important for arch analysis as for beams and frames and may involve influence lines of the reactions, as well as for shear, thrust, and moment, at a given arch cross-section. Figure 8a is a parabolic arch as described before, and Fig. 8b is a free body exposing a cross-section of the arch at E.

With a 1-lb load to the right of the section E, the moment at the section by statics is

$$M_E = V_A x_1 - H y_1$$

and

$$N_E = V_A \sin \theta + H \cos \theta$$

$$V_E = V_A \cos \theta - H \sin \theta$$

The separate influence lines for V_A and H are shown in Figs. 8c and 8d. V_A depends directly on the distance of the 1-lb load from B, while H may be plotted from equation 10-4 previously derived for H. This curve is symmetrical, as can be seen by examining the equation. It is also helpful to have the influence line for V_B available as shown in Fig. 8e.

First construct the influence line for moment at the section E at distance x_1 from A. Multiply the ordinates of the V_A curve by x_1, and the ordinates of the H curve by y_1 as in Fig. 8f. Since the equation $V_A \cdot x_1 - H \cdot y_1$ for M_E is valid only for loads to right of the section, it is necessary to consider a right-hand free body involving V_B when the load is to the left of the section. When this free body is used, $M_E = V_B \cdot x_2 - H \cdot y_1$. $V_B \cdot x_2$ is plotted graphically in Fig. 8f. The graphical subtraction of the effect of H is shown with the resultant areas marked with plus or minus to signify positive or negative moment. Thus, a partial span live loading is required for maximum moment of either sign.

The influence line for thrust or N_E is shown in Fig. 8g. Note the break in curve when the 1-lb load is to the left of the section. Using the right-hand free body, the effect of V_B on N_E is subtracted from effect of H. As may now be determined from influence line, the maximum thrust is always obtained from a full span loading.

The influence line for shear at section E is given in Fig. 8h, and the construction should be checked by the student. It should be noted that a split live loading arrangement is needed for maximum shear at an intermediate section.

While we are on the subject of influence lines, it should be realized that the Maxwell reciprocal deflection relations and Müller-Breslau principles also hold for an arch and that qualitative influence lines can be sketched

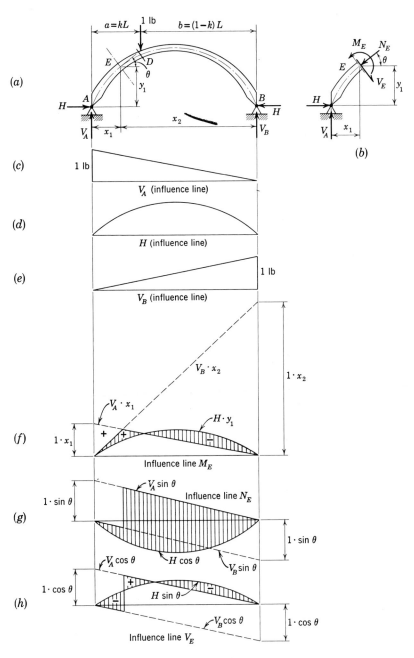

(a)

(b)

(c)

(d)

(e)

(f)

(g)

(h)

Fig. 10–8

for the arch following methods used in beams and frames.   Figure 9*a* demonstrates this procedure for moment at the crown.   The student should be aware of the large distortion in the sketched elastic curve.   The influence line ordinates for moment at the crown depend on the true vertical movement of the arch axis.   Note point *D* moving to point *D'*, where Δ_y represents the vertical deflection and the influence line ordinate.

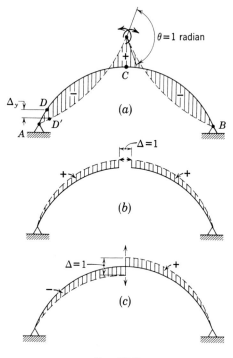

FIG. 10–9

In model analysis, careful measurements of these movements would be required.   The graphical construction is helpful in qualitatively informing the analyst that separate loading conditions for maximum negative and maximum positive moment are required.   Exact load divide points could only be found from a mathematical analysis as in Fig. 8*f* or by model.

The qualitative influence line for thrust at crown can be developed by applying a unit distortion in horizontal direction at the crown as in Fig. 9*b*. Interpretation is not difficult in this case and the conclusion is immediately reached that maximum thrust at the crown is secured by fully loading the span.   The qualitative influence line for shear at the crown also has a definite interpretation.   See Fig. 9*c*.

10-7. FIXED-ENDED ARCHES

Although most steel arches have been constructed as two or three-hinged arches, owing to the difficulty of insuring full fixity at the supports, concrete arches, involving continuity of material and reinforcing, have generally been fixed-ended or restrained. This has the effect of making a single arch span indeterminate to the third degree. The basic theory for solving this problem has already been introduced in Chapter 6 dealing with the elastic center method.

Figure 10a represents a symmetrical fixed-ended parabolic arch with cross-section varying from crown to reactions according to $I = I_c \sec \theta$. The arch is redundant to the third degree, and although M_A, V_A, and H_A could be taken as the redundants, it is more expeditious to employ the elastic center method using redundants M_O, V_O, and H_O.

Assuming that the axis of the arch follows the curve of a second degree parabola, and that the moment of inertia at any section is related to that at the crown by $I = I_c \sec \theta$, then with $ds = dx \sec \theta$, the differential elastic weight is

$$\frac{ds}{EI} = \frac{dx}{EI_c}$$

The total elastic weight of entire arch is then equal to

$$2 \int_0^{L/2} \frac{dx}{EI_c} = \frac{L}{EI_c}$$

With symmetry, as shown in Fig. 10a, the y axis through the elastic center is at the center line of the span. The x axis through the elastic center is determined by finding y_O to the centroid. Moments of the elastic weight about an axis AB are

$$\int_0^L y' \cdot \frac{dx}{EI_c} = \frac{4h}{EI_c} \int_0^L \left[\frac{x'}{L} - \frac{(x')^2}{L^2} \right] dx$$

$$= \frac{2hL}{3EI_c}$$

and
$$y_O = \frac{2hL}{3EI_c} \div \frac{L}{EI_c} = \frac{2}{3} h$$

The moment of inertia of the elastic weights about the y axis is

$$\mathbf{I}_y = \int_{-L/2}^{+L/2} \frac{x^2 \, dx}{EI_c} = \frac{L^3}{12EI_c}$$

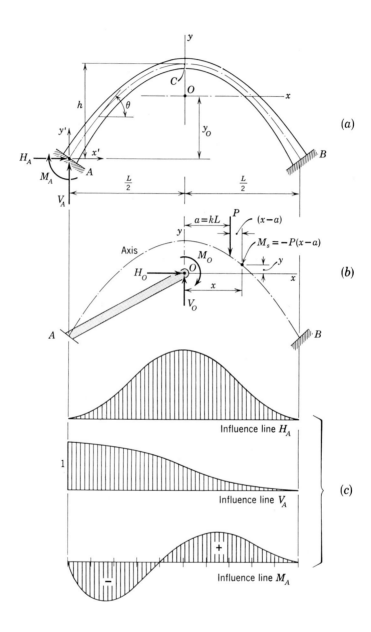

FIG. 10–10

The student should observe, on account of the assumed variation of I, that the total elastic weight and the moment of inertia of the elastic weight are equivalent to the comparable values that would be found for a beam of length L and of uniform cross-section.

The moment of inertia of the elastic weights about the x axis is

$$I_x = \int \frac{y^2\,dx}{EI_c} = \frac{h^2}{9L^4EI_c} \int_{-L/2}^{+L/2} [L^2 - 12x^2]^2 dx$$

$$I_x = \frac{4h^2L}{45EI_c}$$

Due to symmetry $I_{xy} = \int \frac{xy\,ds}{EI} = 0$

Attention may now be given to the load system. In Fig. 10b a single vertical load P is placed on arch and the redundant values of M_O, V_O, and H_O are to be determined. To apply the elastic center theory, certain evaluations must be made. One evaluation required is (see Chapter 6)

$$\int \frac{M_s\,ds}{EI} \quad \text{or} \quad \int \frac{M_s\,dx}{EI_c}$$

From Fig. 10b, M_s has a value different than zero for values of $x = a$ to $x = L/2$, and is expressed as

$$M_s = -P(x - a)$$

$$\int_a^{L/2} \frac{M_s\,dx}{EI_c} = \frac{1}{EI_c} \int_a^{L/2} -P(x - a)\,dx$$

$$= -\frac{P}{8EI_c}(L - 2a)^2$$

$$= -\frac{PL^2}{8EI_c}(1 - 2k)^2$$

An evaluation is also needed of $\int \dfrac{M_s x\,dx}{EI_c}$ (see Chapter 6), or

$$\int \frac{M_s x\,dx}{EI_c} = \frac{1}{EI_c} \int_a^{L/2} M_s x\,dx$$

$$= \frac{1}{EI_c} \int_a^{L/2} -P(x - a)x\,dx$$

$$= -\frac{P}{24EI_c}(L - 2a)^2(L + a)$$

$$= -\frac{PL^3}{24EI_c}(1 - 2k)^2(1 + k)$$

The last evaluation required is that of

$$\int \frac{M_s y\, dx}{EI_c} = -\frac{1}{EI_c} \int_a^{L/2} P(x-a)\left[\frac{h}{3L^2}(L^2 - 12x^2)\right]dx$$

$$= +\frac{PhL^2}{48EI_c}[1 - 8k^2 + 16k^4]$$

The redundants at the elastic center follow from the standard equations 6-4, 6-5, and 6-6 of Chapter 6, and since EI_c appears in both numerator and denominator of each expression it cancels out.

By equation 6-4

$$M_O = -\frac{\left[-\dfrac{PL^2}{8}(1 - 2k)^2\right]}{L} = +\frac{PL}{8}(1 - 2k)^2 \qquad (10\text{-}11)$$

By equation 6-5

$$H_O = +\frac{PhL^2}{48}\frac{[1 - 8k^2 + 16k^4]}{\frac{4}{45}h^2 L} = \frac{15}{64}\frac{PL}{h}(1 - 8k^2 + 16k^4)$$

$$= \frac{15}{64}\frac{PL}{h}(1 - 4k^2)^2 = \frac{15}{64}\frac{PL}{h}(1 - 2k)^2(1 + 2k)^2 \qquad (10\text{-}12)$$

By equation 6-6

$$V_O = -\frac{\left[-\dfrac{PL^3}{24}(1 - 2k)^2(1 + k)\right]}{\dfrac{L^3}{12}} = +\frac{P}{2}(1 - 2k)^2(1 + k) \quad (10\text{-}13)$$

In substituting in the equations 10-11, 10-12, and 10-13, k, varying with position of the load, may range from $-\frac{1}{2}$ to $+\frac{1}{2}$. Influence line values for M_O, H_O, and V_O may be determined by making P equal to unity. Or if influence line values of M_A, H_A, and V_A are desired, the equilibrium conditions of a free body of the rigid arm require that $H_A = H_O$; $V_A = V_O$, and

$$M_A = M_O + \frac{2}{3}hH_O - \frac{L}{2}V_O$$

$$M_A = \frac{PL}{8}(1 - 2k)^2 + \frac{2}{3}h\frac{15}{64}\frac{PL}{h}(1 - 4k^2)^2$$

$$- \frac{L}{2}\left[\frac{P}{2}(1 - 2k)^2(1 + k)\right]$$

$$M_A = \frac{PL}{32}[1 - 2k]^2[1 + 2k][1 + 10k] \qquad (10\text{-}14)$$

The calculations for influence line values may be made in tabular form listing values of k, $(1 + k)$, $(1 - 2k)$, $(1 + 2k)$, $(1 - 2k)^2$, and $(1 + 10k)$, and performing required substitutions in the above equations. It should be noted that H_A is the only value dependent upon the rise of the arch. The influence lines for the redundants are plotted in Fig. 10c. With the influence lines of the redundants established, all other influence lines for the arch may be computed from the principles of static equilibrium.

10–8. UNSYMMETRICAL FIXED-ENDED ARCH

Figure 11a depicts an unsymmetrical arch rib with an assumed variation of structural cross-section. The student should again note that the analysis of an indeterminate structure is based on the prior establishment of structural sections, and many times these initial properties are based on prior designs or past experience. Further discussion of these judgment factors is outside the scope of this text. To use the elastic center method, a rigid arm is installed from A to O, where O is the elastic center, and the new redundants M_O, V_O, and H_O are employed.

Assume that the axis of arch rib follows the curve of a second degree parabola, and that the moment of inertia at any section may be related to that at the crown by $I = I_c \sec \theta$.

To locate the elastic center or centroid of the elastic weights we take moments about the y' axis through A, then

$$x_O \int_A^B \frac{ds}{EI} \quad \text{or} \quad x_O \int_A^B \frac{dx}{EI_c} = \int_A^B x' \cdot \frac{dx}{EI_c}$$

Two integrals must be evaluated to compute x_O

$$\int_0^L \frac{dx}{EI_c} = \frac{L}{EI_c}$$

and

$$\int_0^L \frac{x' \, dx}{EI_c} = \frac{L^2}{2EI_c}$$

giving

$$x_O = \frac{L^2}{2EI_c} \div \frac{L}{EI_c} = \frac{L}{2}$$

The equation of arch axis with origin of the x' and y' coordinates at A is

$$y' = 4h_1 \left[\frac{x'}{\frac{4}{3}L} - \frac{(x')^2}{(\frac{4}{3}L)^2} \right]$$

$$y' = 3h_1 \left[\frac{x'}{L} - \frac{3(x')^2}{4L^2} \right]$$

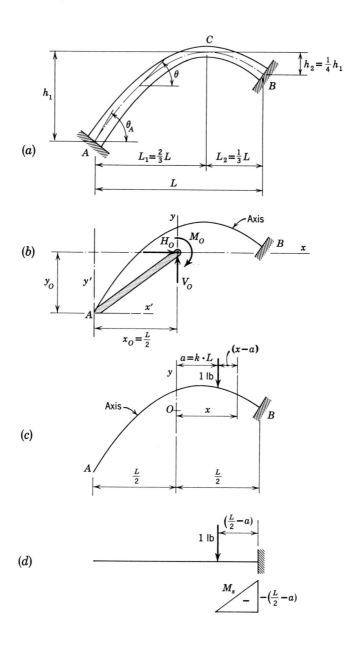

(a)

(b)

(c)

(d)

FIG. 10–11

Then, by moments about the x' axis we write the vertical distance above A to the centroid as

$$y_O = \frac{\int_A^B y' \cdot \frac{dx}{EI_c}}{\int_A^B \frac{dx}{EI_c}}$$

The evaluation of the integral in the numerator follows as

$$\int_0^L y' \frac{dx}{EI_c} = \int_0^L 3h_1 \left[\frac{x'}{L} - \frac{3}{4}\frac{(x')^2}{L^2}\right] \frac{dx}{EI_c}$$

$$= \frac{3h_1}{EI_c}\left(\frac{1}{4}L\right)$$

then

$$y_O = \frac{3}{4}\frac{h_1}{EI_c}L \div \frac{L}{EI_c} = \frac{3}{4}h_1$$

With x_O and y_O determined, the x and y axes through the elastic center or centroid of the elastic weights become the working axes. Now determine the properties of the elastic weights about these axes. In this case it will be more expeditious from the standpoint of the mathematics involved to determine $\mathbf{I}_x$, $\mathbf{I}_y$, and $\mathbf{I}_{xy}$ by working with the x' and y' axes and transferring to the x, y axes by the standard moment of inertia, or product of inertia, transfer formulas. Then

$$\mathbf{I}_y = \int (x')^2 \frac{ds}{EI} - x_O^2 \int \frac{ds}{EI}$$

where

$$\int_0^L (x')^2 \frac{ds}{EI} = \int_0^L (x')^2 \frac{dx}{EI_c} = \frac{L^3}{3EI_c}$$

Therefore

$$\mathbf{I}_y = \frac{L^3}{3EI_c} - \left(\frac{L}{2}\right)^2\left(\frac{L}{EI_c}\right) = \frac{L^3}{12EI_c}$$

Similarly, by the transfer formula

$$\mathbf{I}_x = \int (y')^2 \frac{ds}{EI} - y_O^2 \int \frac{ds}{EI}$$

where

$$\int (y')^2 \frac{ds}{EI} = \int_0^L \left[3h_1\left(\frac{x'}{L} - \frac{3}{4}\frac{(x')^2}{L^2}\right)\right] \frac{dx}{EI_c}$$

$$= \frac{51 \, h'^2 L}{80 \, EI_c}$$

Therefore

$$\mathbf{I}_x = \frac{51\,h'^2L}{80\,EI_c} - \left(\frac{3}{4}h_1\right)^2\left(\frac{L}{EI_c}\right) = \frac{3h_1{}^2L}{40EI_c}$$

The product of inertia of the elastic weights for the x and y axes may be stated as

$$\mathbf{I}_{xy} = \int \frac{(x')(y')\,ds}{EI} - x_o y_o \int \frac{ds}{EI}$$

in terms of the standard transfer formula where $\mathbf{I}_{xy}$ is the product of inertia for a centroidal pair of axes.

First, we obtain

$$\int \frac{(x')(y')\,ds}{EI} = \int_0^L (x')\left[3h_1\left(\frac{x'}{L} - \frac{3}{4}\frac{(x')^2}{L^2}\right)\right]\frac{dx}{EI_c}$$

$$= \frac{7h_1L^2}{16EI_c}$$

Then, by the transfer formula

$$\mathbf{I}_{xy} = \frac{7h_1L^2}{16EI_c} - \left(\frac{L}{2}\right)\left(\frac{3}{4}h_1\right)\frac{L}{EI_c}$$

$$= \frac{h_1L^2}{16EI_c}$$

The basic equations of the elastic center method may now be rewritten as (see previous discussion on elastic center method theory, Chapter 6).

$$\int \frac{M_s\,ds}{EI} + M_O \int \frac{ds}{EI} = 0$$

from which
$$M_O = -\frac{\displaystyle\int \frac{M_s\,ds}{EI}}{\displaystyle\int \frac{ds}{EI}} = -\frac{\displaystyle\int \frac{M_s\,ds}{EI}}{A} \qquad (10\text{-}15)$$

and from equation 6-2 with elastic center redundants

$$-\int \frac{M_s y\,ds}{EI} - M_O \int \frac{y\,ds}{EI} - V_O \int \frac{xy\,ds}{EI} + H_O \int \frac{y^2\,ds}{EI} = 0 \quad (10\text{-}16)$$

and since $\displaystyle\int \frac{y\,ds}{EI}$ equals zero, due to x and y being the centroidal axes, equation 10-16 reduces to

$$-\int \frac{M_s y\,ds}{EI} - V_O \int \frac{xy\,ds}{EI} + H_O \int \frac{y^2\,ds}{EI} = 0 \qquad (10\text{-}17)$$

It should be noted that for a symmetrically framed structure $\int \frac{xy\,ds}{EI}$ would be zero and H_O could be found directly. However, since this is a general case, equation 10-17 must be combined with the third equation, equation 6-3, of the elastic center method (written in terms of elastic center redundants) which, after it is recalled that $\int \frac{x\,ds}{EI} = 0$, becomes

$$+ \int \frac{M_s x\,ds}{EI} + V_O \int \frac{x^2\,ds}{EI} - H_O \int \frac{xy\,ds}{EI} = 0 \qquad (10\text{-}18)$$

A simultaneous solution of equations 10-17 and 10-18 yields the following basic equation

$$H_O = \frac{\int \frac{x^2\,ds}{EI} \int \frac{M_s y\,ds}{EI} - \int \frac{xy\,ds}{EI} \int \frac{M_s x\,ds}{EI}}{\int y^2 \frac{ds}{EI} \int \frac{x^2\,ds}{EI} - \left[\int \frac{xy\,ds}{EI}\right]^2} \qquad (10\text{-}19)$$

which can be rewritten in terms of the shorter notation, where $\mathbf{I}_{xy}$ equals the product of inertia, as

$$H_O = \frac{\mathbf{I}_y \int \frac{M_s y\,ds}{EI} - \mathbf{I}_{xy} \int \frac{M_s x\,ds}{EI}}{\mathbf{I}_x \mathbf{I}_y - \mathbf{I}_{xy}^2} \qquad (10\text{-}20)$$

and the basic equation for V_O, from a solution of equations 10-17 and 10-18, is

$$V_O = \frac{\mathbf{I}_{xy} \int \frac{M_s y\,ds}{EI} - \mathbf{I}_x \int \frac{M_s x\,ds}{EI}}{\mathbf{I}_x \mathbf{I}_y - \mathbf{I}_{xy}^2} \qquad (10\text{-}21)$$

It should be particularly noted that if $\mathbf{I}_{xy} = 0$, the equations for a symmetrically framed structure result. See equations 6-5 and 6-6. We now have equations 10-15, 10-20, and 10-21 for an unsymmetrically framed structure.

The integrals of equations 10-15, 10-20, and 10-21 remain to be evaluated for a given load system. Referring to Fig. 11c, the determinant base structure, and using a simplified general unit load system, the M_s at a section of the arch x distance from the origin will be

$$M_s = -(x - a)$$

for all values of x to the right of the load. Then

$$\int \frac{M_s\,ds}{EI} = -\frac{1}{EI_c} \int_a^{L/2} (x - a)dx = -\frac{1}{2EI_c}\left(\frac{L}{2} - a\right)^2$$

$$= -\frac{L^2}{8EI_c}(1 - 2k)^2 \qquad (10\text{-}22)$$

This result can be easily verified. When $\dfrac{ds}{EI}$ became $\dfrac{dx}{EI_c}$ the base structure could have been thought of as a beam of uniform cross-section cantilevered from B. Then, with the unit load on beam as in Fig. 11d, the M_s diagram is a triangle and

$$\int \frac{M_s \, ds}{EI} = \text{the area of the triangle divided by } EI_c$$

The integral $\displaystyle \int \frac{M_s x \, ds}{EI} = \int \frac{M_s x \, dx}{EI_c}$ and can also be evaluated by taking

the moment of the area just described about the origin in lieu of integration. Thus,

$$\int \frac{M_s x \, dx}{EI_c} = -\frac{1}{2EI_c} \left(\frac{L}{2} - a \right)^2 \left[\frac{L}{2} - \frac{(L/2 - a)}{3} \right]$$

$$= -\frac{1}{6EI_c} \left(\frac{L}{2} - a \right)^2 (L + a)$$

$$= -\frac{L^3}{24EI_c} (1 - 2k)^2 (1 + k) \tag{10-23}$$

To evaluate $\displaystyle \int \frac{M_s y \, ds}{EI} = \int \frac{M_s y \, dx}{EI_c}$, y must be expressed as a coordinate with the elastic center as the origin. For the parabolic arch shown in Figs. 11a and 11c, analytical geometry leads to

$$y = \frac{3h_1}{16L^2} (L^2 + 4xL - 12x^2)$$

Then $\displaystyle \int_a^{L/2} \frac{M_s y \, dx}{EI_c} = -\frac{3h_1}{16EI_c L^2} \int_a^{L/2} [x - a][L^2 + 4xL - 12x^2] dx$

which, with $a = kL$, leads to

$$\int_a^{L/2} \frac{M_s y \, dx}{EI_c} = \frac{-3h_1 L^2}{16EI_c} \left[\frac{5}{48} - \frac{k}{2} + \frac{k^2}{2} + \frac{2}{3}k^3 - k^4 \right] \tag{10-24}$$

All necessary equations are now developed for the determination of the redundants M_O, H_O, and V_O. A general solution could be obtained; however, a numerical evaluation is probably more expeditious.

Example of Unsymmetrical Rectangular Frame

The general equations for M_O, H_O, and V_O just derived are also applicable to an unsymmetrical rectangular frame. Figure 12 shows all necessary work. The elastic center is found as usual, but the x and y axes are not the principal axes. Elementary calculations serve to determine the product of inertia $\mathbf{I}_{xy}$.

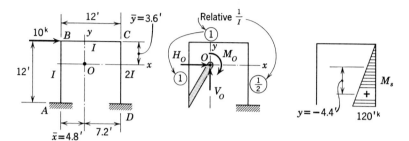

$$\int \frac{ds}{I} = \frac{12}{I} + \frac{12}{I} + \frac{12}{2I} = \frac{30}{I}$$

$$\bar{x} = \frac{12 \times 6 + \frac{1}{2} \times 12 \times 12}{30} = 4.8'$$

$$\bar{y} = \frac{12 \times 6 + \frac{1}{2} \times 12 \times 6}{30} = 3.6'$$

$\mathbf{I}_x$, $\mathbf{I}_y$, $\mathbf{I}_{xy}$ in terms of relative $\frac{1}{I}$

$\underline{\mathbf{I}_x}$

$12 \times 3.6^2 = 156$

$1 \times \frac{12^3}{12} = 144$

$12 \times 2.4^2 = 69$

$\frac{1}{2} \times \frac{12^3}{12} = 72$

$6 \times 2.4^2 = \underline{\quad 35}$

$\overline{476}$

$\underline{\mathbf{I}_y}$

$12 \times 4.8^2 = 276$

$6 \times 7.2^2 = 312$

$1 \times \frac{12^3}{12} = 144$

$12 \times 1.2^2 = \underline{\quad 17}$

$\overline{749}$

$\underline{\mathbf{I}_{xy}}$

$12\,(-4.8)(-2.4) = +\,138$

$6\,(+7.2)(-2.4) = -\,104$

$12\,(+1.2)(+3.6) = +\;\;\underline{52}$

$\overline{+\,86}$

$$\int M_s\,ds = \frac{60 \times 12}{2} \quad \frac{M_s}{\text{Relative } I} = \frac{120}{2} = 60$$

$$= 360$$

$$M_O = -\frac{360}{30} = -12'^k \,\big)$$

$$\int M_s\,y\,ds = 360\,(-4.4) = -1584$$

$$\int M_s\,x\,ds = 360\,(+7.2) = +2592$$

$$H_O = \frac{(749)(-1584) - (86)(2592)}{(476)(749) - (86)^2}$$

$$= -4.05^k \;\longleftarrow$$

$$V_O = \frac{(86)(-1584) - 476\,(2592)}{(476)(749) - (86)^2}$$

$$= -3.92^k \;\downarrow$$

FIG. 10–12

10–9. CIRCULAR FIXED-ENDED ARCH

Figure 13a is a symmetrical arch of constant radius and uniform cross-section. A derivation is to be made for M_A, H_A, and V_A for a single general vertical load P. The elastic center method is employed, and owing to symmetry, the elastic center is easily located vertically from the center of curvature by the standard equation for the centroid of a line. This

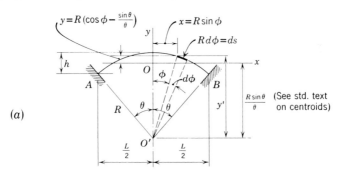

(a)

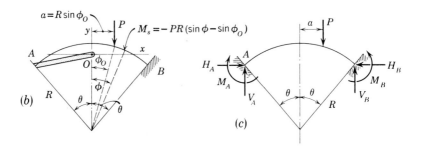

(b)

(c)

FIG. 10–13

distance is $R \sin \theta / \theta$. Since EI is uniform in value it is treated as unity in the following development. The fundamental properties are

$$\int ds = \int_{-\theta}^{+\theta} R\, d\phi = 2R\theta \tag{10-25}$$

$$\int y^2\, ds = \int_{-\theta}^{+\theta} \left[R\left(\cos\phi - \frac{\sin\theta}{\theta} \right) \right]^2 R\, d\phi$$

$$= R^3 \left[\theta + \frac{\sin 2\theta}{2} - \frac{2\sin^2\theta}{\theta} \right] \tag{10-26}$$

$$\int x^2\, ds = \int_{-\theta}^{+\theta} (R\sin\phi)^2 R\, d\phi = \frac{R^3}{2}(2\theta - \sin 2\theta) \tag{10-27}$$

The following integrals are also needed.   See Figs. 13a and 13b for geometry.   Angle ϕ_O locates the point of application of the load P.

$$\int M_s \, ds = -PR^2 \int_{\phi_O}^{\theta} (\sin \phi - \sin \phi_O)d\phi$$

$$= +PR^2[(\cos \theta - \cos \phi_O) + \sin \phi_O(\theta - \phi_O)] \quad (10\text{-}28)$$

$$\int M_s y \, ds = -PR^3 \int_{\phi_O}^{\theta} (\sin \phi - \sin \phi_O)\left(\cos \phi - \frac{\sin \theta}{\theta}\right)d\phi$$

$$= -PR^3\left[\tfrac{1}{2}(\sin^2 \theta - \sin^2 \phi_O) - \sin \phi_O(\sin \theta - \phi_O)\right.$$

$$\left. +\frac{\sin \theta}{\theta}(\cos \theta - \cos \phi_O) + \frac{\sin \phi_O}{\theta}\sin \theta \,(\theta - \phi_O)\right] \quad (10\text{-}29)$$

$$\int M_s x \, ds = -PR^3 \int_{\phi_O}^{\theta} (\sin \phi - \sin \phi_O)(\sin \phi)d\phi$$

$$= -PR^3[\tfrac{1}{2}(\theta - \phi_O) - \tfrac{1}{4}(\sin 2\theta - \sin 2\phi_O)$$

$$+\sin \phi_O(\cos \theta - \cos \phi_O)] \quad (10\text{-}30)$$

Then by the standard equations 6-4, 6-5, and 6-6 of the elastic center method (Chapter 6), we obtain

$$M_O = -\frac{\int M_s \, ds}{\int ds} = -\frac{PR}{2\theta}[(\cos \theta - \cos \phi_O) + \sin \phi_O(\theta - \phi_O)] \quad (10\text{-}31)$$

$$H_O = \frac{\int M_s y \, ds}{\int y^2 \, ds} = \cfrac{-2P[\tfrac{1}{2}(\sin^2 \theta - \sin^2 \phi_O) - \sin \phi_O(\sin \theta - \sin \phi_O) \\ +\dfrac{\sin \theta}{\theta}(\cos \theta - \cos \phi_O) + \sin \phi_O \dfrac{\sin \theta}{\theta}(\theta - \phi_O)]}{2\theta + \sin 2\theta - \dfrac{4\sin^2 \theta}{\theta}}$$

$$(10\text{-}32)$$

$$V_O = -\frac{\int M_s x \, ds}{\int x^2 \, ds}$$

$$= \frac{P[2(\theta - \phi_O) - (\sin 2\theta - \sin 2\phi_O) + 4 \sin \phi_O(\cos \theta - \cos \phi_O)]}{2(2\theta - \sin 2\theta)}$$

$$(10\text{-}33)$$

The reactions and moment at A from equilibrium conditions, wherein all sign conventions are observed, are

$$H_A = H_O; \qquad V_A = V_O$$

and $$M_A = M_O - \frac{V_O \cdot L}{2} + H_O R\left(\frac{\sin \theta}{\theta} - \cos \theta\right) \quad (10\text{-}34)$$

The reactions and moment at B by similar considerations are

$$H_B = H_O; \qquad V_B = P - V_O = P - V_A$$

and

$$M_B = M_O - P\left(\frac{L}{2} - R \sin \phi_O\right) + \frac{V_O L}{2} + H_O R\left(\frac{\sin \theta}{\theta} - \cos \theta\right) \quad (10\text{-}35)$$

Numerical evaluation of the final equations is cumbersome but straightforward. To illustrate, assume a full semicircular arch and place P at the center line. Then $\theta = \pi/2$ and $\phi_O = 0$. By direct substitution in equations 10-31, 10-32, and 10-33,

$$M_O = +\frac{PR}{\pi}; \qquad H_O = \frac{P(4 - \pi)}{(\pi^2 - 8)}; \qquad V_O = \frac{P}{2}$$

$$H_A = H_O = 0.459P$$

From equation 10-34

$$M_A = \frac{PR}{\pi} - \frac{PR}{2} + 0.459 \times \frac{2R}{\pi} = +0.11PR$$

Influence lines for all redundant reactions may be computed from the general equations with P equal to unity and with ϕ_O varying.

ϕ_O is generally varied so that projection of load points onto the span subdivide the span into equal intervals. It is common to use about 10 such intervals.

Influence lines for the redundants may also be calculated from Müller-Breslau principles. To illustrate this approach, a general fixed-ended circular arch of uniform section is assumed as in Fig. 14a. If A is pinned against translation and the tangent at A rotated through an angle of $\theta_A = 1$ radian, Fig. 14b, then by Müller-Breslau principles, the vertical deflection $\varDelta$ at a given point on the arch axis will represent the influence line ordinate for M_A for a vertical loading. Although the simplicity of this principle is intriguing, quantitative results, except as measured from a model, are often complex mathematically. The full semicircular arch will be used to illustrate a solution by the elastic center method.

In Fig. 14c, representing a semicircular arch, free A for rotation and draw the rigid arm from A to O. Rotate the tangent to arch at A, and thus the rigid arm, through a small angle of θ_A. Noting that geometric conditions require zero vertical and horizontal translation at A during this rotation, we see that the end of arm at O must rotate to O', producing the vertical and horizontal translations shown at the elastic center.

With the distance from A to O equal to r, then the arc $OO' = r\theta_A$. By

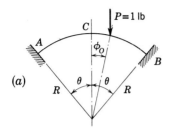

(a)

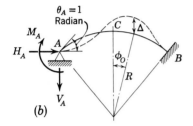

(b)

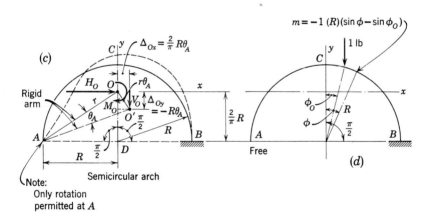

(c)

(d)

Note:
Only rotation
permitted at A

Semicircular arch

FIG. 10–14

proportionality of the sides of similar triangles we determine the translations as

$$\frac{\Delta_{Ox}}{\frac{2R}{\pi}} = \frac{r\theta_A}{r}$$

$$\Delta_{Ox} = \frac{2R}{\pi}\theta_A$$

and
$$\frac{\Delta_{Oy}}{R} = \frac{r\theta_A}{r}$$

$$\Delta_{Oy} = R\theta_A$$

Attention may now be directed to ascertaining the elastic center redundants which would be required to produce this geometrical configuration while only rotating tangent at A. The student should note that M_O, V_O, and H_O, computed by these equations, represent the magnitude of the redundants required to produce the rotation of the rigid arm and the displacements of the elastic center. This is in contrast to the purpose of the redundants for a loaded arch wherein the redundants produce rotations and displacements opposite to those caused by the loading on the statically determinate structure. It follows from the elastic center equations 6-4, 6-5, and 6-6, for a symmetrical arch, when $\theta = \pi/2$, that

$$M_O = \frac{\theta_A}{\int \frac{ds}{EI}}; \qquad H_O = \frac{\frac{2}{\pi}R\theta_A}{\int \frac{y^2\,ds}{EI}}; \qquad V_O = \frac{R\theta_A}{\int \frac{x^2\,ds}{EI}}$$

Equations 10-25, 10-26, and 10-27 enable the properties of the semicircular arch for axes through the elastic center to be found. Substituting $\theta = \pi/2$, we have

$$\int \frac{ds}{EI} = \frac{2R}{EI}\frac{\pi}{2} = \frac{\pi R}{EI}$$

$$\int \frac{y^2\,ds}{EI} = \frac{R^3}{EI}\left(\frac{\pi}{2} + \frac{\sin \pi}{2} - \frac{2\sin^2 \frac{\pi}{2}}{\pi/2}\right) = \frac{0.2976R^3}{EI}$$

$$\int \frac{x^2\,ds}{EI} = \frac{R^3}{2EI}\left(2\frac{\pi}{2} - \sin \pi\right) = \frac{\pi R^3}{2EI}$$

The magnitude of M_O by substitution, directed as shown in Fig. 14c, is

$$M_O = \frac{\theta_A EI}{\pi R}$$

then
$$\theta_A = \frac{\pi R}{EI} M_O$$

Then H_O and V_O, as directed in Fig. 14c, and in terms of M_O, are

$$H_O = \frac{\frac{2}{\pi}R\left(\frac{\pi R}{EI}M_O\right)}{0.2976R^3/EI} = \frac{6.720}{R}M_O$$

$$V_O = \frac{R\left(\frac{\pi R}{EI}M_O\right)}{\pi R^3/2EI} = \frac{2}{R}M_O$$

As a definite illustration of the calculation of influence line values let us take the problem of computing the vertical deflection of the arch at C under the influence of M_O, H_O, and V_O as determined. To carry out Müller-Breslau principles the angle θ_A will be made unity. To compute $\varDelta$ by the virtual work or unit load method, place a unity load at C, or, in Fig. 14d, make

$$\phi_O = 0$$

The general equation is

$$\varDelta = \int \frac{Mm \, ds}{EI}$$

where M at a general section produced by M_O, V_O, and H_O, with θ_A equal to unity is

$$M = \frac{EI}{\pi R} \left[1 - 2\sin \phi - 6.72\left(\cos \phi - \frac{2}{\pi}\right) \right]$$

and where $m = -R \sin \phi$

and $ds = R \, d\phi$

Then

$$\varDelta = -\frac{R}{\pi} \int_0^{\pi/2} \left[1 - 2\sin \phi - 6.72\left(\cos \phi - \frac{2}{\pi}\right) \right] \sin \phi \, d\phi$$

After integration and substitution of limits the deflection produced at C for $\theta_A = 1$ is $\varDelta = -0.11R$ where the negative sign indicates that C has moved up. By Müller-Breslau principles this deflection is interpreted as the influence line ordinate for M_A at C and implies that a moment of $+0.11R$ will be produced at A by a unity vertical load at C. This checks the result previously computed from equation 10-34.

The deflection equation expressing influence line ordinates at any section of the semicircular arch may be found by integration, wherein the section and the location of the unity load is identified by the angle ϕ_O in Fig. 14d. For sections to the right of C ϕ_O is a positive angle, and for sections to the left it is a negative angle. The equation to be integrated is

$$\varDelta = -\frac{R}{\pi} \int_{\phi_O}^{\pi/2} \left[1 - 2\sin \phi - 6.72\left(\cos \phi - \frac{2}{\pi}\right) \right] [\sin \phi - \sin \phi_O] d\phi \quad (10\text{-}36)$$

The integrated result of equation 10-36 is not given. In general, this approach is more difficult than the first method given in this section, but it can be considered as a feasible alternative. Müller-Breslau principles are, however, ideal for model studies wherein the deflections become measured quantities instead of computed quantities.

Influence line values for horizontal reaction by Müller-Breslau principles may be determined by introducing a unit horizontal displacement at A and calculating the vertical deflections of the arch. Figure 15a indicates the displacement and the resulting configuration. Figure 15b demonstrates, for a semicircular arch, that the elastic center moves a unit distance to O'. The elastic center redundants to produce this distortion are

$$M_O = 0; \qquad V_O = 0$$

since the rigid arm does not rotate and the elastic center does not displace vertically. And

$$H_O = \frac{1}{\displaystyle\int \frac{y^2\,ds}{EI}}$$

owing to the unit displacement.

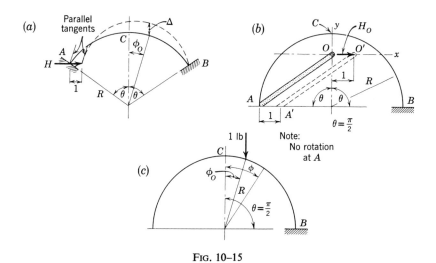

FIG. 10–15

To compute $\varDelta$

$\varDelta = \displaystyle\int \frac{Mm\,ds}{EI}$, where M is produced by H_O. Then,

$$\varDelta = \frac{1}{\int y^2\,ds}\int_{\phi_O}^{\theta}\left(R\cos\phi - \frac{2}{\pi}R\right)(R\sin\phi - R\sin\phi_O)R\,d\phi$$

Finally, in simplified form

$$\varDelta = \frac{\left[\left(\dfrac{\sin^2\theta}{2} - \dfrac{\sin^2\phi_O}{2}\right) + \dfrac{2}{\pi}(\cos\theta - \cos\phi_O) - \sin\phi_O(\sin\theta - \sin\phi_O) + \dfrac{2}{\pi}(\sin\phi_O)(\theta - \phi_O)\right]}{\theta + \dfrac{\sin 2\theta}{2} - \dfrac{2\sin^2\theta}{\theta}}$$

$$(10\text{-}37)$$

Equation 10-37 is valid for $\theta_. = \pi/2$ only. For $\theta = \dfrac{\pi}{2}$ and $\phi_O = 0$, equation 10-37 gives

$$\Delta = -0.459$$

Thus a 1-lb load at centerline produces a horizontal reaction at A to the right of 0.459 lb by Müller-Breslau principles.

The influence line for vertical reaction can be computed by static considerations after the influence lines for moment at A and B have been determined.

10–10. ARCH CALCULATIONS BY NUMERICAL INTEGRATION

Whenever the variations of moment of inertia, or forms of the arch axis, are such as to complicate their expression by analytical equations, then in practical applications, a numerical integration is generally applied. The integration sign is replaced by the Σ sign, and a finite number of ΔS sections used, where ΔS is analogous to the differential ds in prior theoretical expressions. The number of sections used will vary with span of arch, although nominally anywhere from 10 to 20 ΔS sections are sufficient. In some instances, equal ΔS lengths along the arch axis are employed, but equal Δx spaces along a horizontal axis may also be used leading to unequal ΔS lengths. With either practice the ΔS length is divided by the moment of inertia at the midpoint of the ΔS section. This provides $\Delta S/I$ in place of ds/I.

The fundamental quantities identified by their integral designations are

$$\int \frac{ds}{I} = \sum \frac{\Delta S}{I}$$

$$\int \frac{x^2\,ds}{I} = \sum x^2 \frac{\Delta S}{I} \qquad \int \frac{M_s\,ds}{I} = \sum M_s \frac{\Delta S}{I}$$

$$\int \frac{y^2\,ds}{I} = \sum y^2 \frac{\Delta S}{I} \qquad \int \frac{M_s x\,ds}{I} = \sum M_s x \frac{\Delta S}{I}$$

$$\int \frac{xy\,ds}{I} = \sum xy \frac{\Delta S}{I} \qquad \int \frac{M_s y\,ds}{I} = \sum M_s y \frac{\Delta S}{I}$$

In the above finite summations, x and y are the coordinates to the midlength of the ΔS section measured from the elastic center. The elastic center is located at the centroid of the elastic weights which is the centroid of the separately considered $\Delta S/I$ values.

No examples of the numerical method are provided in this book in view of the abundance of illustrative problems in other references such as standard textbooks in concrete and steel design.

REFERENCES

1. Grinter, L. E., *Theory of Modern Steel Structures*, Chapter 8, New York: MacMillan, 1949.
2. Johnson, J. B., C. W. Bryan, and F. E. Turneaure, *The Theory and Practice of Modern Framed Structures*, Part II, tenth edition, Chapter 4, New York: John Wiley, 1929.
3. Maugh, L. C., *Statically Indeterminate Structures*, Chapter 8, New York: John Wiley, 1946.
4. McCullough, C. B., and E. S. Thayer, *Elastic Arch Bridges*, New York: John Wiley, 1931.
5. Parcel, J. I., and R. B. B. Moorman, *Analysis of Statically Indeterminate Structures*, Chapter 10, New York: John Wiley, 1955.
6. Spofford, C. M., *The Theory of Continuous Structures and Arches*, Chapters 2, 5, 6, 7, and 8, New York: McGraw-Hill, 1937.
7. Urquhart, L. C., C. E. O'Rourke, and G. Winter, *Design of Concrete Structures*, Section 10, New York: McGraw-Hill, 1958.

PROBLEMS

10–1. The primary structure supporting a roof over a skating rink is to be constructed on a segmental basis. If the roof loading may be assumed to be uniform per horizontal foot of roof projection, determine the y coordinate to all segmental points by a graphical construction. The rise of the arch must be 30 ft.

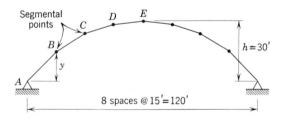

Prob. 10–1

10–2. By graphical construction determine h_1 and h_2 for the best shape of segmental structure to carry the four loads P.

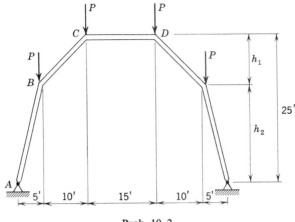

Prob. 10–2

10–3. The axis of the two-hinged arch shown is defined by $y = h \sin \pi x/L$. The moment of inertia at any section of the arch rib is equal to $I_c \sec \theta$, where I_c is the moment of inertia at the crown and θ equals the angle that a tangent to the arch axis makes with the horizontal. Calculate the horizontal reactions.

Ans. $H = \sqrt{2} PL/\pi^2 h$.

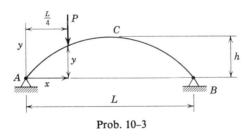

Prob. 10–3

10–4. Calculate the horizontal reactions for the arch of problem 10–3 for a load P at the center line. *Ans.* $H = 2PL/\pi^2 h$.

10–5. Calculate the horizontal reactions for the arch of problem 10–3 when loaded by one load P at the crown, if the moment of inertia is constant throughout instead of varying as stated.

10–6. The axis of the two-hinged arch shown is defined by $Z = mx_1{}^3$. The moment of inertia at any section of the arch rib is equal to $I_c \sec \theta$, where I_c is the moment of inertia at the crown and θ equals the angle that a tangent to the arch axis makes with the horizontal. Calculate the horizontal reactions produced by the central load P.   *Ans. $H = 7PL/40h$.*

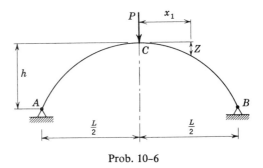

Prob. 10–6

10–7. For the arch shown in problem 10–6, compute the horizontal reactions for a uniform load of w lb per ft of horizontal projection. Also calculate the moment at crown.

10–8. For the arch shown, $x = ky^2$ with origin at A for the left half. The right half is similar. The moment of inertia is constant. Determine the horizontal reactions due to the load P.   *Ans. $H = 0.191P$.*

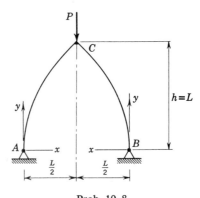

Prob. 10–8

10–9. A parabolic two-hinged steel arch is subjected to a uniform increase in temperature of 40° F. If the properties of the arch are as follows, calculate the horizontal reactions due to this change in temperature: $L = 100$ ft, $h = 20$ ft, $I_c = 2000$ in.⁴, and $E = 30 \times 10^6$ psi. The moment of inertia follows the secant variation.

10–10. Calculate the change in the rise of the arch of problem 10–9 due to the change in temperature.

10–11. Construct the required influence lines for a parabolic two-hinged arch if the moment of inertia follows the secant variation $L = 100$ ft and $h = 20$ ft. Construct influence lines for (*a*) horizontal reactions, (*b*) bending moment at one-fourth point of span, (*c*) thrust at one-fourth point of span, (*d*) shear at one-fourth point of span.

10–12. Construct the following influence lines for the arch of problem 10–11: (*a*) thrust at crown; (*b*) moment at crown; (*c*) shear at crown.

10–13. A fixed-ended arch has its axis defined by $y' = h \sin \pi x'/L$. The

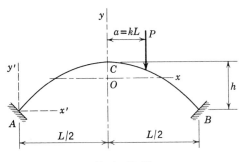

Prob. 10–13

moment of inertia variation follows the secant rule. Determine the value of M_A by the elastic center method when $a = L/4$.

10–14. Construct the required influence lines for the arch of problem 10–13 if $L = 150$ ft and $h = 40$ ft: (*a*) moment at crown, (*b*) moment at one-fourth point of span, (*c*) thrust at one-fourth point of span.

10–15. Construct the required influence lines for a fixed-ended parabolic arch shown in Fig. 10–10 of text for (*a*) moment at crown, (*b*) moment at one-fourth point of span, and (*c*) shear at one-fourth point of span. Let $L = 100$ ft and $h = 20$ ft.

10–16. Develop a general equation for determining the horizontal reaction for a symmetrical fixed-ended parabolic arch produced by a decrease in temperature of ΔT.   *Ans.* $H = 45 C_x \Delta T E I_c / 8h^2$.

10–17. Develop a general equation for moment at a fixed end for the conditions stated in problem 10–16.   *Ans.* $M_A = -15 C_x \Delta T E I_c / 4h$.

10–18. Fix the ends of the arch shown for problem 10–6 and compute the horizontal reactions and fixed-end moments by the elastic center method.
                  Ans. $H_A = 7PL/30h$; $M_A = PL/20$.

10–19. The axis of the arch is defined by $y' = h_1 \sin \pi x'/L$. The moment of inertia follows the secant variation based on the moment of inertia at C. Determine the elastic center and then compute the reactions and moment at A for a load P placed at C.

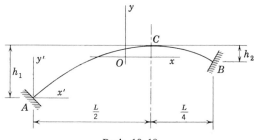

Prob. 10–19

10–20. The axis of the arch is parabolic with the moment of inertia following the secant variation based on moment of inertia at C. Determine the elastic center and then compute H_O, V_O, and M_O for a single load P placed at C. For numerical values let $L = 100$ ft and $h_1 = 20$ ft.

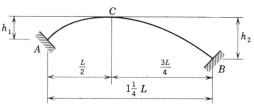

Prob. 10–20

10–21. Solve this unsymmetrical rectangular frame by the elastic center method. Determine M_A. Check by moment distribution.

Ans. -26.5 ft kips.

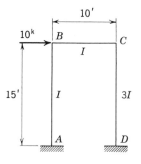

Prob. 10–21

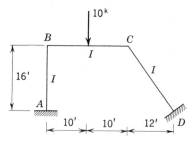

Prob. 10–22

10–22. Solve this unsymmetrical frame by the elastic center method. Determine M_A. Check by moment distribution.

Ans. $H_A = 3.16$ kips; $V_A = 6.16$ kips; $M_A = 20.9$ ft kips.

10–23. Solve this unsymmetrical frame by the elastic center method. Determine the moment at C.

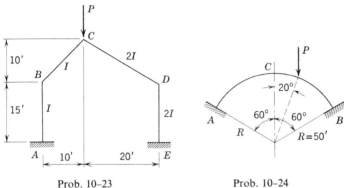

Prob. 10–23   Prob. 10–24

10–24. The circular arch has a uniform section throughout. Working from fundamentals determine the elastic center, and then determine H_O, V_O, and M_O produced by the load P. Calculate the moment at C. Check your results by the established formulas in text.

10–25. Using the circular arch of problem 10–24, and the elastic center method, calculate the moment at A produced by a uniform load of w lb per ft of horizontal projection over the entire span.

10–26. Solve for the moment at A by the elastic center method. The arch has a constant cross-section throughout. Solve by working with fundamentals and check your solution by using the equations in text.

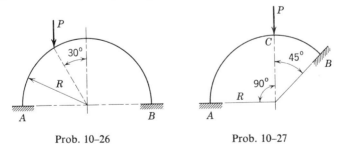

Prob. 10–26   Prob. 10–27

10–27. Solve for H_O, V_O, and M_O by the elastic center method. The arch has a constant cross-section throughout. Determine M_A.

10–28. Solve problem 10–26 by the column analogy method.

10–29. Solve problem 10–27 by the column analogy method.

10–30. By a method and theory of your choice determine the moment and reactions at A. Uniform cross-section. The load is uniform per unit of vertical projection. *Ans.* $V_A = 0.278wR$; $H_A = 0.0304wR$; $M_A = 0.045wR^2$.

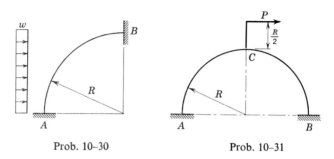

Prob. 10–30   Prob. 10–31

10–31. Determine the moment and reactions at A by a method and theory of your choice. Uniform cross-section.

10–32. Although not discussed in the text, the moment distribution method is applicable to arches and has merit where multiple spans are involved. As an exercise in the fundamentals of arch theory, determine the absolute stiffness and carry-over factors (K_A and C_A). Return to the basic definitions of moment distribution. Place a hinge at A, and determine the moment at A to rotate the tangent through an angle of θ_A. Also compute the moment developed at B by the applied moment at A. These calculations will provide the basic data for formulating K_A and C_A. The arch has constant cross-section.
Ans. $K_A = 2.306EI/R$; $C_A = -0.449$.

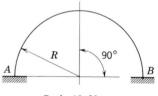

Prob. 10–32

10–33. Twin arches are loaded with a single load P at C. The arch spans are continuous at B, but supported to permit rotation at B as well as at A and D. Solve for the moment in arch at B by the moment-distribution method. Use the factors determined in problem 10–32 and calculate the fixed-end moments for the left-hand span by the formula in text. Constant cross-section.

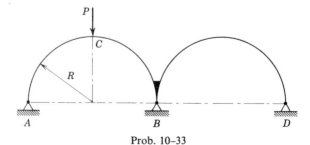

Prob. 10–33

10–34. Determine the vertical deflection at C due to a unit rotation of the tangent at A.   Solve by a method of your choice.   What is the moment at A due to a vertical load at C from Müller-Breslau principles?   Constant cross-section.

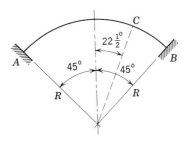

Prob. 10–34

The Cable

11

11–1. INTRODUCTION

Whereas most structural members are able to mobilize resistance to some degree to tension, compression, and bending, the cable can only work in tension. Instead of this limitation working against the use of the cable it works for it, as a high-force resistance is generally possible when materials suitable for cables are working in tension. This feature makes long-span suspension bridges feasible; however, there are many other uses for cables, such as bracing guys for vertical towers.

11–2. CABLE EQUILIBRIUM

Consider a cable suspended between points A and B, at the same elevation, in Fig. 1. Assume that the only load is the weight of the cable and as an approximation assume that this weight effect is uniform per unit of horizontal projection. The horizontal components of the reactions at A and B, termed H, are unknown. Let f equal the sag, the distance of midspan cable point below A and B. An equilibrium analysis will provide the basis for defining the geometry of the suspended cable.

With the origin at A, and a free body of cable as shown in Fig. 1b, a moment equation of static equilibrium, assuming no resisting moment in the cable, may be written about point D. This equation is

$$\frac{wa}{2} x - Hy - \frac{wx^2}{2} = 0$$

or

$$Hy = \frac{wa}{2} x - \frac{wx^2}{2}$$

$$y = \frac{wx}{2H} (a - x) \tag{11-1}$$

The equation 11-1 for y, where y is positive downward and A is the origin, can be identified as the equation of a second degree parabola. Thus, the assumption of uniform load per unit of horizontal projection leads to the equations of the parabolic form of cable so often substituted for the true catenary form. To derive the catenary form the cable weight is treated as uniform per unit length of the cable. The catenary form is not considered in this book.

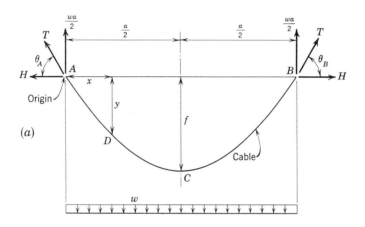

(a)

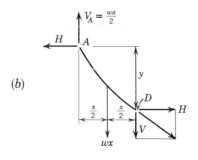

(b)

FIG. 11-1

To evaluate H, the horizontal component of the cable tension, let $y = f$ (the sag) when $x = a/2$; then, from equation 11-1

$$H = \frac{wa^2}{8f} \tag{11-2}$$

This is an important relation as the horizontal component is constant throughout and the total tension in the cable at any point may be easily determined if the slope of cable is known.   For example, at A and B the maximum cable tension would be

$$T_A \quad \text{or} \quad T_B = H \sec \theta_A \qquad (11\text{-}3)$$

To determine the slope at A or B, first take the derivative of y with respect to x from equation 11-1 as follows:

$$\frac{dy}{dx} = \frac{w}{2H} \cdot (a - 2x) \qquad (11\text{-}4)$$

and when $x = 0$

$$\tan \theta_A = \frac{dy}{dx} = \frac{wa}{2H}$$

or with $H = \dfrac{wa^2}{8f};$ then $\dfrac{dy}{dx} = \dfrac{4f}{a^2}(a - 2x)$

and $$\tan \theta_A = \frac{dy}{dx} = \frac{4f}{a}$$

Example

In Fig. 1, let $f = 20$, $a = 200$ ft, and $w = 3$ lb per ft.   By established formulas

$$H = \frac{3 \times 200^2}{8 \times 20} = 750 \text{ lb}$$

$$\tan \theta_A = \frac{4 \times 20}{200} = 0.4$$

$$\theta_A = 21°48'$$

$$T_A = 750 \times \sec 21°48'$$

$$= 810 \text{ lb}$$

The geometric length of the parabolic cable from A to B can be found as follows.   Letting L equal the geometric length of cable then a differential length of the cable dL is expressed as $\sqrt{dx^2 + dy^2}$ or $\left[1 + \left(\dfrac{dy}{dx}\right)^2\right]^{\frac{1}{2}} dx$. The total geometric length, L, in integral form is

$$L = 2 \int_0^{a/2} \left[1 + \left(\frac{dy}{dx}\right)^2\right]^{1/2} dx \qquad (11\text{-}5)$$

which, for an expedient solution, may be rewritten as several terms of a binomial expansion.

By the binomial expansion

$$(1 + Z)^n = 1 + nZ + \frac{n(n-1)}{2} Z^2 + \cdots$$

or with

$$Z = \left(\frac{dy}{dx}\right)^2 = \left[\frac{w}{2H}(a - 2x)\right]^2$$

Using three terms, equation 11-5 expands to

$$L = 2 \int_0^{a/2} \left\{ 1 + \frac{1}{2}\left[\frac{w}{2H}(a - 2x)\right]^2 - \frac{1}{8}\left[\frac{w}{2H}(a - 2x)\right]^4 + \cdots \right\} dx$$

or with $H = \dfrac{wa^2}{8f}$

$$L = 2 \int_0^{a/2} \left\{ 1 + \frac{1}{2}\left[\frac{4f}{a^2}(a - 2x)\right]^2 - \frac{1}{8}\left[\frac{4f}{a^2}(a - 2x)\right]^4 + \cdots \right\} dx$$

$$L = a\left[1 + \frac{8}{3}\left(\frac{f}{a}\right)^2 - \frac{32}{5}\left(\frac{f}{a}\right)^4 \right] \tag{11-6}$$

For the previous example the geometric length of the cable from equation 11-6 is

$$L = 200\left[1 + \frac{8}{3} \times \left(\frac{20}{200}\right)^2 - \frac{32}{5}\left(\frac{20}{200}\right)^4 \right]$$

$$= 205.21 \text{ ft}$$

11-3. INCLINED CABLE

Consider an inclined cable suspended freely between anchored points or fixed points A and B in Fig. 2. The only load is the weight of the cable and as a reasonable approximation it is assumed that this effect is uniform per unit of horizontal projection. T_A is the tension in cable at A, where it is resolved into vertical and horizontal components. Similarly T_B is the tension in cable at B. Horizontal equilibrium requires H_A to equal H_B. Let H signify these horizontal components, where H is also the horizontal component of the cable tension at any point along the length of cable. As a basis for geometrical considerations A and B are joined by a chord making an angle of ϕ with the horizontal.

Since a cable can have no resisting moment, moments of the external forces to the left about any section C must equal zero; or

$$V_A \cdot x - H\left(y' - \frac{x}{a} \cdot b\right) - \frac{wx^2}{2} = 0 \tag{11-7}$$

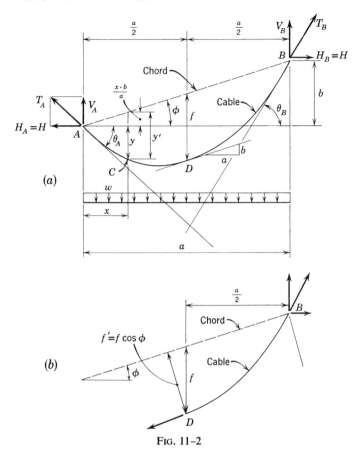

FIG. 11–2

To obtain a second relation involving V_A and H, moments are taken about point B.

$$H \cdot b + V_A \cdot a - \frac{wa^2}{2} = 0$$

giving

$$V_A = \frac{wa}{2} - \frac{H \cdot b}{a} \qquad (11\text{-}8)$$

Solving equations 11-7 and 11-8

$$H \cdot y' = \frac{wa}{2} x - \frac{wx^2}{2} \qquad (11\text{-}9)$$

Equation 11-9 is a general equation for a cable uniformly loaded per horizontal unit of length where y' is the vertical distance from the chord joining A and B to the cable and x is measured from A. y' is a positive quantity as taken in this derivation. The value of y' at the midlength of the chord or cable span is called the sag and is termed f. With y' equal to f when $x = a/2$, equation 11-9 yields

$$H = \frac{wa^2}{8f} \quad \text{or} \quad f = \frac{wa^2}{8H} \tag{11-10}$$

Substituting H from equation 11-10 into equation 11-9, we obtain

$$y' = \frac{4fx}{a^2}(a - x) \tag{11-11}$$

where y' is measured from the chord. If y is the distance to the cable measured vertically from a horizontal line through A, then

$$y = \frac{xb}{a} - y'$$

where y is termed positive when measured upward from a horizontal line through A. Eliminating y' by substituting equation 11-11 leads to an equation for y.

$$y = \frac{xb}{a} - \frac{4fx}{a^2}(a - x) \tag{11-12}$$

which is recognizable as an equation representing a parabolic curve.

From equation 11-12 several important relations may be derived. Taking the derivative of y with respect to x

$$\frac{dy}{dx} = \frac{b}{a} - \frac{4f}{a^2}(a - 2x) \tag{11-13}$$

The value of $\frac{dy}{dx}$ equals $\tan \theta_A$ when $x = 0$, and equals $\tan \theta_B$ when $x = a$, and by substitution into equation 11-13

$$\tan \theta_A = \frac{b}{a} - \frac{4f}{a} \tag{11-14}$$

$$\tan \theta_B = \frac{b}{a} + \frac{4f}{a} \tag{11-15}$$

If it is desired to have the tangent to the cable horizontal at A then $\tan \theta_A = 0$ and $f = \frac{1}{4}b$. This provides an upper limit on sag f equal to one-fourth the slant rise of cable if A is a point at ground level and the cable must just clear the ground at all points.

With θ_A and θ_B known from equations 11-14 and 11-15, the total tension at A and B may be found as

$$T_A = H \sec \theta_A, \qquad T_B = H \sec \theta_B$$

with the maximum tension occurring at the point of greatest slope.

Equation 11-13, if set equal to zero, will locate the low point on the cable which is to one side of the center of span. Letting the distance from A to this low point be x_L, then

$$x_L = \frac{a}{2}\left(1 - \frac{b}{4f}\right)$$

which will be valid for all values of f equal to or greater than $\frac{1}{4}b$, since for values of f less than $\frac{1}{4}b$, x_L would be negative and meaningless as the low point must be at point A.

Another geometric fact of some importance is that the slope of the tangent to the cable is parallel to the chord when $x = a/2$. This may be verified by setting $\frac{dy}{dx}$ of equation 11-13 equal to b/a and solving for x.

Example

In Fig. 2, let $a = 200$ ft and $b = 200$ ft and the uniform load on cable per horizontal foot is assumed to be 3 lb per ft. The sag is 20 ft and the sag ratio is expressed as $f/a = \frac{20}{200} = \frac{1}{10}$. Find the maximum tension in cable, the tension at A and the average tension.

$$H = \frac{wa^2}{8f} = \frac{3 \times 200^2}{8 \times 20} = 750 \text{ lb}$$

$$\tan \theta_B = \frac{b}{a} + \frac{4f}{a} = \frac{200}{200} + \frac{4 \times 20}{200} = 1.400$$

$$\theta_B = 54°28'$$

$$T_B = 750 \times \sec 54° \ 28' = 1290 \text{ lb}$$

$$\tan \theta_A = \frac{b}{a} - \frac{4f}{a} = 1 - 0.4 = 0.6$$

$$\theta_A = 30° \ 58' \text{ (above horizontal)}$$

$$T_A = 750 \times \sec 30° \ 58' = 880 \text{ lb}$$

The average of T_A and T_B is 1085 lb. The tension in the cable, where tangent to cable is parallel to the chord, is

$$T = 750 \sec 45° = 1060 \text{ lb}$$

The average tension, when dealing with small sags, may be satisfactorily approximated by calculating the tension in cable at the point where the tangent is parallel to the chord. This will be of usefulness in calculations dealing with tower guys.

11–4. LENGTH OF INCLINED CABLE

Figure 2b is one-half of the inclined cable extending from B to D. f is the maximum sag measured from the chord and f' is the component of the sag normal to the chord. This component is

$$f' = f \cdot \cos \phi$$

It is considered sufficiently accurate, for most purposes with small sags, to compute the length of the cable based on equation 11-6 previously developed for a horizontal span, taking only the first two terms of equation 11-6 and taking ($a \sec \phi$) the chord length as the total span length. Replacing f by $f \cos \phi$, the approximate geometric length of the cable is

$$L = a\left[\sec \phi + \frac{8}{3}\frac{f^2}{a^2 \cdot \sec^3 \phi}\right] \tag{11-16}$$

Although only approximate, this equation is satisfactory for the general case of inclined cables used as guys where the sag is small.

The geometric length of the cable from A to B for the example problem is

$$L = 200\left[1.41421 + \frac{8 \times 20^2}{3 \times 200^2 \times (1.41421)^3}\right]$$

$$= 200[1.41421 + 0.00943]$$

$$= 284.73 \text{ ft}$$

11–5. CABLE STRETCH

For most work involving guys the elastic stretch of the cable may be computed by using the average tension $\overline{T}$ computed as described in the previous article. This change in length, termed ΔL, would be

$$\Delta L = \frac{\overline{T} \cdot L}{AE}$$

where E and A follow the usual definitions, but where E may be a property of the composite wire rope strand in place of the properties of the basic material composing the cable elements.

A theoretical development for $\overline{T}$ follows. The tension throughout the cable varies, hence unit strain for a given dL varies. Each differential length dL of the cable extends by $\frac{T \cdot dL}{AE}$. This differential change in length of a differential element, referring to Fig. 3, will be

$$d(dL) = \frac{H \sec \alpha \cdot dL}{AE}$$

From Fig. 3

$$\sec \alpha = \frac{dL}{dx}$$

therefore

$$\Delta L = \frac{H}{AE} \int_0^a \frac{dL \cdot dL}{dx} = \frac{H}{AE} \int_0^a \frac{(dL)^2}{dx}$$

Substituting

$$dL = dx\left[1 + \left(\frac{dy}{dx}\right)^2\right]^{1/2}$$

$$\Delta L = \frac{H}{AE} \int_0^a \left[1 + \left(\frac{dy}{dx}\right)^2\right] dx$$

$$\Delta L = \frac{H}{AE} \int_0^a \left[1 + \left(\frac{b}{a} - \frac{4f}{a^2}(a - 2x)\right)^2\right] dx$$

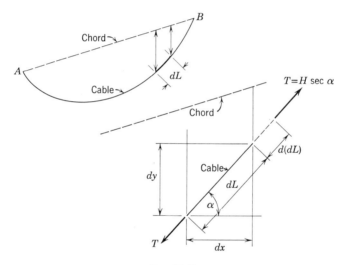

FIG. 11–3

By expanding, integrating, and substituting limits, the following simplified equation results when $s = f/a$

$$\Delta L = \frac{Ha}{AE}\left[1 + \left(\frac{b}{a}\right)^2 + \frac{16}{3}s^2\right] \tag{11-17}$$

equating this to $\Delta L = \dfrac{\overline{T}L}{AE}$

$$\overline{T} = \frac{H \cdot a}{L}\left[1 + \left(\frac{b}{a}\right)^2 + \frac{16}{3}s^2\right] \tag{11-18}$$

For the example previously solved

$$\bar{T} = \frac{750 \times 200}{284.73} \left[1 + (1)^2 + \frac{16}{3} \times \left(\frac{1}{10} \right)^2 \right]$$

$$\bar{T} = 1100 \text{ lb}$$

which may be compared with 1060 lb by the approximate method demonstrated previously.

The change in cable length, if $A = \frac{1}{2}$ sq in. and the effective cable modulus is $E = 20 \times 10^6$ psi, will be

$$\Delta L = \frac{1100 \times 284.73}{\frac{1}{2} \times 20 \times 10^6}$$

$$\Delta L = 0.029 \text{ ft}$$

11–6. GUYED MASTS

For guyed vertical towers and masts, one of the most important problems encountered is the calculation of the stresses in the cable guys. Although the complexity of the analysis varies with the number of guys and the number of sets of such guys, the following example will illustrate the principles involved.

In Fig. 4, a 300-ft high mast is guyed with one set of guys attached at the point B. The plan view locates the three cables involved with anchorages at A, C, and D. The direction of the wind acting on the tower is shown.

The calculation of the wind stress in the separate guys would be relatively simple if the guys were straight members and capable of resisting either tension or compression. In practice the guys are generally initially tensioned to about 20 per cent. of the expected total working tension that will occur under full wind load. To the degree that such pretensioned cables can resist compression through a relief of this initial tension, the guys serve as compression elements. However, the determination of this compressive participation is complexed by the fact that sag of cable is increased by the relief of initial tension.

Since maximum stresses must be used in the design of the cables, the wind force must be considered as acting in the proper direction relative to mast and cables. Figure 4c indicates the horizontal components of the cables in relation to F, the needed resultant external horizontal wind reaction at B where F acts parallel and opposite to the direction of the wind force on mast. *Assuming that there is no initial tension* in the guys, $H_3 = 0$, since cable No. 3 is assumed slack; therefore H_1 and H_2 must together furnish a resisting force equal to F.   The force diagram, Fig. 4d, clearly indicates that the resultant of H_1 and H_2 must equal F.   The student should recognize that the assumption of H_3 equal to zero is a necessary assumption to avoid indeterminacy.   Cable No. 3 must slack

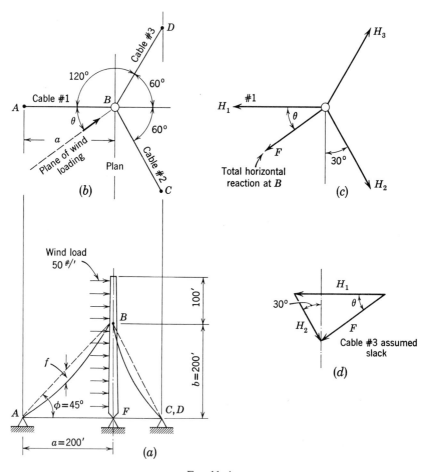

FIG. 11–4

off with a reduction in its initial tension.   This aspect of the guy problem will be considered subsequently.

Resolving F and H_2 into components perpendicular to cable No. 1 in Fig. 4d requires

$$H_2 \cos 30° = F \sin \theta$$

$$H_2 = \frac{F \sin \theta}{\cos 30°}$$

and resolving parallel to H_1 or cable No. 1

$$H_1 - H_2 \sin 30° = F \cos \theta$$

and by substitution

$$H_1 - F\frac{\sin \theta}{\cos 30°} \sin 30° = F \cos \theta$$

from which

$$H_1 = F[\cos \theta + \sin \theta \tan 30°]$$

To determine a maximum value of H_1 take the derivative of H_1 with respect to θ and set equal to 0.

$$\frac{dH_1}{d\theta} = F[-\sin \theta + \cos \theta \tan 30°] = 0$$

$$\tan \theta = \tan 30°$$

$$\theta = 30°$$

This implies that the wind force on the mast must be assumed as making an angle of 30° with cable No. 1 (or at right angles to cable No. 2), for computing maximum wind stress in a single cable. Then, by substitution

$$H_{1(max)} = F[0.866 + 0.5 \times 0.577] = 1.155F$$

and the concurrent value of H_2 is $0.577F$.

A moment equation written about an axis through the base of mast will determine F, the necessary resultant horizontal component of the cable tensions applied at B, owing to wind on the tower. Thus, we obtain

$$F = \frac{50 \times 300 \times 150}{200}$$

$$F = 11,250 \text{ lb}$$

According to the foregoing analysis for maximum cable participation, the maximum horizontal component of H_1, neglecting initial tension, is $1.155 \times 11,250 = 13,000$ lb, and the concurrent H_2 is $0.577 \times 11,250 = 6500$ lb.

Assuming that cable No. 1 is a straight tension member, its approximate stress, neglecting initial tension, is $(13,000) \times 1.414 = 18,400$ lb. If the cable is to be pretensioned to a value equal to the 20 per cent of the working tension, the working tension will be $18,400/0.8 = 23,000$ lb. If a factor of safety of 5 is required, the breaking strength of the cable should be 115,000 lb. As a trial design, a $1\frac{1}{2}$-in. dia. galvanized guy rope weighing 3.6 lb per ft will satisfy this approximate stress requirement.

Although insulators may be required, none are assumed in this case and the cable load per horizontal foot of projection will be taken as

$$\frac{200 \times 1.414 \times 3.6}{200} = 5.1 \text{ lb per ft, which will be rounded off to 5 lb per ft,}$$

for calculation purposes.

The geometry of the cables under initial erection tension and cable dead load must be known. The horizontal component of the initial tension, assuming the cable as a straight member, is

$$(23,000 - 18,400)(0.707) = 3250 \text{ lb}$$

Then, the dead load sag for each cable in the erection condition would be

$$f_E = \frac{5 \times 200^2}{3250} = 6.15 \text{ ft}$$

and the geometric erected length of each cable from anchorage to the center of the mast from equation 11-16 is

$$L = 200\left[1.41421 + \frac{8 \times 6.15^2}{3 \times 200^2} \times (1.41421)^3\right]$$

$$= 200[1.41421 + 0.00087]$$

$$= 283.02 \text{ ft}$$

As the wind load builds up on the mast, making an angle of 30° to cable No. 1, cable stress increases in cables No. 1 and No. 2, and decreases in cable No. 3. Since one of the cables is redundant the problem is to determine the force in each cable compatible with the horizontal deflection of point B. Owing to the complexity of the interaction of the cables, a trial-and-error solution is recommended. Assume that the horizontal component of initial tension in cable No. 3 is reduced 2000 lb by wind action to a value of 1250 lb. Hence, considering equilibrium, the new horizontal components, due solely to initial tension, for cables No. 1 and No. 2 are also 1250 lb.

Based on the foregoing assumption, the approximate total horizontal components in the three guys, when cable No. 1 is carrying its maximum force, are:

$$H_1 = 13,000 + 1250 = 14,250 \text{ lb}$$
$$H_2 = 6500 + 1250 = 7750 \text{ lb}$$
$$H_3 = 3250 - 2000 = 1250 \text{ lb}$$

which leads to the following approximate stress in each guy:

$$S_1 = 14,250 \times 1.414 = 20,200 \text{ lb}$$
$$S_2 = 7750 \times 1.414 = 11,000 \text{ lb}$$
$$S_3 = 1250 \times 1.414 = 1800 \text{ lb}$$

To determine the movement of B, the change in length of each cable from its erection length is required. Since this change in length is associated with the change in stress, the following stress differences are obtained, noting that the initial tension in cables before the assumed relief in cable No. 3 was $3250 \times 1.414 = 4600$ lb. Changes in tensions are

$$\Delta S_1 = 20,200 - 1800 = +18,400 \text{ lb}$$
$$\Delta S_2 = 11,000 - 1800 = +9200 \text{ lb}$$
$$\Delta S_3 = 1800 - 4600 = -2800 \text{ lb}$$

The change in stress is the true wind stress in the cables and could have been found directly by

$$\Delta S_1 = 13,000 \times 1.414 = 18,400 \text{ lb}$$
$$\Delta S_2 = 6500 \times 1.414 = 9200 \text{ lb}$$
$$\Delta S_3 = -2000 \times 1.414 = -2800 \text{ lb}$$

This is an expedient way of determining the incremental forces which produce change in cable length, but either approach is correct.

The change in length of each cable due to ΔS will be computed as $\Delta S L/AE$, where A, the cross-sectional area of wire equals $3.6/3.4 = 1.06$ in.^{2}, and E the effective modulus of elasticity is 15×10^6 psi.

Cable No. 1 $\quad \Delta L = \dfrac{+18,400 \times 283.02}{1.06 \times 15 \times 10^6} = +0.328 \text{ ft}$

Cable No. 2 $\quad \Delta L = \dfrac{+9200 \times 283.02}{1.06 \times 15 \times 10^6} = +0.164 \text{ ft}$

Cable No. 3 $\quad \Delta L = -\dfrac{2800 \times 283.02}{1.06 \times 15 \times 10^6} = -0.050 \text{ ft}$

The lengths of the guys under wind stress, when cable No. 1 is at maximum stress, are

Cable No. 1 $\quad L = 283.02 + 0.328 = 283.348 \text{ ft}$

Cable No. 2 $\quad L = 283.02 + 0.164 = 283.184 \text{ ft}$

Cable No. 3 $\quad L = 283.02 - 0.050 = 282.97 \text{ ft}$

and the sags of cables No. 1 and No. 2 under wind action and the residual initial tension are

$$f_1 = \frac{5 \times 200^2}{8 \times 14250} = 1.75 \text{ ft}$$

$$f_2 = \frac{5 \times 200^2}{8 \times 7750} = 3.22 \text{ ft}$$

The objective of the subsequent calculations is to check the starting assumption for the relief of initial tension in cable No. 3.   With the sags and stretched lengths of cables No. 1 and No. 2 known, the horizontal

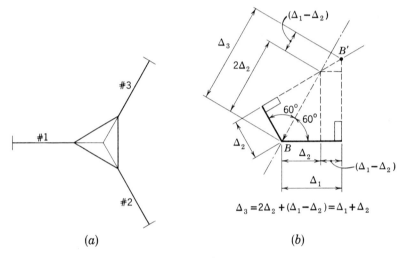

(a)   (b)

F<small>IG</small>. 11-5

displacement of B can be determined.   Figure 5b represents a Williot diagram drawn with Δ_1 and Δ_2 as the horizontal movements of B due to the change in length of cables No. 1 and No. 2.   By the geometry indicated, the horizontal displacement of B in the direction of cable No. 3 will equal $\Delta_1 + \Delta_2$.   The sag of cable No. 3 is now determinable and with this sag known the starting assumption may be checked.

The approximate equation for L (equation 11-16) may be modified by reference to Fig. 5.   The length of cable No. 1, with Δ_1 the horizontal displacement of B due to stretch of cable No. 1 as shown in Fig. 5b, is

$$L_1 = [a + \Delta_1] \left[\sec \phi + \frac{8f_1^2}{3(a + \Delta_1)^2 \sec^3 \phi} \right]$$

which, after transposing and rearrangement of terms, becomes the quadratic equation

$$(a + \Delta_1)^2 - L_1(\cos \phi)(a + \Delta_1) + \frac{8f_1^2}{3 \sec^4 \phi} = 0$$

and, finally, for cable No. 1

$$(200 + \Delta_1)^2 - 283.348 \times 0.707(200 + \Delta_1) + \frac{8 \times 1.75^2}{3 \times 1.414^4} = 0$$

solving $\qquad\qquad 200 + \Delta_1 = 200.348$

$$\Delta_1 = 0.316 \text{ ft}$$

For cable No. 2

$$(200 + \Delta_2)^2 - 283.184 \times 0.707(200 + \Delta_2) + \frac{8 \times 3.22^2}{3 \times 1.414^4} = 0$$

$$\Delta_2 = 0.176 \text{ ft}$$

Δ_3 as a function of Δ_1 and Δ_2 may be obtained from the geometry of Fig. 5b, as $\Delta_3 = \Delta_1 + \Delta_2 = -0.492$ ft. With Δ_3 known, the sag in cable No. 3 may be computed from

$$f_3^2 = \frac{3 \sec^4 \phi}{8} (a + \Delta_3)[L \cos \phi - (a + \Delta_3)]$$

or $\qquad f_3^2 = \frac{3 \times 1.414^4}{8} (200 - 0.492)[282.97 \times 0.707 - (200 - 0.492)]$

$$f_3^2 = 165.19$$
$$f_3 = 12.85 \text{ ft}$$

The computed horizontal component cable tension for cable No. 3 is

$$H_3 = \frac{5 \times (199.51)^2}{8 \times 12.85} = 1930 \text{ lb}$$

Since 1930 lb is not close to the 1250 lb previously assumed for the windward guy, a revised set of calculations must be carried out. This will not be done here, but a value of H_3 of about 2000 lb can now be assumed for the revised calculations.

In the case of a mast with more than one set of guys, the mast becomes an indeterminately supported structure. A first approximation for total horizontal force at each guyed level may be made by assuming the tower as a continuous vertical beam supported by unyielding horizontal supports at the guyed points. After this initial step, cables must be treated as elastic members with calculations similar to those made with the example tower. As the mast is indeterminate, the horizontal deflections at points where guys are attached take on additional importance, since the indeterminate reactions are modified by support movement or sidesway. The total solution is primarily a trial-and-error manipulation.

11–7. STIFFENED SUSPENSION BRIDGES

The monumental suspension bridges of the last three decades emphasize the importance of this type of structure. A complete analysis of all problems associated with suspension bridges is outside the scope of this book. Certain fundamentals of related structural mechanics are presented to form a basis for more intensive study.

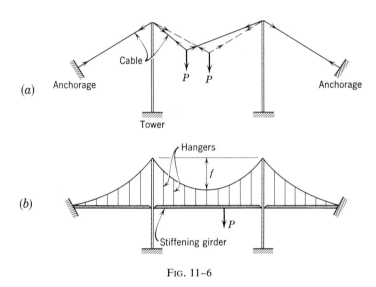

FIG. 11–6

In Fig. 6a, two basic elements of a suspension bridge, the towers and the cable, are shown. The third basic element, the stiffening girder, has been omitted, to demonstrate that if a large load P was applied directly to the cable and its point of attachment considered as movable, a changing cable configuration, as indicated in Fig. 6a, would result as the load moved across the span. If the cable is weightless, the segments of the cable would assume the outline of an equilibrium polygon. This changing configuration would nullify the use of the suspended cable as a traffic carrying bridge. For that reason, the third and fourth elements, the stiffening girder and hangers, are added as in Fig. 6b. A concentrated load P on the roadway will now be transmitted to the cable through the hangers, and extreme changes in cable configuration will not occur as a moving load traverses the bridge. The structure will deflect while retaining its basic form. The dead weight of stiffening girders is suspended from the cable.

There are two well-established theories relating to the analysis of suspended structures. The elastic theory ignores changes in the structural geometry due to deflections, whereas the deflection theory incorporates the changes in the structural geometry. Every structure deflects but, in general, the deflections are too small to effect a substantial change in geometry. However, for long-span flexible arches and suspension bridges, considerable error may result if deflected geometry is not considered.

11–8. ELASTIC THEORY

The elastic theory can only be regarded as an approximate theory and the chief value of its presentation is to clarify some aspects of basic structural action.

In Fig. 7a, either the stiffening girder or the cable can be considered as the redundant structural element, since either system would serve as the determinate *base structure*. The cable will be selected as the redundant element and its horizontal component H selected as the redundant force. To simplify preliminary discussions the effect of the side spans has been eliminated by anchoring the top of towers. Modifications for side span influence will be considered later. Figure 7b indicates the determinate base system, after cutting the cable, in which hanger stresses equal zero and the stiffening girder acts as a simple span hinged at each tower. Figure 7c represents a free body of stiffening girder indicating M_s at a section. Figure 7d is a free body with the redundant cable force in place. The bending moment in the girder equals $M_s + M_i$ where M_i is produced by H.

A unit value for H is introduced in Fig. 7e and, with the assumption that the cable has a parabolic configuration, the axial forces in hangers must all be equal to provide a uniform load on cable per unit of horizontal projection. From parabolic cable theory and with $H = 1$ lb, the equivalent uniform load acting downward on the cable as well as upward on the stiffening girder is

$$p = \frac{8f}{a^2}$$

The force u_n in each hanger is then pb where b is the spacing of hangers and the downward end reaction of stiffening girder is $pa/2$. The bending moment produced in the stiffening girder by the hanger forces is the

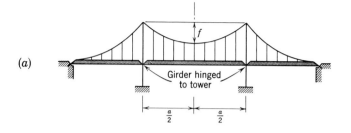

(a)

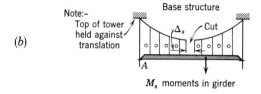

(b)

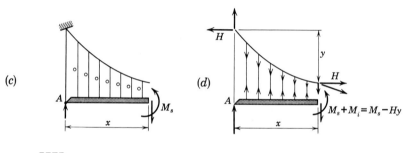

(c)

(d)

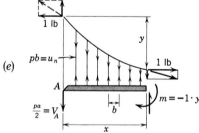

(e)

FIG. 11–7

equivalent of that of a simple span uniformly loaded upward. Letting m equal this moment, where

$$H = 1 \text{ lb}$$

$$m = -\frac{px}{2}(a - x)$$

It may be also noted that $m = -y$ by reference to derivations and equation 11-1 in Art. 11-1 if p is substituted for w and H equals unity.

The final bending moment in the stiffening girder equals $M_s + M_i$, where M_s is the determinate moment in girder caused by a particular live loading and where M_i is the moment caused by H. Since $m = -y$ for H equal to unity, then $M_i = -Hy$. Hence the final moment in the girder or M equals $M_s - Hy$.

We may now return to basic geometrical considerations which are common to the analysis of all indeterminate structures. There can be no relative movement of the two cut ends of the cable at the center line in Fig. 7a, and this geometric condition is the basis for stating that $\Delta_s + \Delta_i = 0$ where Δ_s is the separation of the severed ends of the cable caused by bending in the statically determinate stiffening girder and where Δ_i is the movement of the ends of the cable caused by the redundant force H. If δ equals the sum of all the elastic effects produced by a force H equal to unity, we may write the geometric equation as $\Delta_s + \delta \cdot H = 0$ from which

$$H = -\frac{\Delta_s}{\delta}$$

The equation for H may be formulated as in equation 11-19, wherein the numerator represents Δ_s and the terms in the denominator represent δ. A detailed explanation of all terms follows. The equation for H becomes

$$H = -\frac{\displaystyle\int_0^a \frac{M_s m\, dx}{EI}}{\displaystyle\int_0^a \frac{m^2\, dx}{EI} + \int_0^L \frac{u^2\, ds}{AE} + \sum \frac{u_n{}^2 L_n}{A_n E}} \qquad (11\text{-}19)$$

in which L = length of cable

$\quad L_n$ = length of a given hanger

$\quad dx$ = differential length of girder

$\quad ds$ = differential length of cable

$\quad A$ = cross-sectional area of cable

$\quad A_n$ = cross-sectional area of a hanger

$\quad I$ = moment of inertia of the stiffening girder

$\quad m$ = bending moment in girder due to $H = 1$ lb

$\quad u$ = axial stress in cable due to $H = 1$ lb

$\quad u_n$ = axial stress in a hanger due to $H = 1$ lb

$\quad E$ = modulus of elasticity (Note: If the modulus is the same for cable, hangers, and girder, E can be dropped.)

$\quad b$ = spacing of hangers

The denominator of equation 11-19 for H does not include the effect of the change in length of towers but does include the effect of the hangers. As a first approximation, the last term of the denominator, representing the effect of hangers, will be dropped. Equation 11-19 may also be simplified by canceling E and substituting $m = -y$ to

$$H = \frac{\int_0^a \dfrac{M_s y \, dx}{I}}{\int_0^a \dfrac{y^2 \, dx}{I} + \int_0^L \dfrac{u^2 \, ds}{A}} \qquad (11\text{-}20)$$

The evaluation of the integrals follows, where δ_1/E equals the closure of the cut section at the center line, owing to bending of the stiffening girder from the action of $H = 1$ in the cable.

$$\delta_1 = \int \frac{y^2 \, dx}{I} = \frac{16 f^2}{a^4 I} \int_0^a (ax - x^2)^2 \, dx$$

$$\delta_1 = \frac{8 f^2 a}{15 I}$$

The term $\dfrac{\delta_2}{E} = \displaystyle\int_0^L \dfrac{u^2 \, ds}{AE}$ is the most difficult to evaluate. Physically

interpreted, this term represents the amount that cut ends move toward one another due to the stretch of cable caused by $H = 1$. For a particular differential length the secant of the angle of inclination of the cable is ds/dx, and u representing stress is $1(ds/dx)$.

$$\delta_2 = \int_0^L \frac{u^2 \, ds}{A} = \frac{1}{A} \int_0^L \frac{ds^3}{dx^2}$$

from fundamentals $\overline{ds}^2 = \overline{dx}^2 + \overline{dy}^2$ and

$$ds = \left[1 + \left(\frac{dy}{dx} \right)^2 \right]^{1/2} dx$$

Then $\qquad \delta_2 = \dfrac{1}{A} \displaystyle\int_0^L \dfrac{ds^3}{dx^2} = \dfrac{1}{A} \int_0^a \left[1 + \left(\frac{dy}{dx} \right)^2 \right]^{3/2} dx$

and with $y = (4f/a^2)(ax - x^2)$ this integral is restated by a binomial expansion of three terms as

$$\delta_2 = \frac{1}{A} \int_0^a \left\{ 1 + \frac{3}{2} \left[\frac{4f}{a^2} (a - 2x) \right]^2 + \frac{3}{8} \left[\frac{4f}{a^2} (a - 2x) \right]^4 + \cdots \right\} dx$$

Dropping the last term, the integration gives

$$\delta_2 = \frac{1}{A} \left[\int_0^a x + \frac{24 f^2}{a^2} \left(a^2 x - 2ax^2 + \frac{4x^3}{3} \right) \right]$$

From which we obtain

$$\delta_2 = \frac{a}{A}\left[1 + 8\left(\frac{f}{a}\right)^2\right]$$

δ_2 is thus the approximate length of the main span cable divided by A. Equation 11-20 for H has therefore been reduced to

$$H = \frac{\int \frac{M_s y\, dx}{I}}{\delta_1 + \delta_2} \tag{11-21}$$

wherein the numerator must be evaluated for the given loading on bridge. For a uniform live load of q per unit of length over the entire span, placed on the stiffening girder in its simple span determinate condition, the numerator is

$$\int \frac{M_s}{I} y\, dx = \frac{2fq}{a^2 I}\int_0^a (ax - x^2)^2\, dx$$

$$= \frac{qfa^3}{15I}$$

Then, for a full span live load of q, and equation 11-21, we obtain

$$H = \frac{\frac{qfa^3}{15I}}{\delta_1 + \delta_2} = \frac{\frac{qfa^3}{15I}}{\frac{8f^2 a}{15I} + \frac{a}{A}\left[1 + 8\left(\frac{f}{a}\right)^2\right]} \tag{11-22}$$

11-9. ELASTIC THEORY—SIDE SPANS INCLUDED

The previous article dealt with the suspension bridge as a single span swung between towers anchored at their tops against movement. This was done for simplicity in presenting theory. This theory may now be modified to include the effects of two side spans with cable continuous from anchorage to anchorage but free to move over the top of the towers. Horizontal equilibrium at tower tops requires the horizontal component of cable to be equal in all spans. The expanded equation 11-19 for H for live load, with hanger term eliminated, becomes

$$H = \frac{\int_0^a \frac{M_s y\, dx}{I} + \int_0^{a_1} \frac{M_s y_1\, dx}{I_1} + \int_0^{a_2} \frac{M_s y_2\, dx}{I_2}}{\int_0^a \frac{y^2\, dx}{I} + \int_0^{a_1} \frac{y_1^2\, dx}{I_1} + \int_0^{a_2} \frac{y_2^2\, dx}{I_2} + \int_0^L \frac{u^2\, ds}{A}} \tag{11-23}$$

The nomenclature for the main span is as previously described, and the side span nomenclature may be made clear by reference to Fig. 8. The moment of inertia of the left side span stiffening girder is I_1, and for the right side span it is I_2. The maximum sags for the inclined cables of side spans are measured from inclined chord to cable. If the side spans are identical, let $a_2 = a_1$, $I_2 = I_1$, and $f_2 = f_1$. The last term in the denominator of equation 11-23 represents the effect of change in length of the entire cable in all spans.

If the side span cables follow a parabolic configuration, then after cutting cables at the center line of the main span and introducing $H = 1$ lb, as in Fig. 7e, the side span hangers will impose an upward uniform load on side span stiffening girders. The intensity of this upward load is

$$p_1 = \frac{8f_1}{a_1{}^2}$$

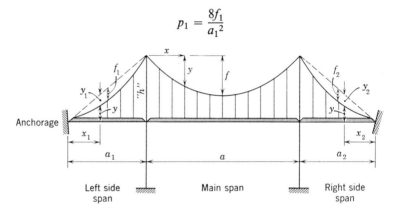

FIG. 11–8

The moment m_1 in side span stiffening girder is

$$m_1 = -\frac{p_1 x_1}{2}(a_1 - x_1)$$

$$m_1 = -\frac{4f_1 x}{a_1{}^2}(a_1 - x_1) = -y_1$$

The integral $\int_0^a \frac{y_1{}^2 \, dx}{EI_1}$ may be termed $\frac{\delta_3}{E}$, or the closure of the cut in the main span due to bending in one side span under the influence of $H = 1$ lb. $\delta_3 = \frac{8f_1{}^2 a_1}{15I_1}$, by comparison with prior development of δ_1.

It is to be noted that the last term in the denominator of equation 11-23 for H takes into account the axial change in the length of cable from anchorage to anchorage. The contribution of the main span is already known to be

$$\delta_2 = \frac{a}{A}\left[1 + 8\left(\frac{f}{a}\right)^2\right]$$

The similar contribution of one side span follows from

$$\delta_4 = \int_0^{L_1} \frac{u_1^2\, ds}{A} = \frac{1}{A}\int_0^{L_1} \frac{ds^3}{dx^2} = \frac{1}{A}\int_0^{a_1}\left[1 + \left(\frac{dy}{dx}\right)^2\right]^{3/2} dx$$

From the equation for inclined cable, using h to represent the slant rise of the cable (see equation 11-12 and Fig. 8),

$$y = \frac{x_1 h}{a_1} - \frac{4f_1 x_1}{a_1^2}(a_1 - x_1)$$

and

$$\frac{dy}{dx_1} = \frac{h}{a_1} - \frac{4f_1}{a_1^2}(a_1 - 2x_1)$$

Then by substitution

$$\delta_4 = \frac{1}{A}\int_0^{a_1}\left\{1 + \left[\frac{h}{a_1} - \frac{4f_1}{a_1^2}(a_1 - 2x_1)\right]^2\right\}^{3/2} dx_1$$

By two terms of a binomial expansion the equation is restated as

$$\delta_4 = \frac{1}{A}\int_0^{a_1}\left\{1 + \frac{3}{2}\left[\frac{h}{a_1} - \frac{4f_1}{a_1^2}(a_1 - 2x_1)\right]^2\right\} dx_1$$

which simplifies to

$$\delta_4 = \frac{a_1}{A}\left[1 + \frac{3}{2}\left(\frac{h}{a_1}\right)^2 + 8\left(\frac{f_1}{a_1}\right)^2\right]$$

For equal side spans and live loads on all spans, equation 11-23 becomes

$$H = \frac{\displaystyle\int_0^a \frac{M_s y\, dx}{I} + \int_0^{a_1} \frac{M_s y_1\, dx}{I_1} + \int_0^{a_2} \frac{M_s y_2\, dx}{I_1}}{\delta_1 + \delta_2 + 2\delta_3 + 2\delta_4} \tag{11-24}$$

The denominator of equation 11-24, when divided by E, represents the movement of cut ends in the cable at the center of main span due to the deformation of all structural elements under the influence of $H = 1$ lb. The sum of the denominator terms will henceforth be termed C. In practical application a term must be added to represent the effect of the deformations of cable extending into both anchorages. This additional term is

$$\delta_5 = 2\int_0^{L_a} \frac{u^2\, ds}{A}$$

where L_a equals length of the cable beyond the parabolic curve of side span at one anchorage. If this section of cable is relatively straight and approaching a horizontal orientation, then the evaluation of the integral, with but small error, is

$$\delta_5 = \frac{2L_a}{A}$$

Maximum cable tension will occur under dead load plus live load over all spans. H for a dead load of w per unit of length is computed by $wa^2/8f$. To calculate H for a uniform live load of q per unit of length over all spans will require the numerator of equation 11-24 for H, by reference to previous evaluations, to be

$$q\frac{fa^3}{15I} + 2\left(\frac{qf_1a_1{}^3}{15I_1}\right)$$

The separate values of H for dead load and live load may be added for total effect.

11–10. INFLUENCE LINES

Influence lines for H, which will in turn lead to influence lines for bending moment in the stiffening girder, may now be developed.

A unit vertical load is placed on the main span girder in determinate condition in Fig. 9a. The M_s moment diagram is illustrated. The

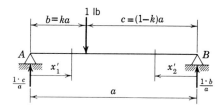

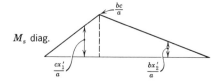

FIG. 11–9

following integration is performed employing two origins, one at A and one at B.

$$\int_0^a \frac{M_s y\, dx}{I} = \frac{1}{I} \int_0^b \frac{cx}{a} y\, dx + \frac{1}{I} \int_0^c \frac{bx}{a} y\, dx$$

$$= \frac{1}{aI} \left[c \int_0^b xy\, dx + b \int_0^c xy\, dx \right]$$

The integral
$$\int xy\, dx = \frac{4f}{a^2} \int (x)(ax - x^2)\, dx$$

$$= \frac{4f}{a^2} \left[\frac{ax^3}{3} - \frac{x^4}{4} \right]$$

Then
$$\int_0^a \frac{M_s y\, dx}{I} = \frac{4f}{a^3 I} \left\{ c \left[\frac{ax^3}{3} - \frac{x^4}{4} \right]_0^b + b \left[\frac{ax^3}{3} - \frac{x^4}{4} \right]_0^c \right\}$$

$$= \frac{fcb}{3a^3 I} [ab^2 + 3abc + ac^2]$$

and if $b = ka$, and $c = (1 - k)a$, we obtain

$$\int_0^a \frac{M_s y\, dx}{I} = \frac{fa^2 k}{3I} (1 - 2k^2 + k^3)$$

Then H, for the one unit vertical load in the main span, by substitution in general equation 11-24 for H where C is the sum of terms in the denominator of equation 11-24, is

$$H = \frac{fa^2 k}{3IC} (1 - 2k^2 + k^3) \qquad (11\text{-}25)$$

and similarly for one unit load in a side span at $b_1 = k_1 a_1$ from anchorage,

$$H = \frac{f_1 a_1^2 k_1}{3I_1 C} (1 - 2k_1^2 + k_1^3) \qquad (11\text{-}26)$$

The influence line for H is diagrammed in Fig. 10b, and can be quantitatively determined by equations 11-25 and 11-26. The maximum live load value for H for calculating cable tension occurs when all spans are loaded. The ordinates to the influence line for the bending moment in the stiffening girder may be generalized from

$$M = M_s - Hy = \left(\frac{M_s}{y} - H \right) y$$

where $-Hy$ represents the effect of hangers on moments.

At the point of application of the 1-lb load

$$M_s = \frac{bc}{a} \quad \text{and} \quad y = \frac{4fb}{a^2} (a - b) = \frac{4f\,bc}{a^2}$$

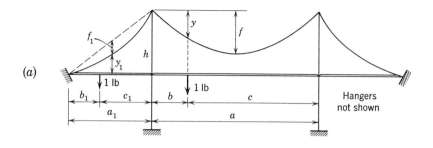

(a)

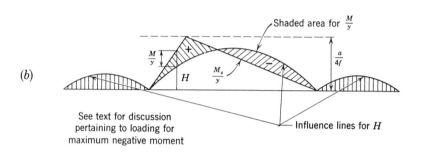

(b)

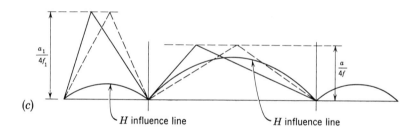

(c)

Fig. 11–10

From the latter expression $bc/a = ay/4f$. Then

$$M_s = \frac{ay}{4f}, \qquad \text{resulting in} \qquad \frac{M_s}{y} = \frac{a}{4f}$$

Thus, M_s/y is equal at every point of application of the unit load in main span to $a/4f$.

This relationship is of great value in construction of influence lines. Furthermore, as an aid to influence line construction, $M = \left(\dfrac{M_s}{y} - H\right)y$

can be restated as $\dfrac{M}{y} = \dfrac{M_s}{y} - H$. Then influence lines may be constructed in a graphical manner as will be illustrated.

The components of the influence line for moment at a given section are shown in Fig. 10b. In Fig. 10c, it is demonstrated that the influence line for moment at any section, after construction of the influence lines for H, is given by a series of simple constructions wherein the altitude of the M_s/y triangle remains constant. It should be noted that the ordinates become true influence line values upon multiplication by the scale factor, which is equal to the y coordinate of cable at the section to which the moment refers. If dealing with areas, scaled from Fig. 10b or 10c, the total areas are to be multiplied by the y coordinate at the loaded point. The main span is loaded by live load over the length indicated by the positive zone of the influence line for maximum positive live load moment. However, for maximum live load negative moment in the main span stiffening girder, the side spans must be fully loaded with live load in addition to the live loads placed over the negative zone of the influence line. The requirement of loading the side spans is based on maximizing H since H contributes a negative effect on the bending moment.

The influence line for shear in the stiffening girder in the main span may be derived from vertical equilibrium conditions. At a general section of the main span the shear is

$$V = V_s - H \tan \theta$$

where
$$V_s = \text{shear in statically determinate } base\ structure$$
$$H = \text{the horizontal component of cable tension}$$
$$\tan \theta = \text{slope of the main span cable at section}$$
$$H \tan \theta = \text{the vertical component of cable tension}$$

From previous development of the parabolic cable, equations 11-2 and 11-4,

$$\tan \theta = \frac{4f}{a^2}(a - 2x) = \frac{4f}{a^2}(c - b)$$

The equation for V may then be restated as

$$V = \left(\frac{V_s}{\tan \theta} - H\right) \tan \theta$$

Figure 11b depicts the graphical construction of the influence line for shear at a given section. The shaded ordinates must be multiplied by $\tan \theta$. The influence line for shear in side spans is constructed by the same principles except that the slope of cable is now $\tan \theta_1$, which equals $(h/a_1) - (4f_1/a_1^2)(a_1 - 2x_1)$. The scaled ordinates must be multiplied by

$\tan \theta_1$.   To determine the maximum values of negative shear in either the main span or a side span, the contribution of H to the value of negative shear must be recognized. H must be maximized by loading the span which contains the given girder section according to the influence line criterion and by also fully loading all other spans.

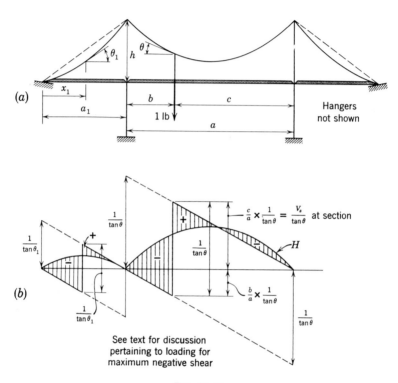

FIG. 11–11

11–11. TEMPERATURE EFFECTS

A change in temperature of ΔT degrees affects the value of H. An increase in temperature will cause a closure of the cut ends of the cable at midspan of main span, resulting in a compressive correction of ΔH to H from other causes. This relief of H will produce a positive moment correction to stiffening girder of $\Delta H \cdot y$. A decrease in temperature will produce opposite effects.

The translations of the cut ends of cable at midspan of the main span for effect of temperature follow from virtual work principles. With C_x

equal to coefficient of expansion, a ds length of cable changes its length by $C_x \, \varDelta T \, ds$, then for the center span

$$\delta_T = \int_0^L (C_x \, \varDelta T \, ds)u = C_x \, \varDelta T \int_0^L u \, ds$$

$$\delta_T = C_x \, \varDelta T \int_0^L \frac{ds}{dx} \cdot ds$$

which, after integrations similar to those used in deriving equation 11-17, evaluates for the center span to

$$\delta_T = C_x \, \varDelta T \left[1 + \frac{16}{3} \left(\frac{f}{a} \right)^2 \right] a$$

The similar effect for a temperature change in a side span (see equation 11-17) is

$$\delta'_T = C_x \, \varDelta T \left[1 + \left(\frac{h}{a_1} \right)^2 + \frac{16}{3} \left(\frac{f_1}{a_1} \right)^2 \right] a_1$$

The temperature effect alone on the horizontal cable component is

$$\varDelta H = -E \frac{(\delta_T + 2\delta'_T)}{{}^{\text{``}} C {}^{\text{''}}} \tag{11-27}$$

The temperature influence on moment and shear in the stiffening girder may then be computed following previously explained techniques.

Example

A suspension bridge employing two cables carries pedestrian traffic across a river. There are many engineering problems associated with a structure of this nature, such as adequate bracing of stiffening system against lateral wind forces, aerodynamic effects, vibrations, and total economy. However, this example problem is associated only with the primary problem of the cable and stiffening system under vertical loads. The data is as follows (see Fig. 12a):

$$a = 300 \text{ ft}; \qquad a_1 = a_2 = 100 \text{ ft}$$
$$f = 30 \text{ ft}; \qquad f_1 = f_2 = 3.33 \text{ ft}$$

The loads for one cable are estimated as follows:

	Dead Load lb per ft	Live Load lb per ft
Cable and hangers	30	
Walkway and fittings	100	150
Stiffening girder	110	
	240	

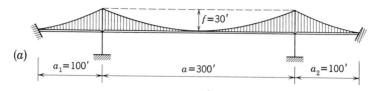

(a)

$a_1 = 100'$ $a = 300'$ $a_2 = 100'$

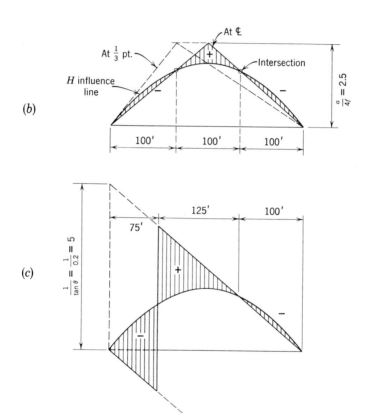

(b)

(c)

FIG. 11–12

If in a zone of severe winters, an additional, and often very significant, ice live load should also be included as a live load.

The dead load effect, the condition after erection of cable, hangers, stiffening girder, and walkway, may be computed directly.

For dead load only,

$$H = \frac{wa^2}{8f} = \frac{240 \times 300^2}{8 \times 30} = 90,000 \text{ lb}$$

For the live load an approximate value for H must first be obtained since the size of cable and stiffening girder is unknown. If δ_2 and $2\delta_4$ are dropped from the denominator of equation 11-24, the remainder of the equation contains only the terms determined from considering deformations of the stiffening girder. Furthermore, assuming that I of girder is the same throughout bridge, equation 11-24 reduces to

$$H = \frac{q(fa^3 + 2f_1a_1{}^3)}{8(f^2a + 2f_1{}^2a_1)} \tag{11-28}$$

An approximate value for H for live load is, by the above equation,

$$H = \frac{150[30(300)^3 + 2 \times 3.33(100)^3]}{8[30^2(300) + 2(3.33)^2(100)]}$$

$$H = 56,000 \text{ lb}$$

The answer may be checked in this instance by recognizing that, owing to the modified form of equation 11-24, the cable has been assumed to be uniformly loaded through the hangers at 150 lb per ft. Hence, H for live load could have been found by multiplying the dead load H by the ratio of the uniform live load to the uniform dead load, or,

$$H = \frac{150}{240} \times 90,000 = 56,000 \text{ lb}$$

The first approximation for the maximum combined value of H may now be found

$$
\begin{aligned}
\text{D.L.} &= 90,000 \text{ lb} \\
\text{L.L.} &= 56,000 \text{ lb} \\
\hline
&146,000 \text{ lb}
\end{aligned}
$$

The maximum tension in cable occurs at the towers and may be simply found as the resultant of the vertical and horizontal cable components, or

$$T \max = \sqrt{146,000^2 + (390 \times 150)^2}$$

$$T \max = 157,000 \text{ lb}$$

This approximate value will be on the high side since the terms dropped in the denominator of equation 11-24 increase the denominator. With the approximate T max equal to 157,000 lb, or 78.5 tons, and an assumed safety factor of 5, the minimum required breaking strength for the cable is 392.5 tons. Although final cable selection and arrangement is a matter of economics, the cable selected is a single 3-in. dia. galvanized plow steel bridge rope, having an approximate weight of 15 lb per ft and a stated breaking strength of 412 tons by the manufacturer. The effective modulus of elasticity of this cable is approximately 20×10^6 psi. The metallic cross-sectional area is 4.25 in.^{2}.

Since E for the stiffening girder is 30×10^6 psi and E for the cable is 20×10^6 psi, equation 11-24 must be modified before a refinement in the live load H can be computed. It is also desirable to clarify the units or dimensions of moment of inertia and area. A choice of units always exists at the option of the analyst. In all integrals involving I and E, where the dimensions of a, a_1, f, f_1, and y are taken in feet and the units of moment are in foot-pounds, I must be in foot^4 units and E must be in pounds per square foot units. For the terms involving the area of the cable, A must be expressed in foot^2 units, and E in pounds per square foot units. With these units and dimensions in mind, the E terms of the numerator and denominator were canceled in equation 11-24 when the E of cable and stiffening girder were identical. But in the present case if E of stiffening girder is identified as E_g and the effective E of the cable is noted as E_c, then the terms δ_2 and $2\delta_3$ must be modified by the ratio of E_g/E_c. On that basis equation 11-24 becomes

$$H = \frac{\int_0^a M_s y \, dx + 2 \int_0^{a_1} M_s y_1 \, dx}{\delta_1 + 2\delta_3 + \dfrac{I \cdot E_g}{E_c}(\delta_2 + 2\delta_4)}$$

or, with all known quantities substituted, we obtain

$$H = \frac{\dfrac{qfa^3}{15} + 2\dfrac{(qf_1 a_1^3)}{15}}{\dfrac{8f^2 a}{15} + 2\dfrac{8f_1^2 a_1}{15} + \dfrac{I \cdot E_g}{E_c}(\delta_2 + 2\delta_4)} \qquad (11\text{-}29)$$

The various terms of equation 11-29 are evaluated as follows.

$$\frac{fa^3}{15} = \frac{30 \times 300^3}{15} = 54 \times 10^6$$

$$\frac{f_1 a_1^3}{15} = \frac{3.33 \times 100^3}{15} = 0.22 \times 10^6$$

$$\frac{8}{15}f^2 a = \frac{8 \times 30^2 \times 300}{15} = 144 \times 10^3$$

$$\frac{8}{15}f_1^2 a_1 = \frac{8 \times 3.33^2 \times 100}{15} = 0.6 \times 10^3$$

$$\delta_2 = \frac{a}{A}\left[1 + 8\left(\frac{f}{a}\right)^2\right]$$

$$\delta_2 = \frac{300}{A}(1 + 8(0.1)^2) = \frac{324}{A}$$

$$\delta_4 = \frac{100}{A}\left[1 + \frac{3}{2}\left(\frac{30}{100}\right)^2 + 8\left(\frac{3.33}{100}\right)^2\right] = \frac{114}{A}$$

$$\frac{E_g}{E_c} = 1.5$$

$$A = \frac{4.25}{144} = 0.026 \text{ ft}^2$$

For purposes of trial calculation, assume the stiffening girder to be a 24 W 100 section, erected with the web oriented vertically.

$$I = \frac{2987.3 \text{ in.}^4}{124 \dfrac{\text{in.}^4}{\text{ft}^4}} = 0.143 \text{ ft}^4$$

All quantities are now available for substitution into the modified equation 11-29, hence,

$$H = \frac{150 \times 10^6[54 + 0.44]}{(144 + 1.2)10^3 + \dfrac{0.143}{0.026}(1.5)(324 + 114)}$$

$$H = \frac{150 \times 54.44 \times 10^6}{145,200 + 3600} = 54,800 \text{ lb}$$

The value of H is only slightly less than that computed neglecting the effect of δ_2 and $2\delta_4$, hence design of cable will not be modified in this case. The denominator is identified as IC equal to 148,800 for future use.

The maximum bending moment in the stiffening girder, due to live load, will occur under a partial span loading.   The influence line of Fig. 12b for the center line of the span illustrates the general problem.

To develop the influence line, the influence line for H is plotted using equation 11-25, which is

$$H = \frac{30(300^2)}{3IC}k(1 - 2k^2 + k^3)$$

and since $IC = 148,800$

$$H = 6.05k(1 - 2k^2 + k^3)$$

The ordinate $a/4f = 300/4 \times 30 = 2.5$.

The intersection of the two curves indicates that the center 100 ft. of span is to be loaded with live load to produce maximum positive moment. The two outer 100-ft segments of the center span and the two side spans must be loaded to produce maximum negative moment in stiffening girder at the center line. For greater precision the intersection points may also be found analytically, but greater precision is seldom warranted.

Equation 11-25 represents the effect on H of a generally placed unity load. To obtain the effect of a partial span loading of uniform load, substitute a differential load $w\,dx$ for the unity load and integrate equation 11-25 over the limits of the load. Let $r = ia$ equal the distance from the left-hand tower to left end of the uniform load, and $s = ja$ equal the distance of the right end of uniform load from the left-hand tower. The total length of the uniform load is $a(j - i)$. With limits established in this manner, any and all conditions of uniform loading in the center span may be examined.

Then equation 11-24 restated, with $k = x/a$, is

$$H = \frac{qfa^2}{3IC} \int_r^s \frac{x}{a}\left(1 - 2\frac{x^2}{a^2} + \frac{x^3}{a^3}\right)dx$$

which integrates and simplifies to

$$H = \frac{qf}{60ICa^2}\left[10a^3(s^2 - r^2) - 10a(s^4 - r^4) + 4(s^5 - r^5)\right]$$

and with $s = ja$ and $r = ia$,

$$H = \frac{qfa^3}{30IC}\left[5(j^2 - i^2) - 5(j^4 - i^4) + 2(j^5 - i^5)\right] \tag{11-30}$$

Equation 11-30 may be easily checked for a full span loading by letting $i = 0$ and $j = 1$, or

$$H = \frac{qfa^3}{15IC}$$

which was found previously in Art. 11-8.

For live load H, relevant to the calculation of positive moment at the center line, use $j = \frac{2}{3}$ and $i = \frac{1}{3}$, then,

$$H = \frac{150 \times 30 \times 300^3}{30 \times 148,800}\left[\frac{5}{9}(4 - 1) - \frac{5}{81}(16 - 1) + \frac{2}{243}(32 - 1)\right]$$

$$H = 27,800 \text{ lb}$$

M_s, the simple span bending moment at the center line for the partial span loading, is

$$M_s = \frac{150 \times 100 \text{ ft}}{2} \times 150 \text{ ft} - \frac{150 \times 50^2}{2}$$

$$M_s = 1,125,000 - 187,500 = 937,500 \text{ ft-lb}$$

The maximum positive moment at the center line is

$$M = M_s - Hy$$
$$M = 937,500 - 27,800 \times 30 = +103,500 \text{ ft-lb}$$

A repetition of the above procedure for maximum negative moment at the center line of the main span follows.

With $i = 0$, and $j = \frac{1}{3}$, in equation 11-30, and including the effect of both loaded segments,

$$H = 2 \times \frac{150 \times 30 \times 300^3}{30 \times 148,800} \left[\frac{5}{9} - \frac{5}{81} + \frac{2}{243} \right] = 27,200 \text{ lb}$$

The live load on the two side spans contributes to H. This contribution is $2(qf_1a_1{}^3)/8(f^2a + 2f_1{}^2a_1)$ and equals 470 lb.

$$M_s = 150 \times 100 \times 50 = +750,000 \text{ ft-lb}$$
$$M = M_s - Hy = 750,000 - 27,670 \times 30 = -80,100 \text{ ft-lb}$$

A repetition of the foregoing will yield data for computation at other sections of the center span. For example, at the one-third point of span the load divide points, Fig. 12b, indicate that the left half of the center span is to be loaded for maximum positive moment, and the right half for maximum negative moment. It may be recalled that the side spans must also be loaded when considering negative moment. The relevant value of y at the one-third point may be computed from the properties of the parabolic curve as

$$y = 30 - (\tfrac{1}{3})^2 \times 30 = 26.67 \text{ ft}$$

The maximum live load shear in the stiffening girder will be investigated at two sections. First at the center line of the main span, and secondly at the one-quarter point of the main span. The maximum shear calculation at the center line is simplified by the fact that $\theta = 0$, and accordingly the cable contributes no vertical resistance to shear. Hence $V = V_s$ where V_s is the shear in statically determinate condition.

For maximum shear at the center line, load one-half of the span as would be done for any simple span.

$$V = V_s = \frac{1}{8} qa = \frac{150 \times 300}{8} = 5600 \text{ lb}$$

To determine the loading conditions providing maximum shear at the one-quarter point, first construct the specific influence line for shear, Fig. 12c. This follows the principles previously discussed and illustrated in Fig. 11b.

The tangent of the slope of cable at the one-third point is

$$\tan \theta = \frac{4 \times 30}{300 \times 300} (225 - 75) = 0.2$$

Figure 12c indicates, for maximum positive shear, that a 125-ft long segment of span must be loaded. The relevant live load H of cable will be calculated with $j = \frac{200}{300} = \frac{2}{3}$ and $i = \frac{75}{300} = \frac{1}{4}$, and is

$$H = \frac{150 \times 30 \times 300^3}{30 \times 148,800} [5(0.383) - 5(0.194) + 2(0.131)]$$

$$H = 33,000 \text{ lb}$$

$$V_s = \frac{150 \times 125 \times 162.5}{300} = 10,300 \text{ lb}$$

$$V = 10,300 - 33,000 \times \tan \theta = +3700 \text{ lb}$$

The student can easily determine the maximum negative shears at the same section by an application of the same principles. Since live loads on the side spans also increase the horizontal component of the cable, the side spans must also be fully loaded with live load. Thus, for maximum negative shear, load the side spans as well as the 75-ft long segment at the left and the 100-ft long segment on the right of the center span to determine the relevant maximum value of H. It is to be recalled that the vertical component of H is $H \tan \theta$ and that it acts downward on the left-hand side of section, thus adding to the negative shear effects.

11–12. DEFLECTION THEORY—DISCUSSION

The elastic theory neglected the effect of changes in the geometry of structure on calculated forces and moments. The vertical deflections of the suspended system changes y and all integrals involving y are more complex mathematically. A presentation of the deflection theory incorporating changes in geometry is outside the scope of this text.

11–13. PRESTRESSED CONCRETE BEAMS—DEFLECTION

The cable is an important component of prestressed concrete beams. Although the complete theory of prestressed concrete cannot be presented in this book the deflections of beams can be considered.

Figure 13a represents a simple span with the stressed cable following a parabolic configuration. The initial prestress in cable is 150 kips. Owing to the parabolic configuration, the cable must exert an upward uniform force per foot of length on the beam, or it is equally correct to say that the concrete beam exerts a downward force of the same intensity on the cable. Since the sag of the cable is small, H, the horizontal cable force component, is essentially equal to the initial prestress. The magnitude of the upward distributed force on beam may be computed from equation 11-2, where $H = 150$ kips, $f = 8$ in. $= \frac{2}{3}$ ft, and $a = 30$ ft. Then

$$w = \frac{8fH}{a^2} = \frac{8(\frac{2}{3})(150)}{30^2} = 0.89 \text{ kips per ft}$$

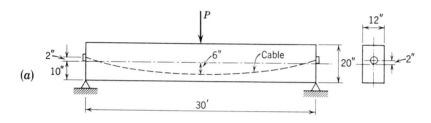

(a)

(b)

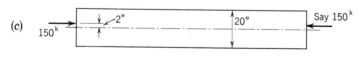

(c)

Fɪɢ. 11–13

Since the dead weight of the beam is 0.25 kips per ft, the net upward force on beam is $(0.89 - 0.25) = 0.64$ kips per ft.

The upward deflection at the center line of beam may be computed from the standard equation

$$\Delta_1 = \frac{5wL^4}{384EI}$$

where $w = 0.64$ kips per ft and I is generally taken as the moment of inertia of the concrete cross-section. Δ_1, however, is not the final deflection of the beam. The eccentricity of the cable at ends produces a uniform moment of M throughout the span equal to H times the end eccentricity measured from the midplane of beam. This moment produces a downward deflection of Δ_2. Δ_2 by moment-area principles is equal to $ML^2/8EI$. The final erected deflection or camber is the difference of Δ_1 and Δ_2. The deflection of the beam due to superimposed loads may then be computed by normal methods and combined with the erected deflection to determine final deflections under working conditions.

There are many additional complications to consider from a practical viewpoint, such as loss of prestress due to shrinkage and plastic flow of the concrete, which are outside the scope of this book.

REFERENCES

1. Johnson, J. B., C. W. Bryan, and F. E. Turneaure, *The Theory and Practice of Modern Framed Structures*, Part II, tenth edition, Chapter 5, New York: John Wiley, 1929.
2. Lin, T. Y., *Design of Prestressed Concrete Structures*, New York: John Wiley, 1955.
3. Morris, C. T., and S. T. Carpenter, *Structural Frameworks*, Chapter 9, New York: John Wiley, 1943.
4. Parcel, J. I., and R. B. B. Moorman, *Analysis of Indeterminate Structures*, Chapter 11, New York: John Wiley, 1955.
5. Steinman, D. B., "A Generalized Deflection Theory for Suspension Bridges," *Trans. Am. Soc. Civil Engineers*, **100**, 1133 (1935).

PROBLEMS

11–1. A cable supported from towers 400 ft apart has a sag of 30 ft. Compute the maximum tension in the cable if the weight per foot of horizontal projection is taken as 2 lb. *Ans.* 1390 lb.

11–2. Calculate the maximum tension for problem 11–1 if the cable weight is 2 lb per lineal foot of cable. (Determine length of the cable and derive the equivalent average weight per foot of horizontal projection.)

11–3. A cable is suspended between two points at the same level. The length of the cable is 300 ft and the horizontal distance between supports is 290 ft. Determine the sag of the cable. *Ans.* 33.5 ft.

11–4. A cable whose weight may be considered to be uniform at 4 lb per ft of horizontal projection is suspended between two points at the same level. The span is 240 ft and the initial sag is 30 ft. (*a*) If it is desired to change the sag to 20 ft, how much should the cable be shortened? (*b*) If it is desired to change the sag of cable to 20 ft, by increasing the span length, what is this increase?

11–5. If a steel cable suspended between fixed points at the same level has a length of 300 ft, and a sag of 30 ft, calculate the new sag if the temperature should increase by 100° F. *Ans.* 30.3 ft.

11–6. A cable 330 ft long is freely suspended and free to slide over support B. Compute the sags in the two spans without adjustment for stretch of cable.

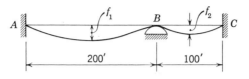

Prob. 11–6

11–7. A cable assumed to have an equivalent weight of 3 lb per ft of horizontal projection hangs as indicated with the sag equal to 70 ft. Determine (*a*) the maximum tension in the cable. (*b*) Locate the low point C by a distance from A and y_C. *Ans.* (*a*) 780 lb; (*b*) 96.5 ft and -28.9 ft.

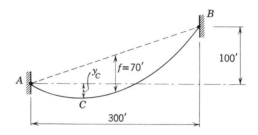

Prob. 11–7

11–8. Compute the necessary magnitude of the counterweight W to make the cable conform to the geometry indicated. The support at A is a small frictionless pulley and the cable weighs 2 lb per ft of horizontal projection.

Ans. $W = 252$ lb.

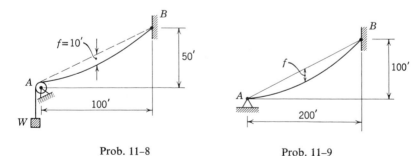

Prob. 11–8   Prob. 11–9

11–9. Calculate the sag and the geometric length of the cable if erected with an initial H of 400 lb. The cable may be taken as weighing 3 lb per ft on the horizontal projection. *Ans.* $f = 37.5$ ft.

11–10. A 1-in. diameter galvanized steel bridge rope has the following properties:

> Gross metallic area = 0.47 in.^{2}
> Approximate weight = 1.67 lb per ft
> Minimum breaking strength = 45.7 tons
> Minimum modulus of elasticity = 20 × 10^6 psi.

This rope is to be used as a guy (refer to sketch) and will be erected with a maximum pretensioning stress equal to 20 per cent of the breaking strength. During an ice storm the cable ice load becomes 1 lb per lineal foot of cable. Considering that the ends of the cable are fixed in position, calculate the maximum tensions in the cable and also compute the total sag of the cable.

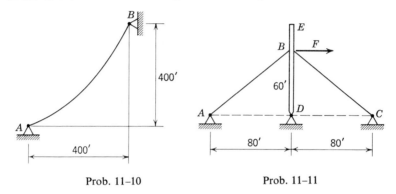

Prob. 11–10   Prob. 11–11

11–11. Assume that vertical tower *DBE* is guyed with only two steel wires (practically unrealistic) *AB* and *BC*. The initial tension in wires equals 2000 lb before the introduction of the external force *F* at *B*. Let *F* equal 3000 lb, and compute the adjusted maximum stress in each cable. The area of the wire is 0.30 in.^{2} and the modulus of elasticity may be used as 30 × 10^6 psi.

11–12. For the guyed tower of Fig. 11–4 in text, assume that the wind loading on the tower is increased to 80 lb per ft. Using the cable data given in text example, recalculate the maximum tensions in each guy.

11–13. A footbridge with main and side spans as shown supports a 6-ft wide walkway. The dead load, excluding the cable, is 60 lb per sq ft of walkway, or 360 lb per ft of span. The live load on one cable is 300 lb per horizontal foot. The cable has a metallic area of 2.90 in.^{2} and weighs approximately 10 lb per ft. The stiffening girders are 15-in. *I* 42.9 and have a moment of inertia of 441.8 in.^{4}

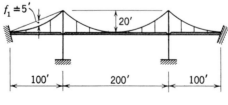

Prob. 11–13

or 0.021 ft⁴. The hangers are steel rods spaced about 8 ft on centers. The breaking strength of the cable is 290 tons. $E_c = 20 \times 10^6$ psi. (a) Calculate the maximum tension in cable. (b) Calculate the maximum positive and negative bending moments in the stiffening girder, for live loading, at the center line of the center span.

11–14. For the suspended footbridge of problem 11–13, assume that once a year, when the local college holds its annual regatta, the live loading on the bridge will be 900 lb per ft over the center half of the main span and 300 lb per ft over the remainder of bridge per cable. (a) Calculate the maximum live load tension in cable. (b) Compute the moment in the stiffening girder at the center line of the center span.

11–15. A suspension bridge is to have a center span of 600 ft, and two side spans of 200 ft each. The dead load, including the weight of the cable, is 2000 lb per ft of bridge per cable. The live load is 1400 lb per ft of bridge per cable. The stiffening girder is of the same stiffness throughout. If the sag in the main span is 70 ft, calculate the maximum tension in cable neglecting contributions of cable deformations to denominator of equation 11–24.

11–16. For the suspended footbridge of problem 11–13, develop the following influence lines: (a) for H; (b) for moment in the stiffening girder at midspan of the center span; (c) for shear at the one-quarter point of the center span.

11–17. A prestressed concrete beam with the initial cable tension equal to 120 kips is as shown. Compute the deflection at midspan after erection. Assume $E_c = 4 \times 10^6$ psi.

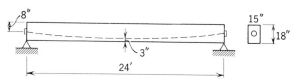

Prob. 11–17

Structural Dynamics

12

12–1. INTRODUCTION

Previously in this book all external loads have been assumed to be slowly or statically applied to the structure and the static equilibrium of structure was considered. However, rapidly applied loads with load varying with time, shock loadings, and induced motion due to earthquakes or other disturbances establish vibrations in elastic structures. The key to analysis lies in describing the motion and behavior of the structure with respect to disturbing forces, time, and structural parameters. $F = ma$ forms the basis of analysis, which is a statement of Newton's second law relating the force acting in the direction of the acceleration to the product of mass times acceleration.

12–2. SIMPLE HARMONIC MOTION

In Fig. 1a, a weight W on rollers is supported on a smooth plane and attached to a rigid support by a spring having a spring factor k. In Fig. 1b the weight is displaced a positive distance x_O by applying an external force P. If the force P is rapidly released, the weight starts moving to the left. Figure 1c is a free body of the weight and spring force at a time t after release.

A clear notion of signs is essential. For a positive x the spring force kx acts to the left in the negative direction of x, hence $F = -kx$. Then a, the acceleration, would also be in the negative direction. The equation of motion, where $m = W/g$, and g equals the acceleration due to gravity, is

$$-kx = ma$$

or

$$a = -\frac{kx}{m}$$

430

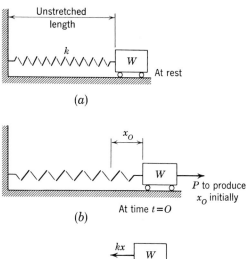

FIG. 12–1

From elementary kinematics, with displacement denoted by x, velocity was mathematically expressed as $\dfrac{dx}{dt}$, and acceleration expressed as $\dfrac{d^2x}{dt^2}$. In standard notation velocity $\left(\dfrac{dx}{dt}\right)$ is written in shorthand form as $\dot{x}$, and acceleration $\left(\dfrac{d^2x}{dt^2}\right)$ as $\ddot{x}$. With this notation the relationship between force and acceleration becomes the differential equation of motion written as follows:

$$-kx = m\ddot{x}$$
$$m\ddot{x} + kx = 0$$
$$\ddot{x} + \frac{k}{m}x = 0$$

where if
$$p^2 = \frac{k}{m} \quad \text{or} \quad \frac{kg}{W}$$

the following basic differential equation results

$$\ddot{x} + p^2x = 0 \tag{12-1}$$

This differential equation is basic for all single degree of freedom problems vibrating as a free and undamped simple system. In Fig. 1 the weight W will vibrate back and forth on a periodic basis identified as simple harmonic motion. A solution of equation 12-1 is obtained by letting $x = C \cos pt$. Equation 12-1 may be seen to be satisfied by substituting $x = C \cos pt$ and $\ddot{x} = -p^2 C \cos pt$ so that

$$-p^2 C \cos pt + p^2 \cdot C \cos pt = 0$$

Hence, $x = C \cos pt$ may be accepted as a solution of equation 12-1, where, if $t = 0$, the constant $C = x_O$, where x_O is the initial starting displacement equal to the amplitude of vibration. Therefore, the fully solved equation of motion is

$$x = x_O \cos pt \qquad (12\text{-}2)$$

where p is the circular frequency in radians per unit of time.

Since the sine and cosine are periodic functions the motion repeats for every angular interval of 2π radians. Inasmuch as motion is time-dependent, an inquiry may be made as to the time period for one complete oscillation by establishing that the angle pt changes by 2π as time changes by τ. This condition may be expressed as

$$p(\tau + t) - pt = 2\pi$$

from which
$$\tau = \frac{2\pi}{p} \qquad (12\text{-}3)$$

τ is termed the natural period of vibration where $p = \sqrt{\dfrac{kg}{W}}$ and may be re-expressed as

$$\tau = \frac{2\pi}{\sqrt{\dfrac{kg}{W}}} = 2\pi \sqrt{\frac{W}{kg}} \qquad (12\text{-}4)$$

where W/k can be physically interpreted as the static deflection of the spring if the weight had been supported by a vertical spring. Terming this static deflection Δ_{st}, equation 12-4 becomes

$$\tau = 2\pi \sqrt{\frac{\Delta_{st}}{g}} \qquad (12\text{-}5)$$

The description of the vibration with respect to time is often stated in terms of the frequency of vibration, or number of complete cycles of vibration per unit of time, where the unit of time is usually one second. Calling the frequency f, the frequency equation is

$$f = \frac{1}{\tau} = \frac{1}{2\pi} \sqrt{\frac{kg}{W}} = \frac{1}{2\pi} \sqrt{\frac{g}{\Delta_{st}}} \qquad (12\text{-}6)$$

The harmonic motion of the weight of Fig. 1 can now be depicted graphically as shown in Fig. 2b, where the displacement is the abscissa and time as the ordinate is increasing in value downward. The period is shown as the time for one complete cycle or oscillation.

The characteristics of simple harmonic motion may also be shown graphically by means of a rotating vector where the vector length represents x_0, the magnitude of the starting displacement at $t = 0$, and p represents the angular velocity or circular frequency of this vector about the center of a circle. This is shown in Fig. 2c, and the projection of the vector on the horizontal diameter, the x direction, represents $x = x_0 \cos pt$.

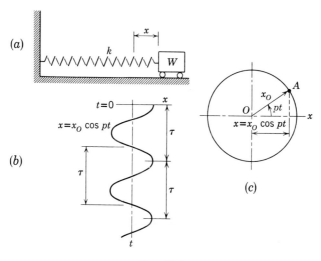

FIG. 12-2

From these basic relations the natural period or natural frequency of vibration for many structural problems may be ascertained.

Examples

If a vertical steel cable 100 in. long has a cross-sectional area of $\frac{1}{2}$ sq in., and supports a weight of 10,000 lb, what is the natural period or natural frequency of the system?

If k is the force required to stretch the cable 1 in.,

$$k = \frac{AE}{L} = \frac{\frac{1}{2} \times 30 \times 10^6}{100} = 1.5 \times 10^5 \text{ lb per in.}$$

Then $\tau = 2\pi \sqrt{\dfrac{10,000}{1.5 \times 10^5 \times 386}}$

$\tau = 0.082$ sec

and $\qquad f = \dfrac{1}{\tau} = 12.2$ cycles per sec (cps)

As a further example of the calculation of natural period and frequency, assume that a vertical cantilever 100 in. long and fixed at the base, supports a concentrated weight of 1000 lb at its upper end. The mass of the vertical pole will be neglected. Figure 3 illustrates the problem wherein the moment of inertia of the vertical cantilever is that of a 6-in. standard steel pipe or $I = 28.14$ in.^{4}. Since vibrations will be horizontal the spring constant k will equal the force required to cause a horizontal deflection Δ of 1 in.

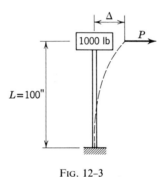

FIG. 12-3

For a cantilever beam

$$\Delta = \frac{PL^3}{3EI}, \text{ and for } \Delta = 1 \text{ in., let } P = k$$

or $\qquad k = \dfrac{3EI}{L^3} = \dfrac{3 \times 30 \times 10^6 \times 28.14}{100^3}$

$$k = 2530 \text{ lb per in.}$$

Then $\qquad \tau = 2\pi \sqrt{\dfrac{1000}{2530 \times 386}}$

$$\tau = 0.20 \text{ sec}$$

and the frequency of free vibrations is

$$f = \frac{1}{0.20} = 5 \text{ cps}$$

As long as the mass of the supporting structure can be neglected, these elementary considerations also apply to other elementary problems. For example, the beam of Fig. 4 is loaded with a concentrated vertical load of W at the center line and the frequency of vertical vibrations is required.

The static deflection due to W may be stated as

$$\Delta_{st} = \frac{WL^3}{48EI}$$

If the beam is put into vibration by causing a movement from this deflected position of static equilibrium and then released for free vibrations, the frequency of vibration will be the frequency of free harmonic vibrations. By equation 12-6

$$f = \frac{1}{2\pi}\sqrt{\frac{g}{\Delta_{st}}}$$

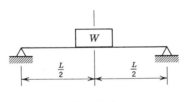

FIG. 12–4

The previous theory and examples refer to vibrating masses, wherein the mass was initially displaced from a position of static equilibrium and then released. Thus the starting conditions were simple, but a general development would assume that at the start of time considerations, or $t = 0$, the mass not only had an initial displacement x_O but also an initial velocity $\dot{x}_O$. The basic differential equation of motion remains the same as equation 12-1 or

$$\ddot{x} + p^2 x = 0$$

This second order homogeneous differential equation has the general solution

$$x = C_1 \cos pt + C_2 \sin pt \qquad (12\text{-}7)$$

in which C_1 and C_2 are constants to be determined from starting conditions.
For the condition $x = x_O$ when $t = 0$, we determine

$$C_1 = x_O$$

Taking the first derivative of equation 12-7 with respect to time

$$\dot{x} = -pC_1 \sin pt + pC_2 \cos pt$$

and when $\qquad t = 0, \quad \dot{x} = \dot{x}_O$

we obtain $\qquad \dot{x}_O = pC_2$

or $\qquad C_2 = \dfrac{\dot{x}_O}{p}$

The equation of motion with determined constants becomes

$$x = x_O \cos pt + \frac{\dot{x}_O}{p} \sin pt \qquad (12\text{-}8)$$

The displacement expressed by equation 12-8 is composed of two parts, the first term representing the vibration which is dependent on the initial displacement and the second term the vibration depending upon the initial velocity. These two terms combine algebraically, and since each

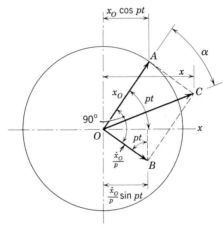

FIG. 12–5

depends on the angle (pt) each separate component will have a period of recurrence equal to τ the natural period. Equation 12-8 may also be thought of as the mathematical combination of two rotating vectors. In Fig. 5, the displacement x is measured horizontally from the center of the circle. The horizontal projection of the vector **OA**, where **OA** is equal to x_O in magnitude, represents $x_O \cdot \cos pt$. The vector **OB** which makes an angle of 90° with **OA** has a magnitude of $\dot{x}_O/p$ and a horizontal projection of $\dfrac{\dot{x}_O}{p} \sin pt$. These two vectors may be combined into the resultant rotating vector **OC** where

$$\mathbf{OC} = \sqrt{x_O{}^2 + \left(\frac{\dot{x}_O}{p}\right)^2}$$

The magnitude of this rotating vector equals the magnitude of the maximum amplitude of vibration since in a periodic fashion this rotating vector will eventually assume a horizontal position either to the left or the right of O. The angle α is termed the phase angle and the displacement may be restated as

$$x = \sqrt{x_O^2 + \left(\frac{\dot{x}_O}{p}\right)^2} \, (\cos{(pt - \alpha)})$$

from which it may be determined that for x_{max}, t equals α/p, $\alpha/p + \tau/2$, $\alpha/p + \tau$, etc.

From Fig. 5 we obtain

$$\tan \alpha = \frac{\dfrac{\dot{x}_O}{p}}{x_O} = \frac{\dot{x}_O}{px_O}$$

Example

As an example involving these relations it is assumed that a small mine car is lowered down an inclined track at a constant velocity of 10 in. per sec, by a steel cable having a cross-sectional area of 2 in.^{2} The track is inclined at an angle of 30°, and the car and contents weigh 20,000 lb. The cable is 500 ft long at the time of this problem and downward motion is suddenly retarded by application of an automatic brake.   Assuming that no friction exists between the track and car the tension in the cable before car is stopped is $T = 10,000$ lb.   The elongation of the cable under this constant tension is $\dfrac{10,000 \times 500 \times 12}{2 \times 30 \times 10^6} = 1.00$ in.   The spring factor k is then 10,000 lb per in.

An inquiry may now be made as to the maximum tension created in the cable by the stopping of the car.   At the time of abrupt braking, taken as $t = 0$, the car's velocity $\dot{x}_O$ was 10 in. per sec, and the initial displacement, or x_O, from the position of static equilibrium is equal to zero.   The circular frequency $p = \sqrt{kg/W}$, or

$$p = \sqrt{\frac{10,000 \times 386}{20,000}} = 13.9 \text{ radians per sec}$$

Since the phase angle equals zero the maximum amplitude of vibration will be

$$x \max = \frac{\dot{x}_O}{p} = \frac{10}{13.9} = 0.718 \text{ in.}$$

This adds to the initial stretch of the cable so that the maximum cable stretch is 1.718 in.   The total tension caused in the cable is increased to

$$10,000\left(\frac{1.718}{1.000}\right) = 17,180 \text{ lb}$$

12–3. HARMONIC VIBRATIONS BY ENERGY ANALYSIS

The principles of conservation of energy may be applied to a vibrating system by an accounting of kinetic and potential energies. The following discussion relates to an ideal system in which there is no loss of energy due to damping, hence, the kinetic energy of the vibrating system plus the potential energy of the system will remain constant during vibration. This is expressed as

$$\text{kinetic} + \text{potential} = \text{constant} \tag{12-9}$$

where kinetic energy of a body is $\dfrac{mv^2}{2}$.

Referring to Fig. 1, the potential energy stored in the spring owing to the initial stretch is $kx_O \cdot x_O/2$ or $kx_O^2/2$.

Before the mass is released or set in vibration the energy is all potential, hence $kx_O^2/2$ will equal the constant on the right-hand side of equation 12-9. Upon release from this initial position the energy equation becomes

$$\frac{m\dot{x}^2}{2} + \frac{kx^2}{2} = \frac{kx_O^2}{2} \tag{12-10}$$

differentiating both sides of equation 12-10 with respect to time where

$$\dot{x}^2 = \left(\frac{dx}{dt}\right)^2$$

we obtain

$$m \cdot \frac{dx}{dt}\left(\frac{d^2x}{dt^2}\right) + kx \cdot \frac{dx}{dt} = 0$$

or

$$m\frac{d^2x}{dt^2} + kx = 0$$

$$\ddot{x} + \frac{k}{m}x = 0$$

$$\ddot{x} + p^2x = 0 \tag{12-11}$$

Since the differential equation 12-11 is the same as equation 12-1, the basic solution is the same as before, showing that the energy viewpoint leads to an identical solution.

12–4. FORCED VIBRATIONS

The structural engineer is confronted with many problems in which forced vibrations occur due to applied disturbing forces or motions. If these disturbing forces or motions follow a periodic pattern the forced vibration is termed the steady state. In this book the transient motion or stage is neglected.

Referring to Fig. 6, let the applied disturbing force Q be represented by $P \sin \omega t$, where ω is the circular frequency in radians per second. The periodic nature of the disturbing force is defined by the period of repetition τ_1 so that $\omega\tau_1 = 2\pi$, or $\tau_1 = 2\pi/\omega$ or its frequency in cycles per second, f_1, is $\omega/2\pi$.

The equation of motion from $F = ma$, where $m = W/g$, is

$$-kx + P \sin \omega t = ma$$

$$ma + kx = P \sin \omega t$$

$$m\ddot{x} + kx = P \sin \omega t$$

simplifying to

$$\ddot{x} + \frac{k}{m}x = \frac{P}{m} \sin \omega t$$

(a)

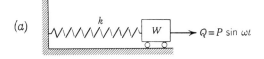

(b)

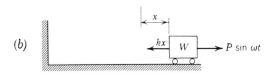

FIG. 12–6

In usual notation, let $p^2 = \dfrac{k}{m}$

and $\qquad q = \dfrac{P}{m}$, where q represents the maximum disturbing

force per unit of mass, then the differential equation of motion, in its standard form, is

$$\ddot{x} + p^2x = q \sin \omega t \qquad (12\text{-}12)$$

Equation 12-12 would be the same as equation 12-1 if the right-hand side was equal to zero. Hence, the solution of equation 12-12 is obtained by combining the general solution of a second order homogeneous equation with a particular solution. Owing to the repetitive nature of $q \sin \omega t$, the particular solution must also be a periodic function. Assume that a particular solution must be in the form $x = C_3 \sin \omega t$, where C_3 is a constant to be determined. Substitution into equation 12-12 gives

$$-\omega^2 C_3 \sin \omega t + p^2 C_3 \sin \omega t = q \sin \omega t$$

$$C_3 = \frac{q}{p^2 - \omega^2}$$

Then the solution of equation 12-12 is

$$x = C_1 \cos pt + C_2 \sin pt + \frac{q \sin \omega t}{p^2 - \omega^2} \tag{12-13}$$

The first two terms represent free harmonic vibrations which have the period $\tau = 2\pi/p$, but the last term represents a periodic forced vibration which has a period of $2\pi/\omega$ equal to the period of the disturbing force Q. For the usual case of vibrations the free vibrations will rapidly damp out and the amplitude of vibration due to the forced vibration prevails. Hence, for our present purposes the steady state of vibration is

$$x = \frac{q \sin \omega t}{p^2 - \omega^2}$$

or
$$x = \frac{P}{mp^2} \frac{\sin \omega t}{(1 - \omega^2/p^2)} = \frac{P}{k} \frac{1}{(1 - \omega^2/p^2)} \sin \omega t \tag{12-14}$$

In equation 12-14, P/k represents the statical deflection of spring if P were applied as a maximum static force. The term $1\big/\left(1 - \dfrac{\omega^2}{p^2}\right)$ modifies this amplitude for the dynamic effect. This factor is usually termed the magnification factor.

The ratio of circular frequencies is

$$\frac{f_1}{f} = \frac{\tau}{\tau_1} = \frac{\omega}{p}$$

Hence, as the ratio of the circular frequencies ω/p varies, the magnification factor varies. For small values of ω/p, or when the natural period is small compared with the period of the disturbing force, the magnification factor is nearly unity. If the two periods are equal, resonance occurs and a vibration of infinite amplitude would occur if it were not for damping.

When the ratio of ω/p exceeds unity, the amplitude of the forced vibration reduces. Figure 7 portrays this variation in magnification graphically. For $\omega/p < 1$ the force and displacement are in phase, but for $\omega/p > 1$ they are 180° out of phase. For maximum effect $\sin \omega t$ in equation 12-14 equals unity.

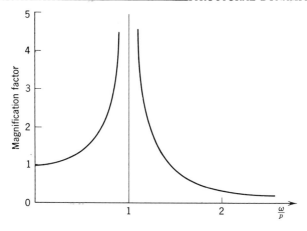

FIG. 12–7

Example

Consider a rotating motor of weight W attached at the center line of a simple beam of negligible mass but of constant section. See Fig. 8. The moment of inertia of the beam cross-section is I. The motor is running at a speed of n revolutions per sec, or at a circular frequency or circular velocity of $\omega = 2\pi n$ radians per sec. The motor is, however, imperfectly balanced and the centrifugal force P is dependent on the amount of unbalance and the radial acceleration. For this particular problem assume that $P = A\omega^2$, where A is the unbalanced force for ω equal to one radian per sec. The unbalanced force delivers a periodic vertical disturbing force to the beam depending on time and ω. This vertical component is $P \sin \omega t$ and the value of the forced vibration at any time t would be computed from equation 12-14 where k is the vertical spring constant of the beam at the center line. The quantity p^2 is associated with the spring constant of the beam and its load W through $p^2 = kg/W$.

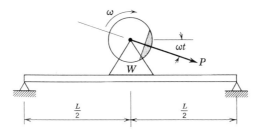

FIG. 12–8

The amplitude of the steady state vibration will be obtained from equation 12-14 as

$$x = \frac{P}{k}\left(\frac{1}{1 - \omega^2/p^2}\right)\sin \omega t$$

where for x max, $\sin \omega t = 1$

$$x \text{ max} = \frac{A\omega^2}{k}\left(\frac{1}{1 - \omega^2/p^2}\right)$$

$$x \text{ max} = \frac{A\omega^2}{k}\left(\frac{p^2}{p^2 - \omega^2}\right) = \frac{Ap^2}{k}\left(\frac{\omega^2}{p^2 - \omega^2}\right)$$

$$x \text{ max} = \frac{Ap^2}{k}\left(\frac{\omega^2/p^2}{1 - \omega^2/p^2}\right)$$

The greater x max the greater is the force imposed on the beam. To minimize x max the ratio of ω/p must become large, causing the second factor of the equation for x max to approach unity. From a structural point of view this may be accomplished by increasing the natural period of vibration τ in order to reduce the circular frequency p.

The effect of the forced vibration is to load the beam dynamically. The dynamic effect is measured by the deflection produced, and in this problem with deflection known, the equivalent dynamic central force on the beam may be determined. When p is small relative to ω, the disturbing force transmitted to the beam is small. The measure of the feasible force reduction may be seen by comparing the maximum dynamic force $P = A\omega^2$, which would apply if the beam were infinitely rigid, to the reduced force $(x \text{ max})k = Ap^2$ if ω/p is made infinitely large. The ratio of the reduced force to P is p^2/ω^2. In mounting machinery on rigid bases, a spring mounting may be employed to achieve a force reduction on the foundation. It is also practicable to use a spring mounting between the beam and the machine, but this will not be dealt with here since the problem would involve a consideration of two degrees of freedom.

Example

A machine weighing 2200 lb is supported on its foundation by a spring mounting. The piston on the machine moves up and down with a harmonic frequency of 10 cps. The piston has a weight of 100 lb, and a total stroke of 20 in. Determine the maximum force transmitted to the foundation if the total k of springs is 24,000 lb per in.

From the given data

$$p^2 = \frac{kg}{W} = \frac{24,000 \times 386}{2200} = 4210$$

and

$$\omega^2 = (10 \times 2\pi)^2 = 400\pi^2$$

Since the piston is undergoing harmonic motion its vertical motion may be represented by

$$y = \tfrac{20}{2} \sin \omega t$$

By two differentiations with respect to time, the acceleration of the piston is

$$\ddot{y} = -10\omega^2 \sin \omega t$$

The maximum acceleration, $\ddot{y}$ max is $\pm 10\omega^2$. The maximum inertia force of piston is

$$F = ma = m\ddot{y} \text{ max}$$
$$= \tfrac{100}{386}(10\omega^2)$$
$$= \tfrac{100}{386} \times 10 \times 400\pi^2 = 10{,}240 \text{ lb}$$

If Q is the periodic disturbing force of $P \sin \omega t$, then with $P = F$

$$Q = 10{,}240 \sin \omega t$$

The vertical displacement of the machine on its supporting springs may be obtained by equation 12-14. The maximum dynamic force transmitted to the foundation through the springs is kx where

$$kx = P \, \frac{1}{1 - \omega^2/p^2}$$
$$= \frac{10{,}240}{1 - \dfrac{400\pi^2}{4210}}$$

Max force transmitted to foundation $= 10{,}240 \ (16.6) = 170{,}000 \text{ lb}$

This large force could have been anticipated by having first computed the ratio of ω/p, which is very close to one. To reduce the dynamic force on the foundation the designer can either add mass or, more practically, decrease the stiffness of the springs to lower the natural frequency. If k is made 12,000 lb per in., then

$$p^2 = \frac{12{,}000 \times 386}{2200} = 2105$$

and the maximum dynamic force is reduced to

$$\frac{10{,}240}{1 - \dfrac{400\pi^2}{2105}} = 11{,}500 \text{ lb}$$

The engineer has many criteria to satisfy in a problem of this nature and the above should be taken as only an example of the basic calculations.

12–5. GROUND MOTION

One of the most important of structural problems is to design and build structures to resist earthquakes. Although this is a subject worthy of extended treatment, suffice it to say for current purposes that a geologic disturbance creates a rapid and periodic displacement of the foundation upon which the structure rests. In attempting to follow the ground motion the structure is set in motion and is subject to forces due to its own mass and acceleration. The most elementary conception of ground motion assumes that there is motion in a horizontal plane only, although it is known that earthquake disturbances may also have a vertical component. The measurement of ground motion due to seismic waves may be described in terms of displacement from the at rest position, period, and acceleration. The instruments used for measuring such characteristics are termed

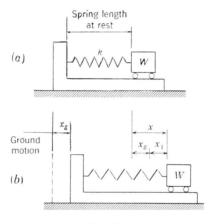

Fig. 12–9

seismometers and their graphic records termed seismographs. These instruments may measure either displacement, velocity, or acceleration. In the following discussion the displacement type of instrument will be theoretically examined.

In Fig. 9a, a weight W is connected to a rigid frame by a spring with a spring factor of k. The frame in turn is supported by the ground and firmly attached thereto. Figure 9a may be termed the static position. Now let the ground, and therefore the base of instrument, be put into horizontal motion. With this ground motion indicated as x_g, the weight will move a distance of x from its original at rest position. The spring, however, changes length by the difference between x and x_g. These displacements are shown in Fig. 9b for a given time t where $x - x_g$ is termed the relative displacement of the spring and is denoted as x_1. The equation of motion, wherein motion is described from the rest position, may now be formulated from $F = ma$.

$$-kx_1 = \frac{W}{g}\ddot{x} = m\ddot{x}$$

$$m\ddot{x} + k(x - x_g) = 0$$

$$\ddot{x} + \frac{k}{m}x = \frac{k}{m}x_g$$

and as before, let $p^2 = k/m$. Then the differential equation of motion becomes

$$\ddot{x} + p^2 x = p^2 x_g \qquad (12\text{-}15)$$

Equation 12-15 is similar to equation 12-12 and has a solution similar to equation 12-13 in the form

$$x = C_1 \cos pt + C_2 \sin pt + \begin{array}{l} \text{a particular solution de-} \\ \text{pendent upon form of ground} \\ \text{motion} \end{array} \qquad (12\text{-}16)$$

If the free vibrations may again be assumed to damp out, and a continuing ground motion assumed upon which the particular solution is based, the motion may be represented as a periodic steady state motion. It may be assumed that the ground motion x_g at a time t may be represented by $x_g = G \sin \omega t$. G is the maximum amplitude of the ground motion and ω is the circular frequency related to the period of the ground motion. A full description of the motion of the weight depends upon its total motion x, and this motion may also be expected to be periodic and dependent upon ω after free vibrations have been damped out. If the weight was at rest at time $t = 0$, its motion may be described by taking $x = A \sin \omega t$ where A is an undetermined constant. This constant may now be evaluated by returning to equation 12-15 and making the proper substitutions as follows:

$$-\omega^2 A \sin \omega t + p^2 A \sin \omega t = p^2 G \sin \omega t$$

solving for A

$$A = \frac{p^2}{p^2 - \omega^2} G$$

or

$$x = \frac{p^2}{p^2 - \omega^2} G \sin \omega t \qquad (12\text{-}17)$$

Equation 12-17 is the equation of motion for a steady and repetitive condition of ground movement with free vibrations damped out.

For a displacement seismograph, a recording drum would be placed on the frame with a stylus attached to the weight. The movement of the stylus relative to the drum would be the relative motion of the weight with respect to the frame. Thus the displacement instrument would record x_1.

With

$$x_1 = x - x_g \text{ and equation 12-17}$$

$$x_1 = x - x_g = \frac{p^2}{p^2 - \omega^2} G \sin \omega t - G \sin \omega t$$

or

$$x_1 = \left[\left(\frac{p^2}{p^2 - \omega^2}\right) - 1\right] G \sin \omega t \qquad (12\text{-}18)$$

Since it is desired to measure the magnitude of the ground motion by such a displacement seismograph, the value of p must be small relative to ω. Since $p^2 = k/m$, k must be small relative to m by construction. Then

$$x_1 = -G \sin \omega t$$

where

$$x_{1(\text{max})} = -G \qquad (12\text{-}19)$$

$x_{1(\text{max})}$ would then be the measured value of ground motion on a recording drum attached to the framework. Theoretically this means that the weight W of Fig. 9b would essentially remain stationary in its static position during the ground motion for the properly constructed displacement type of seismograph.

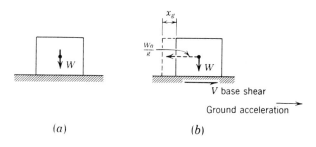

(a)   (b)

FIG. 12–10

The ground motion may be defined by its period τ_1 or by its circular frequency $\omega = 2\pi/\tau_1$ and its amplitude G, or in terms of τ_1 and the acceleration or velocity of the ground movement. These various quantities, as long as a steady state and a simple sinusoidal variation with time are assumed, may be related through derivatives with respect to time of

$$x_g = G \sin \omega t$$

where velocity $\dot{x}_g = \omega G \cos \omega t$, and acceleration $\ddot{x}_g = -\omega^2 G \sin \omega t$.

An example of the effects of ground motion on a structure considered as a rigid block, effectively anchored by its base to the foundation, follows. Figure 10a depicts the rigid block weighing W lb, at rest. Figure 10b

depicts the same block during ground motion. The horizontal ground motion is assumed to be steady and sinusoidal with a stated maximum acceleration of 0.05g where g is the acceleration of gravity. The period of ground motion is also given as 1.5 sec. It is assumed that the acceleration of all parts of the building conform to the ground acceleration. This assumption implies a rigid structure. The maximum inertia force is $(W/g)a$ and if $a = 0.05g$ this lateral force, acting opposite in direction to the acceleration and through the center of mass, equals $0.05W$. This lateral force is resisted by the base shear V. It is usual in discussions of lateral forces due to earthquake disturbances to relate this base shear to the weight of the building by the equation $V = CW$ where C is termed the base shear coefficient. For the example problem $C = 0.05$.

12–6. WATER TANK

The foregoing example of the rigid block lacks a realistic approach to structural problems as they normally exist since most structures possess

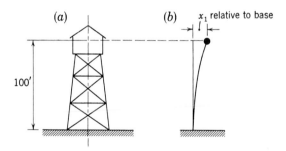

FIG. 12–11

flexibility, with the dynamic response being a result of the coupling of the elastic structure to the ground. To illustrate such a problem, consider the water tank of Fig. 11, which has the bulk of its mass at the centroid of the tank.

The tank weighs 100,000 lb empty and 500,000 lb when filled with water. Since this tank and its tower could be considered as an idealized vertical cantilever the natural period of vibration of the full tank could be computed by methods previously discussed. The horizontal spring factor for the tank and tower is $k = 51,000$ lb per in.

Then τ, the natural period is

$$\tau = 2\pi \sqrt{\frac{W}{kg}} = 2\pi \sqrt{\frac{500,000}{51,000 \times 386}} = 1 \text{ sec}$$

and
$$p = \frac{2\pi}{1} = 6.283 \text{ radians per sec}$$

The steady state ground motion is assumed as having a period of 1.5 sec and a maximum acceleration of 0.05g. From this information the amplitude of maximum horizontal ground motion may be computed from

$$\ddot{x}_g = \omega^2 G \sin \omega t$$

with
$$\omega t = \pi/2 \text{ for maximum conditions}$$

$$G = \frac{\ddot{x}_g}{\omega^2} = \frac{0.05 \times 386}{(2\pi/1.5)^2} = 1.10 \text{ in.}$$

To determine the shear generated at the base of tower by the horizontal inertia force acting through the center of mass, the relative displacement of the top of tower with respect to the base is needed. This relative deflection when multiplied by the spring factor will equal the horizontal shear transmitted to the base.

Equation 12-18 is applicable and is restated

$$x_1 = \left[\left(\frac{p^2}{p^2 - w^2} \right) - 1 \right] G \sin \omega t$$

if
$$x_{1(max)} \text{ is required, then } \sin \omega t = 1$$

and
$$x_{1(max)} = \left\{ \left[\frac{6.283^2}{6.283^2 - \left(\frac{2\pi}{1.5} \right)^2} \right] - 1 \right\} (1.10)$$

$$= \left[\frac{39.4}{21.9} - 1 \right] 1.10 = 0.88 \text{ in.}$$

The maximum base shear is

$$V = 0.88 \times 51,000 = 45,000 \text{ lb}$$

12-7. IMPULSIVE LOADING

Although it is apparent that many cases of periodic disturbing forces exist there are other situations where bodies are subjected to forces varying over a finite time interval. In such cases a steady state of motion is not achieved and a mathematical description of the motion is often impracticable.

In Fig. 12a, an unrestrained body of mass m is subjected to a general disturbing force $Q = F(t)$. The motion of body is to be defined in terms of x measured from a reference position. At the instant that Q is applied the body is assumed to have an initial displacement of x_O and an initial velocity of $\dot{x}_O$. The equation of motion, $m\ddot{x} = $ force, may be restated as

$$m\frac{dv}{dt} = Q$$

and
$$m\,dv = Q\,dt \qquad\qquad (12\text{-}20)$$

(a)

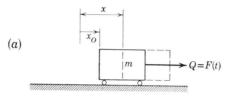

(b)

(c)

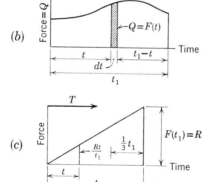

FIG. 12–12

The left-hand side of equation 12-20 represents a differential change in momentum and the right-hand side is the differential impulse. Since $v = \dot{x}$ then $d(v) = d(\dot{x})$ and fundamental expression is restated as $m\,d\dot{x} = Q\,dt$. In Fig. 12b $Q\,dt$ is represented by the area of the elemental strip shown. The force-time curve may thus be thought of as an infinite number of such impulses. Whatever the velocity is at time t, this velocity will be changed by a differential amount $d\dot{x}$. If only one such impulse is involved it will have produced a differential change in displacement by time t_1 of

$$dx = \frac{F(t)}{m} dt(t_1 - t)$$

The total change in displacement, or the sum of all differential displacements, produced in the time interval $t = 0$ to $t = t_1$, by all such differential impulses, is obtained by integration as

$$\varDelta x = \frac{1}{m} \int_{t=0}^{t=t_1} F(t)(t_1 - t) \, dt \qquad (12\text{-}21)$$

If at time $t = 0$ the body had an initial displacement of x_O and an initial velocity of $\dot{x}_O$, then at time t_1 the total displacement is

$$x = x_O + \dot{x}_O t_1 + \varDelta x$$

A geometric or graphical interpretation may be given to the expression for $\varDelta x$. Since $F(t) \, dt$ may be viewed as a differential area, its multiplication by $(t_1 - t)$ is analogous to taking the moment of this area about the right end of the force-time diagram. The integration implies that $m \, \varDelta x$ is the moment of the entire area of Fig. 12b between $t = 0$ and t_1 about t_1. If the force-time relation is linear the diagram is triangular, as in Fig. 12c, with the maximum force designated as equal to R, then at time t_1

$$\varDelta x = \frac{1}{m} \left(\frac{Rt_1}{2}\right)\left(\frac{t_1}{3}\right) = \frac{Rt_1^2}{6m}$$

If x_O and $\dot{x}_O$ are zero, then $x = \varDelta x$. In the absence of restraining or frictional forces, motion would continue beyond time t_1 with a constant velocity equal to

$$\int_0^{t_1} \frac{F(t) \, dt}{m} = \frac{Rt_1}{2m}$$

If the displacement is desired at a time T within the time interval t_1, $F(t)$ is expressible as Rt/t_1 and

$$\varDelta x = \frac{1}{m} \int_0^T \frac{Rt}{t_1} (T - t) \, dt$$

$$\varDelta x = \frac{R}{mt_1} \left(\frac{T^3}{2} - \frac{T^3}{3}\right) = \frac{RT^3}{6mt_1}$$

If the force-time relation is irregular and cannot be expressed analytically, as in Fig. 12b, the time interval may be divided into a series of small finite time intervals of $\varDelta t$. The finite areas are equal to $F(t)\varDelta t$, and their moment about t_1 or T, found by arithmetical integration, will suffice to determine $m \, \varDelta x$.

The body of Fig. 13a is restrained by an elastic spring attached to a fixed support in contrast to Fig. 12a, wherein the body was unrestrained. The body is now subject to the force of the spring in addition to the disturbing force. Nevertheless, basic principles are still valid and the spring force is dependent upon the displacement x from the equilibrium position. Then

$$\ddot{x} = \frac{1}{m} F(x, t) \qquad (12\text{-}22)$$

where $F(x, t)$ is a function describing the resultant horizontal force acting on the body. The momentum-impulse relationship under the action of Q and kx leads to

$$d\dot{x} = \frac{Q \, dt}{m} - \frac{kx}{m} dt \qquad (12\text{-}23)$$

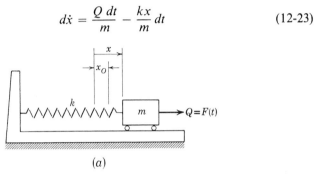

(a)

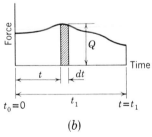

(b)

FIG. 12–13

The derivation of an analytical value for $x = f(t)$ is more complex than when Q was a periodic disturbing force. For that reason a numerical integration procedure based on momentum-impulse principles may be employed with sufficient accuracy for most practical problems. If a finite time interval Δt is substituted for dt, then a finite change in velocity may be computed as

$$\Delta\dot{x} = \frac{(Q - S)}{m} \Delta t \qquad (12\text{-}24)$$

where $S = kx$ is the resisting force in spring.   Designating the incremental impulse $(Q - S)\Delta t$ as I

$$\Delta\dot{x} = \frac{I}{m} \tag{12-25}$$

The velocity at any time t, after $i\,\Delta t$ time intervals, is

$$\dot{x} = \dot{x}_O + \sum_i \Delta\dot{x} \tag{12-26}$$

If the velocity at any time t is $\dot{x}$, then the change in displacement during the next Δt time interval, where $\bar{x}$ is the average velocity within the interval, is

$$\Delta x = \bar{x}\,\Delta t \tag{12-27}$$

Application of the numerical procedure requires an estimate of the displacement at a given time to determine S.   Since this displacement must often be intelligently guessed, it is easily noted that the numerical procedure becomes a trial-and-error process.   The preferred procedure deals with average values of Q and S within the Δt time intervals, designated as $\bar{Q}$ and $\bar{S}$, since both vary over a given Δt interval.   The Δt interval must be made small enough so that changes in direction of motion may be detected and maximum displacement determined.

An example of numerical integration is shown in Fig. 14a.   A body at rest weighing 980 lb, attached to a support by a long spring having $k = 100$ lb per in., is subjected to a force $Q = F(t)$.   Just prior to the application of the force, the conditions are $x_O = 0$; $\dot{x}_O = 0$; $\ddot{x}_O = 0$. The natural period of the restrained body is

$$\tau = 2\pi\sqrt{\frac{980}{100 \times 386}} = 1 \text{ sec}$$

The disturbing force, Fig. 14b, varies linearly from zero to a maximum of 500 lb in 0.5 sec.   The time scale is divided into five Δt intervals of 0.1 sec each, as well as several 0.05 sec intervals beyond 0.5 sec.   The notation t_0, t_1, t_2, etc., indicates the intervals on the time scale.   The value of the resisting force is determined by the displacement x and displacement at time t_1 equals x_1, for time t_2 equals x_2, etc.

Calculations for the example problem are made in table of Fig. 15. Starting with first time interval, an assumption of a trial value is made for average resisting force $\bar{S}$ of 4 lb and all calculations are made leading to $x_1 = 0.09$ in.   The calculated S at this displacement is $0.09 \times 100 = 9$ lb and the calculated average $\bar{S}$ for first time interval is 4.5 lb, which agrees very closely with assumed value.   Calculations for the second time interval, based on assumed displacement of 0.3 in., make the trial value of

$\bar{S}$ equal to 30 lb.   The calculations proceed for all intervals in the same manner; however, the table does not indicate that one or more revisions of assumed magnitude of $\bar{S}$ is often necessary before computed $\bar{S}$ and assumed $\bar{S}$ closely agree.   In problems of this type extreme precision is

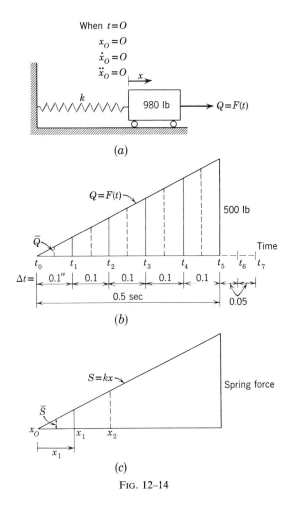

FIG. 12–14

seldom warranted.   If great precision is required it is recommended that a trial solution be made with $\Delta t = 0.1$ sec and then, with this solution as an aid to judgment, a revised solution using smaller time intervals carried out. It is to be noted that the body of this example continues to increase in displacement after the termination of disturbing force, hence calculations

Time	Δt, sec	$\bar{Q}$, lb	Assumed $\bar{S}$, lb	$\bar{Q} - \bar{S}$, lb	I	$\Delta \dot{x}$	$\dot{x}$	$\bar{\dot{x}}$	Δx, in.	x, in.	$S = kx$	Calculated $\bar{S}$, lb
t_0	0.1	50	4	46	4.6	1.80	0	0.90	0.090	0	0	4.5
t_1	0.1	150	30	120	12.0	4.72	1.80	4.16	0.416	0.090	9	29.8
t_2	0.1	250	100	150	15.0	5.90	6.52	9.47	0.947	0.506	50.6	98.0
t_3	0.1	350	220	130	13.0	5.12	12.42	14.98	1.498	1.453	145.3	220.1
t_4	0.1	450	390	60	6.0	2.36	17.54	18.72	1.872	2.951	295.1	388.7
t_5	0.05	0	530	-530	-26.5	-10.4	19.90	+14.7	0.735	4.823	482.3	519.0
t_6	0.05	0	570	-570	-28.5	-11.2	+9.5	+3.9	0.200	5.558	555.8	565.8
t_7	0.05	0	550	-550	-27.5	-10.8	-1.7	-7.1	-0.355	5.758	575.8	558.0
							-12.5			5.403	540.3	

$m = \frac{980}{386} = 2.54.$

$I = (\bar{Q} - \bar{S})\Delta t.$

$\Delta \dot{x} = I/m.$

$p^2 = k/m = \frac{100}{2.54}.$

$p = 6.28$ radians per sec.

Fig. 12-15

are carried out beyond $t = 0.5$ sec in order to determine the maximum displacement of 5.76 in.   The dynamic effect has produced a maximum force in spring of 576 lb at a time 0.6 sec after the introduction of the load. After the time of 0.5 sec the response of body is governed solely by the restraining force of spring, and after reaching maximum displacement the body will vibrate as an elastically supported body and eventually come to rest.

It is instructive to consider that at time t_5 the body has a velocity of 19.90 in. per sec and a displacement of 4.823 in.   These quantities may be viewed as initial values of $\dot{x}_O$ and x_O at a starting time of zero and maximum displacement computed.   From previous theory, the equation 12-8 describing the motion of a vibrating body becomes

$$x = 4.823 \cos pt + \frac{19.90}{6.28} \sin pt$$

Since we desire to determine the maximum x, take the derivative dx/dt and set it equal to zero, deriving

$$\tan pt = 0.658$$

$$pt = 33° 20' = 0.582 \text{ radian}$$

$$t = \frac{0.582}{6.28} = 0.093 \text{ sec}$$

This means that x max will occur at time $t = 0.593$ sec instead of at 0.600 sec as determined in tabular solution.

$$x \text{ max} = 4.823 \times \cos 33° 20' + \frac{19.90}{6.28} \sin 33° 20'$$

$$= 4.04 + 1.74 = 5.78 \text{ in.}$$

This value compares with the calculated value of 5.76 in. by tabular solution.   A similar computed value can be obtained by energy principles. The student should realize that after the body reaches its maximum displacement it will return under the action of the spring force and pass through $x = 0$.   The time for the body to return to its starting point will equal one-fourth of the natural period and free vibrations will continue until damped out.

It is instructive to graph the results as in Fig. 16 against time.   Fig. 16 represents the forcing and resisting functions.   The difference of $Q - S$ $= F$, and a differential impulse $F \, dt$ is equal to $m \, d\dot{x}$.   The total impulse, the area between the two curves from $t = 0$ to $t = 0.5$ sec, divided by mass, equals the velocity at $t = 0.5$ sec.

If the natural period of vibration is increased by increasing the weight of the body, the maximum displacement will be less as $\Delta \dot{x}$ is less for each time interval, and the dynamic effect on force in spring becomes small. The converse, a decrease in the natural period of vibration by decreasing the mass, increases $\Delta \dot{x}$ and displacement increases. The Δt time intervals must be reduced in magnitude near the end of maximum displacement to define explicitly maximum displacement, since the body within the final finite time intervals may have started its return to equilibrium position.

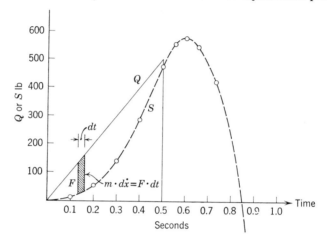

FIG. 12–16

12–8. ONE-STORY FRAME—IMPULSIVE LOADING

Buildings are subjected to gust wind loads, blast pressures; piers are subject to lateral impacts and shock loadings from approaching vessels. The effect on simple one-story frames may be analyzed following procedures of previous article.

Figure 17 represents a one-story bent with columns hinged at the bottom. The girder is assumed to be infinitely rigid. For purposes of this discussion all the mass of structure is concentrated at the top and taken as W/g. The spring factor is computed as the lateral force at the upper level to produce a unit deflection Δ. With a rigid girder, and where F is the total force at the top, the deflection equation for an end loaded cantilever beam may be used, leading to

$$\Delta = \frac{FL^3}{2(3EI)}$$

I represents the moment of inertia of a single column about the bending axis.   Thus

$$F = \frac{6\Delta EI}{L^3}$$

and if $\Delta = 1$ in., $k = F$, or the spring constant is

$$k = \frac{6EI}{L^3}$$

The natural period of vibration is then computed as

$$\tau = 2\pi \sqrt{\frac{W}{kg}} = 2.56L \sqrt{\frac{WL}{EIg}}$$

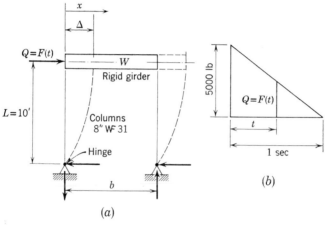

(a)

(b)

FIG. 12–17

For 8WF 31 steel columns, the spring constant in units of pounds/foot

$$k = \frac{6 \times 3 \times 10^7 \times 12^2 \times 109.7/(12)^4}{10^3}$$

$$k = 137,000 \text{ lb/ft}$$

With $W = 28,000$ lb, and $m = 28,000/32.2 = 870$ slugs, we obtain the natural period as

$$\tau = 2\pi \sqrt{\frac{870}{137,000}} = 0.50 \text{ sec}$$

The disturbing force $Q = F(t)$ is represented in Fig. 17b.   The calculations for determining motion of structure with time are given in the table of Fig. 18. S and $\bar{S}$ always refer to the sum of horizontal shears in

Table of Calculations: One-Story Frame of Fig. 17

Time	Δt, sec	$\bar{Q}$, lb	Assumed $\bar{S}$, lb	$\bar{Q} - \bar{S}$	I	$\Delta\dot{x}$	$\dot{x}$	$\bar{\dot{x}}$	Δx, ft	x, ft	$S = kx$	Calculated $\bar{S}$, lb
t_0	0.02						0			0	0	
		4950	100	4850	97	0.112		0.056	0.0011			75
t_1	0.02						0.112			0.0011	150	
		4850	350	4500	90	0.103		0.163	0.0033			375
t_2	0.02						0.215			0.0044	600	
		4750	1000	3750	75	0.086		0.258	0.0052			960
t_3	0.04						0.301			0.0096	1320	
		4600	2400	2200	88	0.101		0.352	0.0141			2280
t_4	0.04						0.402			0.0237	3240	
		4400	4300	+100	−4	0.005		0.400	0.0160			4350
t_5	0.04						0.397			0.0397	5460	
		4200	6400	−2200	−88	−0.101		0.346	0.0138			6400
t_6	0.04						0.296			0.0535	7340	
		4000	8800	−4800	−192	−0.211		0.190	0.0076			8600
t_7	0.04						0.085			0.0711	9750	
		3975	10,000	−6025	−60.2	−0.069		0.050	0.0005			9780
t_8	0.01						0.016			0.0716	9800	
		3925	10,000	−6075	−60.7	−0.071		−0.020	−0.0002			
t_9	0.01						−0.055			0.0714	⋯	⋯

$m = 870$ slugs.

Maximum displacement at approximately $t = 0.24$ sec.

Maximum column shears $= 9800/2 = 4900$ lb per column.

Fig. 12-18

both columns. The average unbalanced force on mass is $(\bar{Q} - \bar{S})$ as in previous calculations. Owing to stiffness of structure, small time intervals are used at the beginning of calculations and near the time of maximum displacement. Displacements are in feet. The maximum displacement occurs at approximately 0.24 sec after shock impinged on structure. Maximum column shears are about 4900 lb per column.

12–9. ENERGY SOLUTIONS

From basic principles, and for rectilinear motion, we again write that

$$F = ma = m\frac{dv}{dt} \tag{12-28}$$

where F is the resultant force on body and v is velocity. This fundamental equation may be restated, by multiplying the right-hand side by ds/ds and recognizing that $ds/dt = v$, as

$$F = m\frac{dv}{dt}\cdot\frac{ds}{ds} = m\frac{ds}{dt}\cdot\frac{dv}{ds} \tag{12-29}$$

or $$mv\,dv = F\,ds$$

Integration leads to

$$\frac{mv^2}{2} = \int F\,ds + \text{constant} \tag{12-30}$$

where $mv^2/2$ is termed kinetic energy and $\int F\,ds$ is the external work of the forces acting on body.

An earlier article presented the work and energy solution for harmonic motion of Fig. 1a. The external work of the spring force action on the body was

$$-\int_0^x \frac{kx\,dx}{2} = -\frac{kx^2}{2}$$

and the equal force acting on the spring does the work of $kx^2/2$ on the spring. This work is stored in the spring as potential energy termed V. The external work of the spring force on the body numerically equals the potential energy stored in the spring but is opposite in sign. For the spring problem

$$\frac{mv^2}{2} - \int F\,ds = \text{constant, where} -\int F\,ds = V$$

Using T for kinetic energy and V for potential energy, the basic equation expressing the law of conservation of energy is

$$\frac{mv^2}{2} + \frac{kx^2}{2} = \text{constant}$$

or
$$T + V = \text{constant} \tag{12-31}$$

in all configurations.

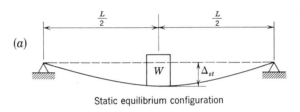

(a)

Static equilibrium configuration

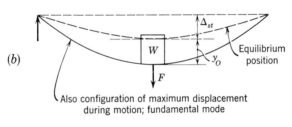

(b)

Also configuration of maximum displacement during motion; fundamental mode

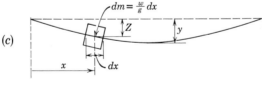

(c)

Fig. 12–19

Figure 19a is a beam of uniform weight and stiffness supporting a weight W. W is assumed to be large compared to weight of beam and beam weight will be neglected. The motion of the weight is to be measured from the static equilibrium position. If W is displaced y_0, in Fig. 19b, from the static equilibrium position by slowly applying an external force F, the net gain in potential energy is $ky_0^2/2$. That is, if k is the force required to deflect the center of the beam 1 in., ky_0 is the maximum force

F required to displace the beam through a distance of y_O. Since a linear relationship exists between force and deflection the work done on the beam by this force is $(ky_O)(y_O/2) = ky_O^2/2$.

A further clarification of the energy accounting may be desirable. In Fig. 19b, the loss in potential energy of the gravity force W, during displacement from equilibrium position, is $-Wy_O$. The external work done by the gravity force W is $+Wy_O$ and the external work done by the uniformly increasing downward force is $ky_O^2/2$. All of this increase in external work is stored in the beam as potential elastic energy. Hence the net gain in potential energy during the y_O displacement is

$$- Wy_O + Wy_O + \frac{ky_O^2}{2} = \frac{ky_O^2}{2}$$

Then, for the deflected static configuration of Fig. 19b, T_1 equals zero, since the beam is at rest, and

$$V_1 = \frac{ky_O^2}{2}$$

If F is quickly removed W will start its vibrating motion. Noting that displacements of W are measured from the original static equilibrium position by a y coordinate, then $\dot{y}$ represents the velocity of this motion. While in motion, the kinetic energy and potential energy are T_2 and V_2, respectively.

Since the sum of kinetic plus potential energy must remain constant

$$T_2 + V_2 = T_1 + V_1$$

We may then state that

$$\frac{W\dot{y}^2}{2g} + \frac{ky^2}{2} = \frac{ky_O^2}{2} \tag{12-32}$$

which has been previously demonstrated in an earlier article to become, after differentiation with respect to time,

$$\ddot{y} + p^2 y = 0 \tag{12-33}$$

from which the fundamental period of vibration has been shown to be

$$\tau = 2\pi \sqrt{\frac{W}{kg}}$$

A simplified view for determining the fundamental period is derived by considering only two configurations of the vibrating system. The initial displaced condition from equilibrium position shown in Fig. 19b defines the maximum amplitude, and since velocity is zero at this time all energy is potential energy. This maximum potential energy in the form of strain energy in the beam is

$$V_1 = \frac{k y_0^2}{2}$$

As the vibrating beam returns and passes through the static equilibrium position of Fig. 19b at maximum velocity all energy must be kinetic and must be equal to the maximum potential energy. If harmonic motion is assumed and described by $y = y_0 \cos pt$, then the maximum velocity, $\dot{y}$ max, by differentiation is $\pm y_0 p$, where p is the circular frequency. The maximum kinetic energy is

$$T \max = \frac{W}{2g}(y_0^2 p^2)$$

Equating T max and V_1

$$p^2 = \frac{kg}{W}; \qquad p = \sqrt{\frac{kg}{W}}$$

and natural period is

$$\tau = \frac{2\pi}{p} = 2\pi \sqrt{\frac{W}{kg}}$$

and if Δ_{st} is the static deflection at the center line, owing to W, then

$$\Delta_{st} = \frac{W}{k} \qquad \text{and} \qquad \tau = 2\pi \sqrt{\frac{\Delta_{st}}{g}}$$

An additional simplification may now be made by assuming that the vibrating beam will always have a configuration similar to the deflected configuration of the statically loaded beam. Letting n equal the scale factor, we assume $y_0 = n\Delta_{st}$.

In simple harmonic motion

$$y = y_0 \cos pt$$

or

$$y = n\Delta_{st} \cos pt$$

Then

$$\dot{y} = -pn\Delta_{st} \sin pt$$

$$\dot{y} \max = \pm pn\Delta_{st} \tag{12-34}$$

and kinetic energy of the beam in motion when passing through the equilibrium configuration is

$$T \max = \frac{W}{2g} p^2 n^2 \Delta_{st}^2 \tag{12-35}$$

This kinetic energy must be equal to the maximum potential energy. In terms of the scale factor, the maximum potential energy is

$$V = \frac{k(n\Delta_{st})^2}{2}$$

which may be restated as

$$V = \frac{k\Delta_{st}(n^2\Delta_{st})}{2}$$

and since $W = k\Delta_{st}$

$$V = \frac{n^2 W\Delta_{st}}{2} \qquad (12\text{-}36)$$

Equating T max and V

$$\frac{W}{2g}p^2n^2\Delta_{st}^2 = \frac{n^2 W\Delta_{st}}{2} \qquad (12\text{-}37)$$

$$p^2 = \frac{g}{\Delta_{st}}$$

from which the natural period is

$$\tau = 2\pi \sqrt{\frac{\Delta_{st}}{g}}$$

The elimination of the scale factor in the final result, since it appears on both sides of equation 12-37, indicates that the solution could have been written directly by representing the potential strain energy in the static configuration as

$$V = \frac{W\Delta_{st}}{2} \qquad (12\text{-}38)$$

and writing the maximum kinetic energy when the beam is in motion as

$$T = \frac{p^2 W\Delta_{st}^2}{2g} \qquad (12\text{-}39)$$

This leads to a direct computation for p^2. Compare this answer with the solution wherein a scale factor was employed. This simplification leads to the Rayleigh method.

The Rayleigh method enables an approximate solution to be obtained by assuming that displacements during motion are similar to some static deflection curve. This may be demonstrated by the example of Fig. 20, employing the simplified procedure, where Δ_1 and Δ_2 are the static deflections.   The potential energy term may be written as (see equation 12-38)

$$V = \frac{W_1\Delta_1}{2} + \frac{W_2\Delta_2}{2} \tag{12-40}$$

and the maximum kinetic energy term as (see equation 12-39)

$$T = \frac{p^2}{2g}(W_1\Delta_1^2 + W_2\Delta_2^2)$$

The equation for p^2 is

$$p^2 = \frac{g(W_1\Delta_1 + W_2\Delta_2)}{W_1\Delta_1^2 + W_2\Delta_2^2} \tag{12-41}$$

and the natural period of the fundamental mode of vibration is

$$\tau = \frac{2\pi}{p}$$

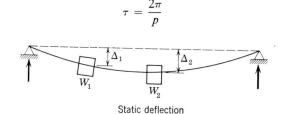

Static deflection

FIG. 12–20

If a correction is to be made for weight of the beam, an assumed form of elastic curve during vibration is generally involved.   In the beam of Fig. 19c, displacements may be assumed to be proportional to the elastic curve of a center loaded beam, in conformance with the Rayleigh method, since the inertia effect of the weight at center is dominant during motion. The deflection Z, in terms of the center line displacement, from deflection principles, may be shown to be

$$Z = y\left(\frac{3xL^2 - 4x^3}{L^3}\right) \tag{12-42}$$

Each elemental mass of $\dfrac{w}{g}\,dx$ vibrates with a velocity proportional to $\dot{y}$, the velocity at the center of span, since each elemental mass of the vibrating beam is in the harmonic motion.

Letting this velocity equal

$$\dot{Z} = \dot{y}\left(\frac{3xL^2 - 4x^3}{L^3}\right) \tag{12-43}$$

The kinetic energy of the beam weighing w per unit of length is

$$T_b = 2 \int_0^{L/2} \dot{Z}^2 \frac{w \, dx}{2g} \qquad (12\text{-}44)$$

$$= \frac{2}{2g} \int_0^{L/2} \dot{y}^2 \left(\frac{3xL^2 - 4x^3}{L^3} \right)^2 dx$$

$$T_b = \frac{17}{35} \frac{wL}{2g} \dot{y}^2 \qquad (12\text{-}45)$$

We may then write the fundamental equation relating kinetic and potential energy corrected for the effect of the weight of the beam as

$$\frac{\dot{y}^2}{2g} \left(W + \frac{17}{35} wL \right) + \frac{ky^2}{2} = \frac{ky_o^2}{2}$$

Since harmonic motion has been assumed, we may write the equation of motion for the center line element of Fig. 19c as $y = y_o \cos pt$.

A derivative with respect to time gives $\dot{y} = -py_o \sin pt$, or $\dot{y} \max = \pm py_o$. Considering $\dot{y}$ as $\dot{y} \max$ the energy equation is

$$\frac{(py_o)^2}{2g} \left(W + \frac{17}{35} wL \right) = \frac{ky_o^2}{2}$$

and

$$p^2 = kg \Big/ \left(W + \frac{17}{35} wL \right)$$

By comparing with the results which omitted the weight of the beam, it can be noted that effect of beam weight is to add $\frac{17}{35} wL$ to W and that the natural period becomes

$$\tau = 2\pi \sqrt{\frac{(W + \frac{17}{35} wL)}{kg}} \qquad (12\text{-}46)$$

The value of $17wL/35g$ is often spoken of as the reduced mass of the beam.

A further example of the energy method may be illustrated by the uniform cantilever beam of Fig. 21. Figure 21 depicts the elastic curve of the beam at maximum amplitude. As the beam vibrates in harmonic motion, in its fundamental mode in the vertical plane, the maximum velocity of all elements occurs as the beam passes through the equilibrium position taken as horizontal in Fig. 21. At this instant, kinetic energy of the beam is at a maximum. Each elemental length of dx has a maximum velocity of $y(p)$, where p is the circular frequency of the vibration in radians per second and where y is the maximum deflection of the elemental length.

For a length of dx the maximum kinetic energy is $\dfrac{mv^2}{2} = \dfrac{w\,dx}{2g}(py)^2$ and for the entire beam we determine the maximum kinetic energy as

$$\max T_b = \frac{wp^2}{2g}\int_0^L y^2\,dx \tag{12-47}$$

In the maximum displacement condition shown in Fig. 21, the velocity is zero, hence all energy is potential strain energy caused by bending. From the previous theory of Art. 2-6 this energy is

$$V_1 = \int_0^L \frac{M^2\,dx}{2EI} = \frac{EI}{2}\int_0^L \left(\frac{d^2y}{dx^2}\right)^2 dx \tag{12-48}$$

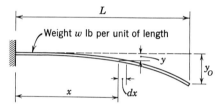

FIG. 12–21

The energy equation for the two positions is

$$\max T_b = V_1$$

or

$$\frac{wp^2}{2g}\int_0^L y^2\,dx = \frac{EI}{2}\int_0^L \left(\frac{d^2y}{dx^2}\right)^2 dx \tag{12-49}$$

from which the circular frequency relation of the free vibration may be found as

$$p^2 = \frac{EIg}{w}\frac{\int \left(\dfrac{d^2y}{dx^2}\right)^2 dx}{\int y^2\,dx} \tag{12-50}$$

The Rayleigh method of solution, equation 12-50, is based on assuming the dynamic deflection curve as being similar to the static deflection curve, or an approximation thereof. Small errors will result, depending on degree of approximation, but will be of no great consequence in most cases. In this example of the cantilever beam of Fig. 21, we may assume the static deflection curve as that which can be represented by a cosine curve since this curve fits the boundary condition of zero slope at the fixed end.   Based on this assumption

$$y = y_0\left(1 - \cos\frac{\pi x}{2L}\right)$$

and
$$\frac{d^2y}{dx^2} = \frac{y_0\pi^2}{4L^2}\cos\frac{\pi x}{2L}$$

Then by substitution into equation 12-50 we obtain

$$p^2 = \frac{EIg}{w}\frac{\displaystyle\int_0^L \left(\frac{y_0\pi^2}{4L^2}\cos^2\frac{\pi x}{2L}\right)^2 dx}{\displaystyle\int_0^L y_0{}^2\left(1 - \cos\frac{\pi x}{2L}\right)^2 dx}$$

Integration and substitution leads to

$$p = 3.65\sqrt{\frac{EIg}{wL^4}}$$

The exact solution found in many standard textbooks provides a factor of 3.52 instead of 3.65.

12–10. TWO DEGREES OF FREEDOM

Previously in this chapter the simple case of free or forced harmonic motion has been emphasized for systems possessing one degree of freedom.

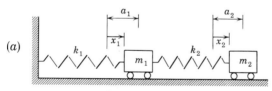

(a)

Fundamental mode

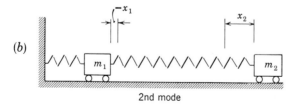

(b)

2nd mode

FIG. 12–22

If systems have more than one degree of freedom the motion is more complex and the vibratory motion is considered to possess as many separate harmonic motions as there are degrees of freedom.   The degree of freedom may be defined as the number of independent coordinates which are required to describe the configuration or displacement of the system with respect to time.   For example, in Fig. 22, two separate but interconnected masses, m_1 and m_2, may move horizontally with x_1 describing the displacement of m_1 and with x_2 representing the displacement of m_2.   This system then has two degrees of freedom since two independent coordinates are required to define the motion.

As many natural periods or frequencies of harmonic vibration exist as there are degrees of freedom, and a natural mode of vibration is associated with each of these natural periods or frequencies.   The lowest natural frequency is spoken of as the first or fundamental frequency, the next highest as the second natural frequency, etc.   The modes of vibration are spoken of in a corresponding manner, namely the fundamental mode, the second natural mode, etc.

A mode of vibration refers to the configuration of the moving masses relative to one another when the motion of the masses is conforming to a harmonic pattern at a specific frequency.   This may be made clear by reference to Fig. 22.   In the fundamental mode, masses m_1 and m_2 oscillate in phase, with x_1 reaching a positive maximum at the same time x_2 reaches a positive maximum.   In the second mode, Fig. 22b, masses m_1 and m_2 move to and fro out of phase by 180°, so that when x_1 is at its positive maximum value, x_2 is at its maximum negative value.   For any given mode of vibration the masses pass through the equilibrium position at the same instant.

To clarify these concepts refer to the elastic beam of Fig. 23.   Three equal masses concentrated at the one-quarter points of the span may oscillate in a vertical plane in the three different modes shown in Fig. 23, since three vertical coordinates are necessary to define the configuration of motion.   This is an example of three degrees of freedom.

A given mode represents a configuration which repeats itself periodically, with the motion of all masses undergoing a harmonic oscillation of the same frequency.   At all times the motion is such as to maintain accelerations proportional to maximum displacement.

A simple harmonic motion of a single mass was shown to be

$$x = a \cos pt \tag{12-51}$$

where a was the initial displacement, as well as the maximum displacement, and the acceleration of the mass by differentiation of 12-51 is

$$\ddot{x} = -ap^2 \cos pt$$

If the two masses of Fig. 22a are oscillating in a natural mode at the same circular frequency, in phase or out of phase, then assigning subscript one for quantities referring to mass one, and subscript two for quantities referring to mass two, we may designate their accelerations by

$$\ddot{x}_1 = -a_1 p^2 \cos pt$$
$$\ddot{x}_2 = -a_2 p^2 \cos pt \tag{12-52}$$

and the ratio of their accelerations as

$$\frac{\ddot{x}_1}{\ddot{x}_2} = \frac{a_1}{a_2} \tag{12-53}$$

where a_1 and a_2 represent maximum displacements.

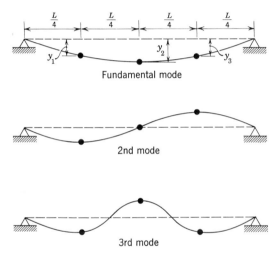

Fundamental mode

2nd mode

3rd mode

Fig. 12–23

Thus the relationships of equation 12-53 demonstrate that the accelerations remain proportional to maximum displacements during a given mode of vibration. With accelerations maintaining this proportionality during motion, then the unbalanced forces acting on the two masses are also proportional to the product of mass times displacement.

In Fig. 24, two weights W_1 and W_2 are supported on a simple beam span. The beam is assumed to be vibrating in a vertical plane with each mass m_1 and m_2 undergoing harmonic motion. The lowest frequency of vibration (the fundamental mode) may be easily found by the energy method previously presented or by equation 12-41. The configuration of the fundamental mode is as shown in Fig. 24, with m_1 and m_2 moving up and down in phase as required by the fundamental mode. The co-ordinates y_1 and y_2 will define the configuration from the position of static equilibrium as a function of time. If a_1 and a_2 represent the maximum amplitudes of the harmonic motion of m_1 and m_2, respectively, the harmonic motion of each mass is expressible as

$$y_1 = a_1 \cos pt$$
$$y_2 = a_2 \cos pt$$

(12-54)

By successive differentiations the maximum accelerations are

$$\ddot{y}_1 = -a_1 p^2$$
$$\ddot{y}_2 = -a_2 p^2$$

(12-55)

with the direction of acceleration acting towards the origin.

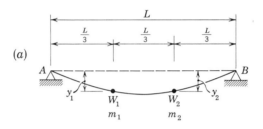

(a)

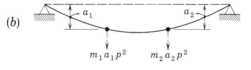

(b)

FIG. 12–24

The maximum inertia forces acting on the beam are shown in Fig. 24b. To explore or obtain all frequencies for all possible modes of vibration we now view the displacements as a deflection problem, where a_1 and a_2 are a result of the inertia forces. Using the deflection coefficients for unit loads the maximum deflections of m_1 and m_2 are expressible as

$$\delta_{11}(m_1 a_1 p^2) + \delta_{12}(m_2 a_2 p^2) = a_1$$
$$\delta_{21}(m_1 a_1 p^2) + \delta_{22}(m_2 a_2 p^2) = a_2$$

(12-56)

Equations 12-56 can be rearranged in homogeneous form as

$$(\delta_{11} m_1 p^2 - 1)a_1 + \delta_{12} m_2 p^2 a_2 = 0$$
$$\delta_{21} m_1 p^2 a_1 + (\delta_{22} m_2 p^2 - 1)a_2 = 0$$

(12-57)

Since equations 12-57 are homogeneous equations, their determinantal equation must be equal to zero for a nontrivial solution.*   This determinantal equation is formed by first writing the coefficients of a_1 and a_2 of the two homogeneous equations (equations 12-57), as a determinant

$$\begin{vmatrix} (\delta_{11}m_1p^2 - 1) & \delta_{12}m_2p^2 \\ \delta_{21}m_1p^2 & (\delta_{22}m_2p^2 - 1) \end{vmatrix} = 0$$

Then the equation formed by the determinant rule is

$$(\delta_{11}m_1p^2 - 1)(\delta_{22}m_2p^2 - 1) - (\delta_{12}m_2p^2)(\delta_{21}m_1p^2) = 0 \quad (12\text{-}58)$$

which will simplify to the frequency equation

$$(\delta_{11}\delta_{22} - \delta_{12}\delta_{21})m_1m_2p^4 - (\delta_{11}m_1 + \delta_{22}m_2)p^2 + 1 = 0 \quad (12\text{-}59)$$

Since the beam of Fig. 24 is a two degree of freedom problem there will be two circular frequencies which are relevant.   These two values of p^2 will be secured by solving for the roots of the frequency equation 12-59 and the roots denoted as

$$p_1{}^2 \text{ for the first fundamental mode}$$

$$p_2{}^2 \text{ for the second mode}$$

As an example, assume in Fig. 24 that W_1 and W_2 equal 3860 lb, making m_1 and m_2 equal to 10 lb-sec^2 per in., and that these masses are located at the one-third points of the span.   From standard deflection equations or by a moment-area solution we obtain

$$\delta_{11} = \delta_{22} = \frac{4L^3}{243EI}$$

and since $\qquad \delta_{12} = \delta_{21}$ (by Maxwell's reciprocal theorem)

we also obtain $\qquad \delta_{12} = \delta_{21} = \dfrac{7L^3}{486EI}$

To obtain numerical values, let

$$L = 12 \text{ ft}, \qquad E = 30 \times 10^6 \text{ psi, and } I = 500 \text{ in.}^4$$

By substitution

$$\delta_{11} = \delta_{22} = 3.27 \times 10^{-6} \text{ in.}$$

$$\delta_{12} = \delta_{21} = 2.86 \times 10^{-6} \text{ in.}$$

The frequency equation 12-59 for the example problem of Fig. 24 is

$$[3.27^2 - 2.86 \times 2.86][10 \times 10]p^4 10^{-12} - [32.7 + 32.7]p^2 \times 10^{-6} + 1 = 0$$

* See Art. 13–4 for a complete explanation.

Multiplying through by (10^{12}) and simplifying

$$p^4 - 261.6 \times 10^3 p^2 + 4 \times 10^9 = 0$$

Solving by the quadratic formula we obtain

$$p^2 = \frac{261.6 \times 10^3 \pm \sqrt{261.6^2 \times 10^6 - 16 \times 10^9}}{2}$$

$$p^2 = \frac{261.6 \times 10^3 \pm 229 \times 10^3}{2}$$

giving the roots

$$p_1{}^2 = 16.3 \times 10^3$$
$$p_2{}^2 = 245.3 \times 10^3$$

and finally

$$p_1 = 128 \text{ radians per sec}$$
$$p_2 = 496 \text{ radians per sec}$$

In terms of cycles per second, the corresponding frequencies of vibration are

$$f_1 = \frac{p_1}{2\pi} = 20.4 \text{ cps} \qquad f_2 = \frac{p_2}{2\pi} = 79.0 \text{ cps}$$

A check may be obtained on the fundamental frequency from the basic equation 12-41 obtained by the energy method where Δ_1 and Δ_2 represent static deflections produced by W_1 and W_2. Equation 12-41 yields

$$p_1{}^2 = g\left(\frac{W_1\Delta_1 + W_2\Delta_2}{W_1\Delta_1{}^2 + W_2\Delta_2{}^2}\right)$$

and due to symmetry of loading

$$p_1{}^2 = \frac{g}{\Delta_1}$$

Using deflection coefficients, we calculate the deflection as

$$\Delta_1 = 3860(3.27 + 2.86)10^{-6} = 2.37 \times 10^{-2} \text{ in.}$$

Then

$$p_1{}^2 = \frac{386}{2.37 \times 10^{-2}} = 16.3 \times 10^3 \text{ sec}^{-2}$$

This checks with previously obtained value for the fundamental mode. Now returning to the basic problem, the two configurations of the fundamental and second mode of free vibrations are as shown in Fig. 25, and will vibrate in these modes if the beam is disturbed periodically by an external influence whose frequency corresponds to the frequency of these natural modes.   Naturally, resonance would occur if damping is absent. From a structural point of view such a prospect of resonance should be avoided by properly designing the beam to control its natural frequencies in relation to the frequency of expected disturbing forces.

Although it should be clear that in the fundamental mode y_1 and y_2 have the same sign and that in the second mode they are of opposite sign as shown in Fig. 25, this fact may be proved by substituting $p_1{}^2$ or $p_2{}^2$ in the basic deflection equations 12-56 and solving for a_2 in terms of a_1.   For the fundamental mode $a_2/a_1 = +1$, and for the second mode $a_2/a_1 = -1$.

Fundamental mode

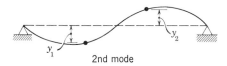

2nd mode

FIG. 12–25

For a beam carrying three masses, three degrees of freedom are involved with three relevant frequencies.   Procedures are identical to that of the example problem except that the frequency equation will yield three different values of p^2.   From a practical viewpoint no more than the fundamental frequency may be required and in that event the energy solution will be found to be more expeditious.

Example: Free Vibrations of a Mast

A mast carrying two concentrated masses is shown in Fig. 26a and, as a preliminary to studying its action under a forced vibration, the two principal frequencies of free vibration must be determined.   The solution neglects the mass of the mast itself.   As for the beam, the relative inertia forces in any mode are equal to mass times initial displacement.   The deflection coefficients for unit loads by moment-area theory are

$$\delta_{11} = \frac{L^3}{3EI}; \qquad \delta_{22} = \frac{8L^3}{3EI}$$

and by Maxwell's reciprocal relation and deflection theory

$$\delta_{12} = \delta_{21} = \frac{5L^3}{6EI}$$

The horizontal deflection equations for m_1 and m_2 are

$$\delta_{11}m_1p^2a_1 + \delta_{12}m_2p^2a_2 = a_1$$
$$\delta_{21}m_1p^2a_1 + \delta_{22}m_2p^2a_2 = a_2$$

The homogeneous equations are formulated as

$$(\delta_{11}m_1p^2 - 1)a_1 + \delta_{12}m_2p^2a_2 = 0$$
$$\delta_{21}m_1p^2a_1 + (\delta_{22}m_2p^2 - 1)a_2 = 0$$

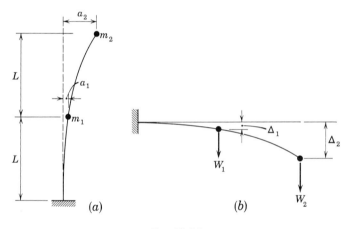

FIG. 12–26

The determinantal equation is identical in form to that of the beam example and the frequency equation is again

$$(\delta_{11}\delta_{22} - \delta_{12}\delta_{21})m_1m_2p^4 - (\delta_{11}m_1 + \delta_{22}m_2)p^2 + 1 = 0 \qquad (12\text{-}60)$$

For example purposes let m_1 and m_2 equal unity.   Then the frequency equation 12-60 is

$$\left(\frac{8}{9} - \frac{25}{36}\right)\left(\frac{L^3}{EI}\right)^2 p^4 - 3\frac{L^3}{EI}p^2 + 1 = 0$$

$$7\left(\frac{L^3}{EI}\right)^2 p^4 - 108\frac{L^3}{EI}p^2 + 36 = 0$$

and by the quadratic formula

$$p^2 = \frac{L^3}{EI}\left(\frac{+108 \pm \sqrt{108^2 - 4 \times 7 \times 36}}{14(L^3/EI)^2}\right)$$

$$p^2 = \frac{EI}{L^3}\left(\frac{+108 \pm 103.3}{14}\right) = 0.336\frac{EI}{L^3} \text{ and } \frac{15.1EI}{L^3}$$

where p_1^2 equals the smallest value and p_2^2 equals the largest value.

Again the two modes may be separately excited by periodic disturbing influences applied to the mast. A check on the fundamental frequency by work and energy may be made by using a static deflection curve obtained by treating the mast as a horizontal cantilever as in Fig. 26b. Since masses $m = W/g$ in the example problem were assumed equal and of unity value, W_1 and W_2 will be equal to g. The fundamental frequency p_1^2 from equation 12-41 is

$$p_1^2 = g\left(\frac{g\Delta_1 + g\Delta_2}{g\Delta_1^2 + g\Delta_2^2}\right) = g\left(\frac{\Delta_1 + \Delta_2}{\Delta_1^2 + \Delta_2^2}\right)$$

With

$$\Delta_1 = \frac{gL^3}{3EI} + \frac{5}{6}\frac{gL^3}{EI} = \frac{7}{6}\frac{gL^3}{EI}$$

and

$$\Delta_2 = \frac{8}{3}\frac{L^3}{EI} + \frac{5}{6}\frac{gL^3}{EI} = \frac{21}{6}\frac{gL^3}{EI}$$

we obtain

$$p_1^2 = 0.344\frac{EI}{L^3}$$

This again indicates the simplicity of the energy method in obtaining the fundamental frequency.

12–11. FREE VIBRATIONS OF A TWO-STORY BENT

Although special methods exist for solving problems of the type shown in Fig. 27a, only the fundamentals of the analysis will be indicated in this text. The fundamentals are the same as for the mast example of the previous article. The primary difficulty is in determining the deflection coefficients for the indeterminate structure. A simple approach is to perform the required moment distribution solutions for unit loads as shown in Figs. 27b, 27c, and then to compute the deflection coefficient by routine moment-area procedures. It should also be noted that the masses involved are generally concentrated at the floor levels.

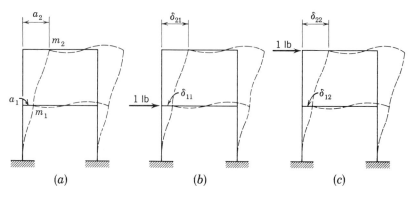

FIG. 12–27

12–12. FREE VIBRATIONS OF A BEAM

To examine the modes of vibration of an elastic beam, consider the beam AB of weight w per unit of length. Figure 28 depicts the beam vibrating

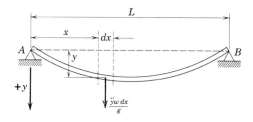

FIG. 12–28

in the vertical plane. If $\ddot{y}$ is equal to the acceleration, the inertia loading per unit of length is

$$q = \frac{\ddot{y}w \, dx}{g} \div dx = \frac{\ddot{y}w}{g} \tag{12-61}$$

For a given differential element the acceleration is

$$\ddot{y} = -\frac{d^2y}{dt^2} \tag{12-62}$$

where the minus sign indicates that the acceleration is oppositely directed to deflection which is taken as positive down. From deflection theory

$$EI\frac{d^2y}{dx^2} = -M$$

and by two successive differentiations

$$\frac{d^2}{dx^2}\left(EI\frac{d^2y}{dx^2}\right) = -\frac{d}{dx}\left(\frac{dM}{dx}\right) = -\frac{dV}{dx} = q$$

or
$$EI\frac{d^4y}{dx^4} = q \tag{12-63}$$

Since y will be a function of x and time, or $f(x, t)$, partial derivative notation must be employed, hence

$$q = -\frac{w}{g}\frac{\partial^2 y}{\partial t^2} \tag{12-64}$$

and by equation 12-63

$$EI\frac{\partial^4 y}{\partial x^4} = -\frac{w}{g}\frac{\partial^2 y}{\partial t^2} \tag{12-65}$$

or
$$\frac{\partial^2 y}{\partial t^2} + \frac{EIg}{w}\frac{\partial^4 y}{\partial x^4} = 0 \tag{12-66}$$

and letting
$$a^2 = \frac{EIg}{w}$$

the controlling differential equation becomes

$$\frac{\partial^2 y}{\partial t^2} + a^2\frac{\partial^4 y}{\partial x^4} = 0 \tag{12-67}$$

If motion is harmonic then, at any general section along the beam span, the displacement with respect to time can be considered as $y = y_x \cos p_n t$. Furthermore, y_x, the maximum amplitude at a given general section for a simple span, may be considered as

$$y_x = b_n \sin\frac{n\pi x}{L} \tag{12-68}$$

where b_n equals maximum amplitude of the deflection and n denotes the number of one-half waves, as in a sine series representing deflection. Thus, while y_x of equation 12-68 defines the maximum displacement of a given section of the vibrating beam, $\cos pt$ modifies this maximum displacement to describe a harmonic motion with respect to time. Since n is any integer an infinite number of degrees of freedom exist and each will possess a natural frequency. In Fig. 29 the configuration of the first, second, and third modes are shown.

Then y_x as a function of x and time is

$$y_x = b_n \sin\frac{n\pi x}{L}\cos p_n t \tag{12-69}$$

From which, by partial differentiation, we obtain

$$\frac{\partial^2 y}{\partial t^2} = -p_n^2 b_n \sin \frac{n\pi x}{L} \cos p_n t$$

and

$$\frac{\partial^4 y}{\partial x^4} = \frac{n^4 \pi^4}{L^4} b_n \sin \frac{n\pi x}{L} \cos p_n t$$

Upon substitution of the above derivatives in the fundamental controlling differential equation 12-67, the frequency equation becomes

$$p_n^2 = \frac{n^4 \pi^4}{L^4} a^2$$

and

$$p_n = \frac{n^2 \pi^2}{L^2} a$$

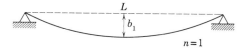

$n = 1$

$n = 2$

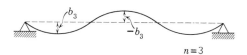

$n = 3$

FIG. 12–29

For the fundamental mode with $n = 1$,

$$p_1 = \frac{a\pi^2}{L^2}$$

for the second mode with $n = 2$,

$$p_2 = \frac{4a\pi^2}{L^2}; \quad \text{etc.}$$

Corresponding periods of vibration are

$$\tau_1 = \frac{2\pi}{p_1}, \qquad \tau_2 = \frac{2\pi}{p_2}, \quad \text{etc.}$$

12–13. CONCLUSION

Structural dynamics is a subject of significant importance and this chapter should be considered as no more than a bare introduction to the subject.   The chapter has established many of the principles and the mathematics of the general problem, but damping has purposely been omitted to simplify the presentation.   Damping, on the other hand, is of extreme importance and it is recommended that each student be required to study this aspect in the many excellent textbooks on vibrations.

The catastrophic collapse of the Tacoma Narrows Bridge in 1940 and the subsequent comprehensive studies of this failure have done much to stimulate the engineer's interest in structural dynamics.   The vibration of stacks and towers is a constant problem to those working in this field. The student should refer to the references cited.

REFERENCES

1. Bernhard, R. K., *Mechanical Vibrations*, Chapter 11, New York: Pitman, 1943.
2. Bleich, F., C. B. McCullough, R. Rosecrans, and G. S. Vincent, *The Mathematical Theory of Vibration in Suspension Bridges*, Washington, D.C.: Dept. of Commerce, Bureau of Public Roads, U.S. Government Printing Office, 1950.
3. Dickey, W. L., and G. B. Woodruff, "The Vibrations of Steel Stacks," *Trans. Am. Soc. Civil Engineers*, **121**, 1054–1112 (1956).
4. Edgerton, R. C., and G. W. Beecroft, "Dynamic Stresses in Continuous Plate-Girder Bridges," *Trans. Am. Soc. Civil Engineers*, **121**, 266–292 (1958).
5. Goldberg, J. E., "Natural Period of Vibration of Building Frames," *Journal of Am. Concrete Institute*, **36**, 81–95 (September 1939).
6. Hoskins, L. M., and J. D. Galloway, "Earthquakes and Structures," *Trans. Am. Soc. Civil Engineers*, **105**, 269–322 (1940).
7. Jacobsen, L. S., "Natural Periods of Uniform Cantilever Beams," *Trans. Am. Soc. Civil Engineers*, **104**, 402 (1939).
8. Morrill, B., *Mechanical Vibrations*, Chapters 1, 2, and 3, New York: Ronald Press, 1957.
9. Newmark, N. M., "An Engineering Approach to Blast Resistant Design," *Trans. Am. Soc. Civil Engineers*, **121**, 45–65 (1956)
10. Timoshenko, S., *Vibration Problems in Engineering*, Chapter 1, New York: Van Nostrand, 1955.
11. Timoshenko, S., and D. H. Young, *Advanced Dynamics*, Chapters 1 and 2, New York: McGraw-Hill, 1948.
12. Williams, H. A., "Dynamic Distortions in Structures Subjected to Sudden Earth Shock," *Trans. Am. Soc. Civil Engineers*, **102**, 838 (1937).

PROBLEMS

12–1. The profile of a railroad track, which is badly out of vertical alignment, may be represented as a sinusoidal curve with the variation in elevation from crest to hollow as 24 in.   If a rigidly mounted car moves along the track, so that the time taken to travel from crest to crest is 5 sec, determine the maximum and minimum pressures on track.   The car weighs 30,000 lb.

Ans. 31,470 lb, 28,530 lb.

12–2. A weight of 2000 lb is suspended vertically by an erection cable having a metallic area of 0.5 in.^{2} and a modulus of elasticity of 20×10^6 psi.   The cable is 50 ft long.   Neglect weight of cable.   (*a*) Calculate the natural frequency of free vibrations.   (*b*) The crane operator is lowering the weight at 60 ft per min and suddenly applies the brake.   What is the maximum amplitude of vibration and what is the maximum stress in cable?

Ans. (*a*) 9.1 cps;   (*b*) .212 in., 5,530 lb.

12–3. A weight W is restrained by two springs as shown.   Set up the differential equation of motion and determine the natural frequency of vibration.

Prob. 12–3

12–4. A steel 12-in. I 31.8-lb beam supports a weight W of 1000 lb at the one-third point of an 18-ft simple span.   Neglecting the mass of the beam, determine the natural frequency.   Use $E = 30 \times 10^6$ psi.

Ans. $f = $ 19.6 cps.

12–5. A beam ABC supports a weight of W at C.   Considering the beam as weightless and of uniform EI, compute the natural period of vibration.

$$Ans. \ \tau = 2\pi \sqrt{\frac{WL^3}{8EIg}}.$$

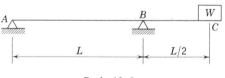

Prob. 12–5

12–6. A machine weighing W lb is supported on four springs.   Determine the natural frequency of vibration if $W = 2000$ lb.   The static deflection of the springs due to W is 1 in.   *Ans.* 3.12 cps.

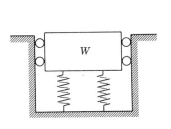

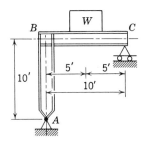

Prob. 12–6   Prob. 12–7

12–7. The frame supports a weight of W as shown. Neglecting the weight of the frame, compute the natural period of horizontal vibration in the plane of the frame. $W = 10,000$ lb, and the frame is composed of 8 in. W⁻ 31 beam sections.

12–8. Determine the natural period of vertical vibration of W if all of the mass is concentrated at the center line of BC. Assume that EI is constant for all members.

$$Ans. \ \tau = 2\pi \sqrt{\frac{Wa^3}{9.6EIg}}.$$

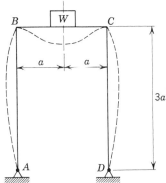

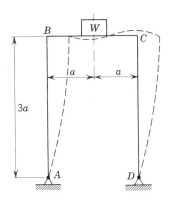

Prob. 12–8   Prob. 12–9

12–9. Determine the natural frequency of horizontal vibrations if all of the mass is concentrated at the BC level. Assume that EI is constant for all members.

$$Ans. \ f = \frac{1}{2\pi} \sqrt{\frac{EIg}{6Wa^3}}.$$

12–10. A rod having negligible weight is hinged at *A* and has a weight of *W* at *B*. Determine the natural frequency for small displacement from the equilibrium position. *W* = 5 lb, and *k* = 25 lb per in. *Ans. f* = 1.75 cps.

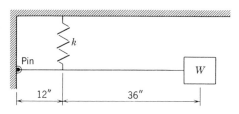

Prob. 12–10

12–11. A steel 12-in. I 31.8-lb beam has a simple span of 18 ft. A rolling wheel load weighing 2000 lb passes over a 4-in. high brick at the center line of the beam and drops onto the top flange of the beam. Compute the maximum deflection produced by this impact.

12–12. A motor is supported on a foundation which in turn is supported by four springs. The foundation and motor together have a weight of 2000 lb. The motor operates at a speed of 1500 revolutions per min. Some centrifugal un-balanced force may normally be expected, owing to an unbalance of the rotating parts. What would be the required spring factor to insure that the transmitted dynamic force to the foundation was no greater than one-fifth of the centrifugal force?

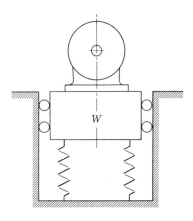

Prob. 12–12

12–13. A machine rests on a concrete block which in turn is supported on springs. A piston weighing 400 lb moves up and down in harmonic motion, at 12 cps. The total stroke of the piston is 36 in. The piston is concentric with the center of gravity of the machine and center of gravity of the foundation. From the standpoint of the operator the base of the machine should not move up and down more than one-fourth of an inch. If eight springs can be used in support-ing the block and machine on its foundation, what is the required *k* value of each spring and the required weight of the machine and the concrete block on which it rests if the ratio of w/p = 2. Also determine the total force acting on the foundation. *Ans. W* = 38,500 lb; *k* = 17,700 lb per in.

12–14. A bridge girder has a known natural frequency of about 2 cps. A simple instrument is to be designed to record the vertical motion at the center line of the bridge. The instrument should permit the direct measurement of the deflections on the recording drum within ± 5 per cent. If the instrument has the form shown, determine the spring factor for the instrument. $W = 2$ lb.

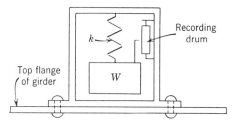

Prob. 12–14

12–15. The frame of problem 12–9 rests on a concrete foundation. Assume that a steady state horizontal ground motion in the plane of the frame, may be stated as having a period of 2 sec and a maximum acceleration of 30 in. per sec². Also assume that the frame and its mass have a natural frequency of 80 cycles per min. Determine the dynamic horizontal shearing force acting on the frame as a percentage of W if W equals 10,000 lb. *Ans.* 9.1 per cent.

12–16. The frame of problem 12–9 is subjected to a steady state horizontal ground motion having a maximum amplitude of 1.5 in., and a period of 1.5 sec. If the frame and its mass have a natural frequency of 80 cycles per min, determine the dynamic horizontal shearing force acting on the frame as a percentage of W if $W = 10,000$ lb.

12–17. This problem is identical to the example problem of Fig. 12–14 in text except for the following changes in given data. Let $k = 50$ lb per in. and $W = 1960$ lb. The force Q is defined as indicated in Fig. 12–14. Assuming finite time intervals, calculate the maximum displacement of the weight and also plot the results in a manner similar to Fig. 12–15 in text.

12–18. This problem is identical to the example problem of Fig. 12–17 in text except for the following changes in given data. Let the disturbing force have a duration of 0.6 sec and vary linearly as indicated. Assuming finite time intervals, calculate the maximum displacement of structure and the maximum shearing force transmitted to the foundations. Plot the displacement results versus time.

12–19. A structure is subjected to a horizontal disturbing force of 2 sec duration, at the level of mass concentration. If the effective mass of the structure is 1000 slugs and its natural frequency is 1 sec, determine the maximum horizontal displacement of the structure. The disturbing force is maximum at $t = 0$ and has a maximum value of 10,000 lb. Its variation with respect to time is parabolic.

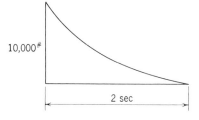

Prob. 12–19

12–20. A ship approaches a pier at a velocity of 1.5 ft per sec. The ship is moving broadside and it is assumed that it contacts all elastic pier bumpers at the same instant. If the ship has a weight of 10,000 long tons, determine the total maximum force transmitted to bumpers if the motion of the ship is arrested after a bumper compression of 0.75 ft.   *Ans.* 1868 long tons.

12–21. Solve problem 12–5 by the energy method.

12–22. An elastic beam supports three concentrated weights as shown. Determine the natural fundamental frequency by the energy method. The beam has constant *EI*. Neglect weight of the beam.

$$Ans.\ f = \frac{1}{2\pi} \sqrt{\frac{16.3EIg}{WL^3}}.$$

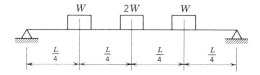

Prob. 12–22

12–23. A beam has a simple span of *L* and is uniform in cross-section. The beam has a weight of *w* lb per unit of length. By assuming the vibrating configuration to have the shape of one-half of a sine curve, determine the proportion of the mass of the beam that should be concentrated at the center line of the span to determine the natural frequency. Solve by the energy method.
$$Ans.\ 0.5wL/g.$$

12–24. A cantilever beam of uniform weight *w* per unit of length and length *L* is set into harmonic vibration. Assuming that the elastic curve is a second degree parabola, determine the equivalent reduced mass of the beam, which, when concentrated at the free end, will suffice for approximating the natural frequency.

12–25. Determine the frequencies of the fundamental mode and second mode. Check your result for the fundamental frequency by the energy method. Constant *EI* and neglect weight of the beam. Let $W = 7720$ lb.

$$Ans.\ f_1 = \frac{1.57}{2\pi} \sqrt{\frac{EI}{L^3}};\ f_2 = \frac{5.76}{2\pi} \sqrt{\frac{EI}{L^3}}.$$

Prob. 12–25

12–26. Determine the frequency of all modes of vibration.   Constant EI and neglect weight of the beam.   Let $W = 3860$ lb.

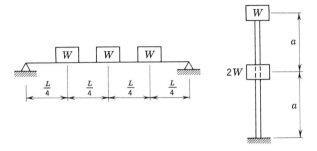

Prob. 12–26   Prob. 12–27

12–27. Determine the frequency of the modes of vibration.   The mast has a constant EI.   Neglect weight of the mast.   Let $W = 3860$ lb.

$$Ans.\ f_1 = 0.0278\ \sqrt{EI/a^3};\ f_2 = 0.1445\ \sqrt{EI/a^3}.$$

Axially Loaded Members
and Beam Columns

13

13–1. INTRODUCTION

The general theoretical views of column stability and buckling were dealt with in mechanics of materials, and the purpose of this chapter is to enlarge on these theoretical concepts and add analytical methods for handling the general problem. In addition, the effect of axial load on moment distribution procedures is considered.

13–2. EULER CONCEPTS

As a matter of review consider the axially loaded pin-ended column of Fig. 1a. If a perfectly straight member is postulated, with complete alignment of the line of action of end loads P with center line of column, the column would remain straight with increasing load. This ideal state is probably never reached and minor imperfections would cause the column to bend under load. However, it may also be assumed that a small chance lateral force, say F of Fig. 1b, would displace the column by Δ_1, causing bending moment in column due to F and also to P. An increase in either F or P will increase Δ_1. If maximum stresses in the column are below the proportional limit of the material, the column will return to its straight form of Fig. 1a upon the removal of F, provided P is below the Euler critical load P_{cr}. The critical load by definition is that axial load which would just maintain a deflected equilibrium configuration. Thus, with P_{cr} on the column, the column would remain in a deflected equilibrium configuration with Δ indeterminate. That is to say, as long as Δ is small, loads of less than P_{cr} would not maintain the configuration, and end loads greater than P_{cr} would cause Δ to increase.

486

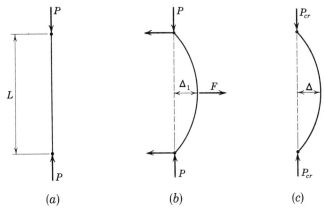

FIG. 13-1

Significantly, this increase in deflection for loads greater than P_{cr} implies that P_{cr} is the ultimate static load condition, or failure load. The phenomenon of failure by buckling is unique in contrast to tearing and fracture in tension and crushing in compression.

Euler was the first to note these concepts. The indeterminateness of Δ may be explained by an approximate approach. In Fig. 2a, assume that the equilibrium configuration is defined by a second degree parabola. The bending moment curve is also parabolic, Figure 2b, where moment

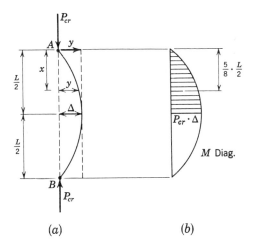

FIG. 13-2

values equal $P_{cr}y$. If the column has a uniform value of EI, then by moment-area principles

$$EI\Delta = \frac{2}{3}(P_{cr}\Delta)\left(\frac{L}{2}\right)\left(\frac{5}{8}\frac{L}{2}\right)$$

or

$$P_{cr} = \frac{48EI}{5L^2} = \frac{9.6\ EI}{L^2}$$ (13-1)

It should be noted that Δ appears on both sides of the initial deflection equation and cancels in the final expression, showing that the derivation is valid for any small initial value of Δ and that Δ is indeterminate. The critical load $9.6EI/L^2$ is approximate, but indicates that the only property of the material which enters is the modulus of elasticity.

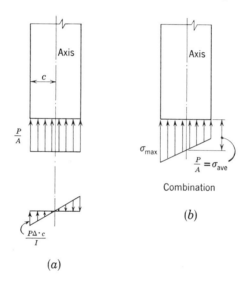

Fig. 13–3

The second notable observation is that nothing is implied by these abstractions concerning the stress level or state of deformation. It may be recalled, however, that one of the restrictions must be that the proportional limit is not exceeded. Figure 3a shows the detail of stress analysis at the column's critical section by combining axial stress and bending stress. The maximum bending stress depends on P and Δ as well as section modulus, but since Δ is indeterminate the bending stress is indeterminate. In Fig. 3b the average stress at the center line of the column will continue to be equal to P/A for all values of Δ, while σ max depends on Δ. It may therefore be noted that if P/A is limited to proportional limit values,

σ max will be greater than this limit if the column deflects, and when the end loads are removed the column would retain a small set in deflection.

In order to determine additional relationships let $I = Ar^2$ where r is the radius of gyration of the column cross-section about the bending axis. Then

$$P_{cr} = \frac{9.6EAr^2}{L^2} \qquad (13\text{-}2)$$

and

$$\sigma_{cr} = \frac{P_{cr}}{A} = \frac{9.6E}{(L/r)^2} \qquad (13\text{-}3)$$

σ_{cr} is the average stress at the center line, which, of course, tells us nothing more than that if σ_{cr} is present any small deflection $\varDelta$ tends to produce a σ max greater than average stress. The basic formulation for σ_{cr}, equation 13-3, also indicates that σ_{cr} is inversely proportional to $(L/r)^2$, where L/r is termed the slenderness ratio.

We may expand these basic views by developing Euler's formulation for critical load for an end loaded column as follows:

From Fig. 2a, the bending moment, with A as an origin for x and y positive as shown, is $+Py$.

Since
$$EI\frac{d^2y}{dx^2} = -M = -Py$$

$$EI\frac{d^2y}{dx^2} + Py = 0$$

or letting $k^2 = P/EI$ we determine the fundamental differential equation

$$\frac{d^2y}{dx^2} + k^2y = 0$$

The general solution of this differential equation is

$$y = A \cos kx + B \sin kx$$

where the two constants A and B may be evaluated from the following conditions.

When
$$x = 0, \qquad y = 0; \ A = 0$$

$$x = L/2, \quad y = \varDelta \ ; \ B = \frac{\varDelta}{\sin (kL/2)}$$

The equation for y then becomes

$$y = \frac{\varDelta}{\sin (kL/2)} \sin kx \qquad (13\text{-}4)$$

The condition of zero slope at center line first involves taking the derivative

$$\frac{dy}{dx} = \frac{k\varDelta}{\sin kL/2} \cdot \cos kx$$

and by substituting $\dfrac{dy}{dx} = 0$ and $x = L/2$ it may be seen that cos $(kL/2)$ must equal zero for a nontrivial solution to exist, and this will occur when

$$\frac{kL}{2} = \frac{\pi}{2}, \ \frac{3}{2}\pi, \ \frac{5}{2}\pi, \ \text{etc.}$$

Since $\pi/2$ is the solution that provides the half-sine wave configuration and the smallest critical load, we may write

$$\frac{kL}{2} = \frac{\pi}{2}, \ \text{ or } \ kL = \pi$$

as the solution for the present problem; and since

$$k^2 = \frac{P}{EI} \ \text{ and } \ P = P_{cr}$$

we obtain

$$P_{cr} = \frac{\pi^2 EI}{L^2} \tag{13-5}$$

Equation 13-5 is the standard form of the Euler equation. Note that the deflected configuration is a sine curve instead of a parabola as was assumed in approximate solution and that the value of $\varDelta$ does not enter the final formulation. Equation 13-5 may be restated in terms of the average critical stress σ_{cr} in pounds per square inch as

$$\sigma_{cr} = \frac{\pi^2 E}{(L/r)^2} \tag{13-6}$$

Equation 13-6 is graphically interpreted in Fig. 4 in the conventional manner. Observe that the Euler curve is invalid for values of σ_{cr} above the proportional limit. The valid limits for slenderness ratio in the elastic range extends from $(L/r)'$ to higher values and this upper range of slenderness ratio is defined as the long column range. The lower range, the inelastic range, from slenderness ratio of zero to $(L/r)'$, is the range of short and intermediate column. For very short columns failure is by deformation or localized overstrain.

Theoretical considerations for the ideal solution for σ_{cr} in the intermediate range have varied from many empirical solutions to the currently accepted Engesser solution. The concern for a correct theoretical solution

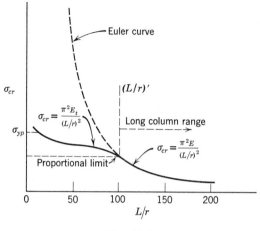

Fig. 13–4

represents one of the most important problems in applied mechanics and it is beyond the scope of this book to discuss its complete ramifications. Engesser modified the Euler equation for the inelastic range by introducing E_t, the tangent modulus, and restating equation 13-6 as

$$\sigma_{cr} = \frac{\pi^2 E_t}{(L/r)^2} \tag{13-7}$$

In fact, equation 13-7 now becomes the general solution for the pin-ended column of Fig. 2a, since for a long column $E_t = E$, and for the intermediate

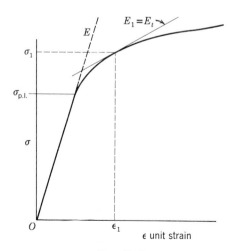

Fig. 13–5

column range the stress-strain diagram determines the tangent modulus. Figure 5 is a stress-strain diagram indicating the changing value of the modulus. For every value of σ_1 a finite value for $E_t = E_1$ exists. The value of L/r from equation 13-7 is determined, employing this modulus, and σ_1 and a point on the curve of Fig. 4 located. For low-carbon structural steels, with a sharp yield point knee, the upper limit of the Engesser curve may be taken as the yield point of the material. The theoretical validity of equation 13-7 has been verified by many tests.

13–3. EFFECT OF END CONDITIONS

The free standing column of Fig. 6, may be treated as a pin-ended column with effective length equal to $2L$ instead of L. The critical load from equation 13-5 would be $\pi^2EI/4L^2$, or one-fourth as much as for a pin-ended column of length L.

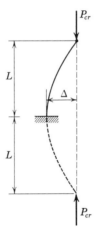

FIG. 13–6

In Fig. 7, the column is restrained against rotation at A and B, but A is free to move towards B. An assumed deflected equilibrium configuration of the column with critical load P is shown and the critical load is to be determined. The deflection Δ is general and indeterminate.

With A as the origin and y positive as shown, we may write

$$EI\frac{d^2y}{dx^2} = -Py + M_A$$

or
$$\frac{d^2y}{dx^2} + k^2y = \frac{M_A}{EI} = \frac{k^2M_A}{P}$$

The solution of this differential equation is the general solution of the homogeneous equation plus a particular solution, which may be shown to be

$$y = A \cos kx + B \sin kx + \frac{M_A}{P}$$

The end condition $x = 0$, $y = 0$, leads to the determination of

$$A = -\frac{M_A}{P}$$

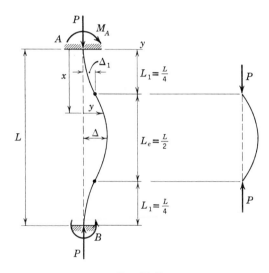

FIG. 13–7

The slope condition at A may be used to determine constant B

$$\frac{dy}{dx} = -Ak \sin kx + Bk \cos kx$$

and with
$$x = 0, \quad \frac{dy}{dx} = 0$$

$$B = 0$$

or finally
$$y = -\frac{M_A}{P} \cos kx + \frac{M_A}{P}$$

$$y = \frac{M_A}{P}(1 - \cos kx) \qquad (13\text{-}8)$$

but $\dfrac{dy}{dx}$ must also equal zero when $x = \dfrac{L}{2}$

With this condition substituted in the $\dfrac{dy}{dx}$ equation we obtain

$$-\frac{M_A}{P} k \sin \frac{kL}{2} = 0$$

and for a nontrivial solution $\sin kL/2$ equals zero, or

$$\frac{kL}{2} = 0, \pi, 2\pi$$

The smallest valid value of $kL/2$ equals π, hence

$$k = \frac{2\pi}{L}$$

and

$$k^2 = \frac{4\pi^2}{L^2} = \frac{P}{EI}$$

from which, with $P = P_{cr}$

$$P_{cr} = \frac{4\pi^2 EI}{L^2} \tag{13-9}$$

Thus, the fixed-ended column in a deflected equilibrium configuration for a long column has a critical load four times that for a pin-ended column of equal length. It may be rationalized from equation 13-9 that the distance between inflection points is $L/2$, or through further mathematical analysis the points of inflection or points of zero bending moment of Fig. 7 may be located as follows.

Since bending moment is $EI\dfrac{d^2y}{dx^2}$, two differentiations of equation 13-8 set equal to zero provide the basic equation

$$EI\frac{M_A}{P} k^2 \cos kx = 0$$

where, if a deflected equilibrium configuration exists, $\cos kx$ must equal zero, requiring kx to equal $\pi/2$. Then, since

$$k = \frac{2\pi}{L}$$

$$L_1 = x = \frac{\pi}{2}\left(\frac{L}{2\pi}\right) = \frac{L}{4}$$

Owing to symmetry, a second point of inflection exists at $L/4$ from B. With inflection points known the central part of column is similar to a pin-ended column of effective length L_e equal to $L/2$. Had this been known, or surmised, the critical load could have been obtained directly from the Euler load equation as expressed by equation 13-5. All critical load formulations may be stated as

$$P_{cr} = \frac{\pi^2 EI}{(L_e)^2} \qquad (13\text{-}10)$$

where the effective length L_e is dependent upon the restraining end conditions of the column, or if $n = L/L_e$, an alternate formulation would be

$$P_{cr} = \frac{n^2 \pi^2 EI}{L^2} \qquad (13\text{-}11)$$

13–4. EULER APPROXIMATIONS BY FINITE DIFFERENCES

For many special cases a differential equation analysis to obtain the critical load is laborious and is not warranted. Recourse may be had to finite difference approximations. The finite difference analysis starts with

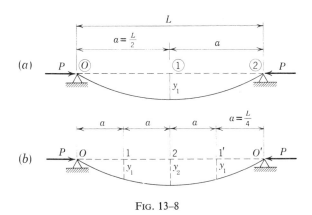

FIG. 13–8

the assumption of a deflected configuration. It is then rationalized that this configuration is held in a state of static equilibrium under the action of the critical axial loads.

In Fig. 8a a pin-ended member is assumed under critical axial compression with a central deflection of y_1 and with two finite intervals of a.

From basic deflection theory

$$\frac{d^2y}{dx^2} = -\frac{M}{EI}$$

and specifically for point 1 M equals Py_1 requiring

$$\frac{d^2y}{dx^2} = -\frac{Py_1}{EI}$$

A finite difference substitution similar to equation 2-25 for the second derivative and in terms of the notation of Fig. 8a gives

$$y_0 - 2y_1 + y_2 = -\frac{a^2Py_1}{EI}$$

and since

$$y_0 = y_2 = 0$$

we determine that

$$y_1 = \frac{a^2Py_1}{2EI}$$

and since y_1 cancels out and P may be interpreted as P_{cr}, the critical load is

$$P_{cr} = \frac{2EI}{a^2} = \frac{8EI}{L^2} \tag{13-12}$$

The value of P_{cr} by equation 13-12 is in error by 19 per cent compared with the exact Euler solution as a result of the assumed configuration being too approximate. The result can be improved by assuming the shape of Fig. 8b, and using four intervals.

The formulation of $\dfrac{d^2y}{dx^2} = -\dfrac{M}{EI}$ at point 1 of Fig. 8b by finite differences is

$$-2y_1 + y_2 = -\frac{a^2Py_1}{EI}$$

and at point 2

$$-2y_2 + 2y_1 = -\frac{a^2Py_2}{EI}$$

Letting $k^2 = P/EI$ the above equations may be restated as

$$(a^2k^2 - 2)y_1 + y_2 = 0 \tag{13-13}$$

$$2y_1 + (a^2k^2 - 2)y_2 = 0 \tag{13-14}$$

which represent linear homogeneous equations linking y_1 and y_2. By algebraic determinants the solution of the two general nonhomogeneous equations

$$a_1 x + b_1 y = c_1$$

$$a_2 x + b_2 y = c_2$$

was

$$x = \frac{\begin{vmatrix} c_1 b_1 \\ c_2 b_2 \end{vmatrix}}{\begin{vmatrix} a_1 b_1 \\ a_2 b_2 \end{vmatrix}} \qquad y = \frac{\begin{vmatrix} a_1 c_1 \\ a_2 c_2 \end{vmatrix}}{\begin{vmatrix} a_1 b_1 \\ a_2 b_2 \end{vmatrix}}$$

and if constants c_1 and c_2 are zero, the determinants in the numerator each evaluate to zero providing a trivial solution of $x = 0$ and $y = 0$. The nontrivial solution for x and y, when c_1 and c_2 equal zero, must be found by setting the determinant of the coefficients equal to zero. This requirement is represented by

$$\begin{vmatrix} a_1 b_i \\ a_2 b_2 \end{vmatrix} = 0$$

since

$$x \begin{vmatrix} a_1 b_1 \\ a_2 b_2 \end{vmatrix} = 0$$

and x is not zero for a nontrivial solution.

Similarly,

$$y \begin{vmatrix} a_1 b_1 \\ a_2 b_2 \end{vmatrix} = 0$$

where the determinant must equal zero.

For the two homogeneous equations, equations 13-13 and 13-14, formulated for Fig. 8b, the determinant of the coefficients is

$$\begin{vmatrix} (a^2 k^2 - 2) & 1 \\ 2 & (a^2 k^2 - 2) \end{vmatrix} = 0$$

which, when evaluated by determinant rules as

$$a_1 b_2 - a_2 b_1 = 0$$

becomes the determinantal equation

$$(a^2 k^2 - 2)^2 - 2 = 0 \qquad\qquad (13\text{-}15)$$

An algebraic solution of this equation is

$$a^2 k^2 - 2 = \pm \sqrt{2}$$
$$a^2 k^2 = \pm \sqrt{2} + 2$$
$$a^2 k^2 = 0.586 \quad \text{or} \quad 3.414$$

To determine the smallest load to maintain the equilibrium configuration, take

$$a^2k^2 = 0.586$$

$k^2 = P/EI$ and P may be interpreted as P_{cr}.

Thus,
$$P_{cr} = \frac{0.586EI}{a^2}$$

and with
$$a = \frac{L}{4}$$

$$P_{cr} = \frac{9.38EI}{L^2} \tag{13-16}$$

The error decreases with the increase in number of intervals, approaching zero as the number of intervals approaches infinity.

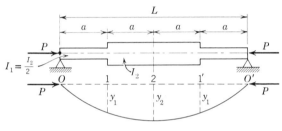

FIG. 13–9

The critical load for a column of variable moment of inertia may be determined by a similar approach.

Figure 9 represents such a problem. The analysis now requires the use of an average value for M/EI at changes in cross-section.

Using I_2 as the base, $k^2 = P/EI_2$, or $k^2 = P/2EI_1$. The formulation of the finite difference equations follows.

At point 1, Fig. 9

$$-2y_1 + y_2 = \left(\frac{-a^2k^2 - 2a^2k^2}{2}\right)y_1$$

At point 2, Fig. 9

$$2y_1 - 2y_2 = -a^2k^2y_2$$

Therefore in homogeneous form the equations become

$$(1.5a^2k^2 - 2)y_1 + y_2 = 0$$
$$2y_1 + (a^2k^2 - 2)y_2 = 0$$

For the nontrivial solution the determinant is formed as

$$\begin{vmatrix} (1.5a^2k^2 - 2) & 1 \\ 2 & (a^2k^2 - 2) \end{vmatrix} = 0$$

and the determinantal equation is

$$(1.5a^2k^2 - 2)(a^2k^2 - 2) - 2 = 0$$

and when solved yields $a^2k^2 = 0.467$ as the smaller root.   Then

$$P_{cr} = \frac{0.467EI_2}{a^2}$$

$$P_{cr} = \frac{7.47EI_2}{L^2}$$

13–5. EULER APPROXIMATIONS BY NEWMARK METHOD

Newmark has established a numerical procedure for determination of critical load which eliminates the mathematics of either the exact differential equation approach or the finite difference approximations.   In brief, the procedure is to estimate a reasonable trial deflected configuration and calculate the trial bending moments produced by the axial load.   The deflections are then computed as produced by these bending moments and compared with the estimated trial deflections.   The ratio of the trial deflections to the computed deflections must equal one for the stable configuration to exist.   This must be true since it has been pointed out that the actual deflection, as long as it is small, is immaterial insofar as the logic of defining the critical load is concerned.

Figure 10a represents an axially loaded pin-ended member of uniform cross-section.   For this example four intervals or panels are employed with the trial configuration based upon assuming the deflected panel points as falling on a parabolic curve.   Letting y_a equal these original trial deflections, at any order of magnitude, and where for convenience the center deflection is taken equal to 4, the bending moment may be computed at each panel point.   Figure 10b is the M/EI diagram with the calculated concentrated angle changes.   It is to be particularly noted that the deflected configuration is considered as being represented by a series of chords, hence, the moment diagram is also assumed to have a linear variation between panel points.   The linear variation simplifies the calculation of the concentrated angle changes.   It is outside the scope of this book to present corrections entailed by considering curved moment

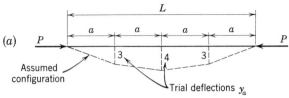

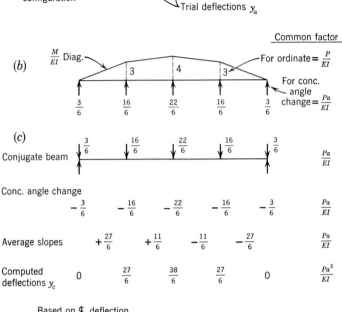

FIG. 13–10

diagrams.   Newmark, in the reference previously cited in Chapter 2, develops these corrections.   In any event, the errors in final values of critical load based on linear variation tend to become smaller as the number of intervals increase.

   The standard solution for the computed deflections is given in Fig. 10c and needs no explanation since the procedures of Art. 2-5 are followed. The computed deflections are designated as y_c.   The common factors are noted at the right of Fig. 10c.   The assumed deflections and computed

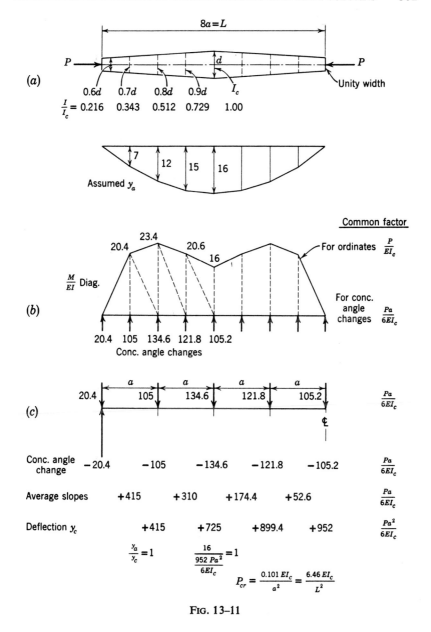

FIG. 13–11

deflections may now be introduced into the stability criterion of $y_a/y_c = 1$. Utilizing center line deflections, it is shown in Fig. 10 that $P_{cr} = 10.1EI/L^2$. It is considered the more correct practice to give weight to all deflections and utilize the criterion $\dfrac{\sum y_a}{\sum y_c} = 1$.   This practice yields $P_{cr} = 10.4EI/L^2$. These values for the critical load are larger than the Euler value.   If the assumed configuration had been taken to follow a sine curve variation, much greater precision would result, although in practice the answer based on the parabolic shape may be considered to be sufficiently precise.   An improved value would be secured by increasing the number of intervals.

As a second example of this approach to critical load, consider the shaped column of Fig. 11*a* where buckling is considered to occur in the vertical plane.   The calculations are referenced to I_c, the moment of inertia at the center line.   A parabolic deflection configuration is assumed with center line deflection y_a taken arbitrarily at 16.   The M/EI diagram is drawn and concentrated angle changes computed.   Standard procedures in Fig. 11*c* lead to the computed deflections.   All calculations are given in Fig. 11.   The critical load based solely on the center line deflection is $6.46EI_c/L^2$.

13–6. ENERGY METHOD FOR COLUMNS

As discussed in the preliminary section of this chapter, a perfect column may remain straight under axial load or it may bend.   Either configuration is an equilibrium configuration if the end loads are equal to the buckling load.   If a critically loaded straight column is deflected by an outside influence it will pass into equilibrium in a bent form.   Figure 12*a* represents the loaded straight column and Fig. 12*b* the loaded bent column.   In changing from the straight form to the curved form the ends of columns approach one another by λ and the external loads P equal to P_{cr} do external work equal to $P\lambda$ or $P_{cr}\lambda$.   The external work done must be stored in the column as potential internal strain energy due to bending.

From previous considerations internal strain energy is

$$U = \int_0^L \frac{M^2\, dx}{2EI} = \frac{EI}{2} \int_0^L \left(\frac{d^2 y}{dx^2}\right)^2 dx$$

The value of λ is determined from differential geometrical considerations of Fig. 12*c*, namely,

$$d\lambda = ds - dx = \sqrt{dx^2 + dy^2} - dx$$

$$= dx\left[1 + \left(\frac{dy}{dx}\right)^2\right]^{1/2} - dx$$

and for small values of $\dfrac{dy}{dx}$

$$\left[1 + \left(\frac{dy}{dx}\right)^2\right]^{1/2} \approx 1 + \frac{1}{2}\left(\frac{dy}{dx}\right)^2$$

by two terms of a binomial series expansion, or

$$d\lambda = \frac{1}{2}\left(\frac{dy}{dx}\right)^2 dx$$

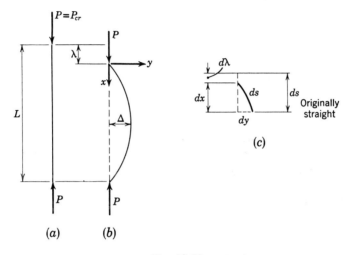

FIG. 13–12

Then λ for the entire column by integration is

$$\lambda = \int_0^L d\lambda = \frac{1}{2}\int_0^L \left(\frac{dy}{dx}\right)^2 dx$$

with $P_{cr}\lambda = U$

$$P_{cr} = \frac{EI \int_0^L \left(\dfrac{d^2y}{dx^2}\right)^2 dx}{\int_0^L \left(\dfrac{dy}{dx}\right)^2 dx} \tag{13-17}$$

Since only the configuration of the column axis enters into the evaluation for critical load by equation 13-17, we are free to assume the shape. To illustrate, assume the configuration of Fig. 12b to be

$$y = \Delta \sin \frac{\pi x}{L}$$

By successive differentiations we obtain

$$\frac{dy}{dx} = \frac{\Delta\pi}{L}\cos\frac{\pi x}{L}$$

$$\frac{d^2y}{dx^2} = -\frac{\Delta\pi^2}{L^2}\sin\frac{\pi x}{L}$$

The evaluations of the integrals of equation 13-17 are made as follows

$$\int_0^L \left(\frac{dy}{dx}\right)^2 dx = \frac{\Delta^2\pi^2}{L^2}\int_0^L \cos^2\frac{\pi x}{L}\ dx = \frac{\Delta^2\pi^2}{2L}$$

and

$$\int_0^L \left(\frac{d^2y}{dx^2}\right)^2 dx = \frac{\Delta^2\pi^4}{L^4}\int_0^L \sin^2\frac{\pi x}{L}\ dx = \frac{\Delta^2\pi^4}{2L^3}$$

Then, from equation 13-17, the critical load is

$$P_{cr} = \frac{(\Delta^2\pi^4/2L^3)\ EI}{\Delta^2\pi^2/2L} = \frac{\pi^2 EI}{L^2} \tag{13-18}$$

The value of P_{cr} is exact owing to the fact that the sine curve assumption is exact. In general, the true curve will not be known and the result will be approximate. Timoshenko* states that greater accuracy will result when employing approximate assumed curves if the strain energy is computed from

$$U = \int \frac{M^2\ dx}{2EI}$$

This is true since any approximation in y will introduce a larger approximation in the second derivative. In any event, the critical load will be greater than the theoretical exact value since only the theoretical configuration is compatible with minimum load. This situation may be demonstrated by assuming that the configuration of Fig. 12b follows a second degree variation. The first solution is made by using equation 13-17.

With $y = \dfrac{4\Delta}{L^2}(xL - x^2)$

and by successive differentiations

$$\frac{dy}{dx} = \frac{4\Delta}{L^2}(L - 2x)$$

$$\frac{d^2y}{dx^2} = -\frac{8\Delta}{L^2}$$

---

* See Reference 4.

The denominator of equation 13-17 is

$$\int_0^L \left(\frac{dy}{dx}\right)^2 dx = \frac{16\Delta^2}{L^4}\int_0^L (L-2x)^2\, dx = \frac{16\Delta^2}{3L}$$

and the integral in the numerator evaluates to

$$\int_0^L \left(\frac{d^2y}{dx^2}\right)^2 dx = \frac{64\Delta^2}{L^4}\int_0^L dx = \frac{64\Delta^2}{L^3}$$

Finally, by equation 13-17

$$P_{cr} = EI\frac{64\Delta^2/L^3}{16\Delta^2/3L} = \frac{12EI}{L^2}$$

which is a very approximate answer when compared with the exact Euler equation. However, returning to the basic equation

$$P_{cr}\lambda = U$$

and letting

$$U = \int_0^L \frac{M^2\, dx}{2EI} = \frac{P_{cr}^2}{2EI}\int_0^L y^2\, dx$$

$$U = \frac{P_{cr}^2}{2EI}\left(\frac{16\Delta^2}{L^4}\right)\int_0^L (xL - x^2)^2\, dx$$

$$U = \frac{4P_{cr}^2\,\Delta^2 L}{15EI}$$

With

$$\lambda = \frac{1}{2}\times\frac{16\Delta^2}{3L} = \frac{8\Delta^2}{3L}$$

we write the equality of external work and internal strain energy as

$$P_{cr}\left(\frac{8\Delta^2}{3L}\right) = \frac{4P_{cr}^2\Delta^2 L}{15EI}$$

from which

$$P_{cr} = \frac{10EI}{L^2}$$

This indicates a closer approximation to the exact Euler equation for the reason stated, namely the elimination of the error due to the use of the second derivative.

13–7. COLUMNS AS COMPONENTS OF FRAMES

In general, most columns exist as component members of frames or trusses and are elastically restrained at the ends by the adjoining connected members.   Figure 13a indicates the nature of the problem, where the upper end A of AB is restrained by beam AC.   If AC is long and I_1 is small, the restraining condition at A would tend to vanish and AB would revert to a pin-ended column.   For analysis purposes assume that joint A rotates through an angle θ_A as P approaches a critical value.   The moment at the top of the column in terms of θ_A, since C is fixed, may be stated by equation 4-2 as

$$M_A = \frac{4EI_1}{L_1} \theta_A$$

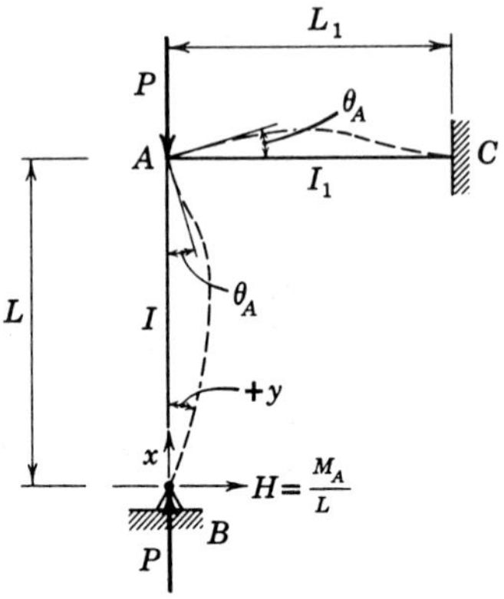

FIG. 13–13

The horizontal reaction acting to the right at B is M_A/L by a free body analysis of the column.   The differential equation of the deflection curve with origin at B is written as

$$EI \frac{d^2y}{dx^2} = -Py + \frac{M_A}{L} x$$

or
$$\frac{d^2y}{dx^2} + k^2y = \frac{k^2 M_A}{PL} x \qquad (13\text{-}19)$$

for which the general and particular solution combined is

$$y = A \cos kx + B \sin kx + \frac{M_A}{PL} x \qquad (13\text{-}20)$$

To determine the constants, let

$$x = 0 \text{ when } y = 0, \text{ requiring } A = 0$$

Letting
$$x = L \text{ when } y = 0$$

$$0 = B \sin kL + \frac{M_A}{P}$$

we obtain
$$B = -\frac{M_A}{P \sin kL}$$

Therefore, the final equation for deflection is

$$y = -\frac{M_A}{P \sin kL} \sin kx + \frac{M_A}{PL} x \qquad (13\text{-}21)$$

To use the slope condition at A we take the first derivative of equation 13-21 as

$$\frac{dy}{dx} = -\frac{kM_A}{P \sin kL} \cos kx + \frac{M_A}{PL}$$

At A, where $x = L$,

$$\frac{dy}{dx} = -\theta_A$$

for the positive direction of y chosen in Fig. 13, hence

$$-\frac{M_A L_1}{4EI_1} = -\frac{kM_A}{P} \cot kL + \frac{M_A}{PL}$$

and canceling common factors

$$-\frac{L_1}{4EI_1} = -\frac{k \cot kL}{P} + \frac{1}{PL}$$

Letting $k^2 = P/EI$, this equation becomes

$$\frac{1}{EI}\left(\frac{1 - kL \cot kL}{k^2 L}\right) = -\frac{L_1}{4EI_1}$$

whereupon
$$(kL \cot kL - 1) = \frac{k^2 L L_1}{4}\left(\frac{I}{I_1}\right) \qquad (13\text{-}22)$$

Equation 13-22 is the controlling equation for calculating the critical load.

If I_1 is infinitely large, θ_A will approach zero and I/I_1 will also approach zero, or from equation 13-22, we obtain

$$kL \cot kL - 1 = 0$$

$$\cot kL = \frac{1}{kL}$$

or
$$\tan kL = kL$$

The resulting transcendental equation may be solved by trial and error for smallest value of kL, and this is found to be $kL = 4.493$. The value of the critical load is then found as

$$P_{cr} = k^2 EI = \frac{20.16EI}{L^2} = \frac{\pi^2 EI}{(0.7L)^2} \qquad (13\text{-}23)$$

It is to be noted that the effective length is about 0.7 of L, owing to end restraint.

Another example is provided by letting

$$I_1 = I, \quad \text{and} \quad L_1 = L$$

then, from equation 13-22,

$$kL \cot kL - 1 = \frac{k^2 L^2}{4}$$

and the smallest kL by trial-and-error solution is $kL = 3.84$, making

$$P_{cr} = \frac{3.84^2 EI}{L^2} = \frac{14.8EI}{L^2}$$

or

$$P_{cr} = \frac{\pi^2 EI}{(0.81L)^2}$$

The effective length increased from $0.7L$ to $0.81L$ by decreasing rigidity of the beam. Thus it is demonstrated that a full analysis of a column must take into account frame action. This refinement in analysis is more important for long columns than for intermediate and short columns. Designers usually ignore these effects and use the effective length as L. This practice in light structures fails to take advantage of the benefits of restraints imposed by frame action.

13–8. BEAM COLUMNS

The structural combination of a beam supporting a transverse load while subjected to an axial compression is termed a *beam column*. The simultaneous action of the two load systems produces an interacting effect on deflections and the principle of superposition is not applicable. In Fig. 14a, the beam has deflected under the action of Q, and in Fig. 14b, when the beam is loaded solely with an axial load P, no beam deflection is produced. The deflected configuration for Q and P simultaneously applied cannot be represented by the superposition of these separate deflections.

The lateral load deflection provides a lever arm at every general cross-section by which the axial load P produces additional bending moment. This additional bending moment increases the transverse load deflections, and for the final equilibrium position the deflection may be considerably greater than that due to Q alone. Only elementary combinations will be

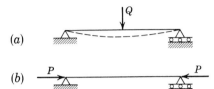

FIG. 13–14

dealt with in this book, and as a first demonstration refer to Fig. 15. The problem is first analyzed from a bending moment point of view. The bending moment at any section with origin of coordinates at A is

$$M = \frac{wL}{2} x - \frac{wx^2}{2} + Py$$

where M is a function of y as well as of x. Since both x and y enter the equation a differential equation solution is preferable to integration procedures. As the first step in transforming the moment equation to a differential equation, the equation for moment is differentiated once with respect to x, and the shear equation is derived as

$$V = \frac{dM}{dx} = \frac{wL}{2} - wx + P\frac{dy}{dx}$$

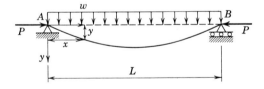

FIG. 13–15

and after a second differentiation the load equation is derived as

$$q = \frac{d^2M}{dx^2} = -w + P\frac{d^2y}{dx^2}$$

The last term of the expressions for shear and load represent the augmentations to the usual transverse load effects due to the axial load.

Furthermore, from $EI \dfrac{d^2y}{dx^2} = -M$ (note: positive y downward), the last term of the load equation may be restated as $-PM/EI$ and the desired differential equation becomes

$$\frac{d^2M}{dx^2} + \frac{PM}{EI} = -w \qquad (13\text{-}24)$$

The general solution of equation 13-24, letting $k^2 = P/EI$, is

$$M = A \cos kx + B \sin kx - \frac{w}{k^2} \qquad (13\text{-}25)$$

The constant A equals w/k^2 from the condition $M = 0$ at $x = 0$. The constant B from the condition $M = 0$ at $x = L$ is

$$B = \frac{w}{k^2} \left[\frac{1 - \cos kL}{\sin kL} \right]$$

The final simplified equation for moment, by substituting the constants in equation 13-25, is

$$M = \frac{w}{k^2} \left[\frac{1 - \cos kL}{\sin kL} \sin kx + \cos kx - 1 \right] \qquad (13\text{-}26)$$

The maximum bending moment at the center line is obtained from equation 13-26 when $x = L/2$.

$$M \max = \frac{w}{k^2} \left[\frac{1 - \cos kL}{\sin kL} \sin \frac{kL}{2} + \cos \frac{kL}{2} - 1 \right] \qquad (13\text{-}27)$$

Letting
$$\sin kL = 2 \sin \frac{kL}{2} \cos \frac{kL}{2}$$

and
$$\cos kL = 2 \cos^2 \frac{kL}{2} - 1$$

$$M \max = \frac{w}{k^2} \left(\sec \frac{kL}{2} - 1 \right) \qquad (13\text{-}28)$$

If limit rules are applied, M max may be shown to be $wL^2/8$ as the axial load approaches zero, or as k approaches zero, as it should for beam action alone. For increasing ratios of P to P_E the Euler critical load, the maximum moment increases. For studying this effect the equation 13-28 for maximum moment may be algebraically and trigonometrically transformed to

$$M \max = \frac{wL^2}{8} \cdot \frac{8[1 - \cos (kL/2)]}{k^2L^2 \cos (kL/2)}$$

By definition,
$$k^2 = \frac{P}{EI} \quad \text{and} \quad P_E = \frac{\pi^2 EI}{L^2}$$

Then
$$k^2 L^2 = \pi^2 \frac{P}{P_E} \quad \text{and} \quad \frac{kL}{2} = \frac{\pi}{2}\sqrt{\frac{P}{P_E}}$$

and M max can be restated as

$$M \text{ max} = \frac{wL^2}{8}\left\{ \frac{8P_E}{\pi^2 P} \frac{\left[1 - \cos\left(\frac{\pi}{2}\sqrt{\frac{P}{P_E}}\right)\right]}{\cos\left(\frac{\pi}{2}\sqrt{\frac{P}{P_E}}\right)} \right\} \tag{13-29}$$

where the term in brackets is the factor expressing the magnification of moment due to P. As an example, take $P = 2EI/L^2$, then,

$$\frac{P_E}{P} = \frac{\pi^2}{2} = 4.93, \quad \text{and} \quad \frac{P}{P_E} = 0.203$$

$$\cos\frac{\pi}{2}\sqrt{\frac{P}{P_E}} = \cos(0.706) = \cos 40.5° = 0.76$$

Then
$$M \text{ max} = \frac{wL^2}{8}\frac{8}{\pi^2}(4.93)\left(\frac{1 - 0.76}{0.76}\right)$$

$$= \frac{wL^2}{8}(1.26)$$

which indicates a 26 per cent increase due to P. In many instances when w is small, or equal only to the weight of the member, such increase may be small and inconsequential. However, in studying the ultimate load capacity for a structure where P is a direct function of the load w, it is imperative to study these effects. The chart of Fig. 16 also indicates that the magnification factor of equation 13-29 increases rapidly as the ratio of P/P_E increases.

The previous considerations led directly to bending moment. A complete study of beam column action also involves a knowledge of slopes and deflections. Once the equation for M is known the transitional theoretical step is

$$EI\frac{d^2y}{dx^2} = -M$$

and by substituting equation 13-26

$$EI\frac{d^2y}{dx^2} = -\frac{w}{k^2}\left[\frac{1 - \cos kL}{\sin kL}\sin kx + \cos kx - 1\right] \tag{13-30}$$

by integration

$$EI\frac{dy}{dx} = -\frac{w}{k^3}\left[-\frac{1-\cos kL}{\sin kL}\cos kx + \sin kx\right] + \frac{wx}{k^2} + C_1 \quad (13\text{-}31)$$

Constant C_1 is evaluated from the condition $\frac{dy}{dx} = 0$
when $x = L/2$, or

$$C_1 = -\frac{wL}{2k^2}$$

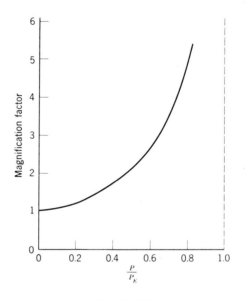

FIG. 13–16

Thus,

$$EI\frac{dy}{dx} = -\frac{w}{k^3}\left[-\frac{1-\cos kL}{\sin kL}\cos kx + \sin kx\right] + \frac{wx}{k^2} - \frac{wL}{2k^2} \quad (13\text{-}32)$$

A second integration provides

$$EIy = -\frac{w}{k^4}\left[-\frac{1-\cos kL}{\sin kL}\sin kx - \cos kx\right] + \frac{wx^2}{2k^2} - \frac{wLx}{2k^2} + C_2$$

$$(13\text{-}33)$$

From the condition $y = 0$ when $x = 0$

$$C_2 = -\frac{w}{k^4}$$

The final equation for y is

$$EIy = \frac{w}{k^4}\left[\frac{1 - \cos kL}{\sin kL}\sin kx + \cos kx - 1\right] + \frac{w}{2k^2}(x^2 - xL) \quad (13\text{-}34)$$

If $k^2 = P/EI$ is substituted in the equations for slope and deflection then the maximum end slope when $x = 0$, from equation 13-32, is

$$\left(\frac{dy}{dx}\right)_{max} = \frac{wL}{2P}\left[\frac{2\tan (kL/2)}{kL} - 1\right] \quad (13\text{-}35)$$

and maximum deflection when $x = L/2$, from equation 13-34, is

$$y\,max = \frac{w}{Pk^2}\left[\frac{1}{\cos (kL/2)} - 1 - \frac{(kL)^2}{8}\right] \quad (13\text{-}36)$$

and by algebraic manipulation the maximum deflection can be restated in the following form for comparison with laterally loaded beams.

$$y\,max = \frac{5wL^4}{384EI}\left[\frac{\dfrac{1}{\cos kL/2} - 1 - \dfrac{(kL)^2}{8}}{\dfrac{5}{384}(kL)^4}\right] \quad (13\text{-}37)$$

An alternate differential equation approach, preferred by many authors, for the beam column of Fig. 15 starts with the differential equation of the elastic curve.   Referring to Fig. 15

$$EI\frac{d^2y}{dx^2} = -M = -\frac{wLx}{2} + \frac{wx^2}{2} - Py \quad (13\text{-}38)$$

or

$$\frac{d^2y}{dx^2} + k^2y = -\frac{wLx}{2EI} + \frac{wx^2}{2EI} \quad (13\text{-}39)$$

The solution of equation 13-39 is $y = C\cos kx + D\sin kx$ plus the particular solutions.   The first term of the particular solution is derivable by assuming y as $m(wLx/2EI)$ and substituting in equation 13-39

$$0 + k^2\frac{mwLx}{2EI} = -\frac{wLx}{2EI}$$

Then $m = -1/k^2$ and the first term of the particular solution is $-wLx/2k^2EI$.

The second part of the particular solution of equation 13-39 may be taken as $(wx^2/2k^2EI - w/k^4EI)$.   If taken only as $wx^2/2k^2EI$ a residual of w/k^2EI exists due to the $\dfrac{d^2y}{dx^2}$ term, hence the particular solution must be taken as indicated to eliminate the residual.

The deflection equation becomes, when $EI = P/k^2$,

$$y = C \cos kx + D \sin kx + \frac{w}{2P}\left(x^2 - xL - \frac{2}{k^2}\right) \qquad (13\text{-}40)$$

To evaluate the constant C, let $y = 0$ when $x = 0$

$$C = \frac{w}{Pk^2}$$

To evaluate the constant D let $y = 0$ when $x = L$

$$D = \frac{w}{Pk^2}\frac{(1 - \cos kL)}{\sin kL}$$

Then equation 13-40 becomes

$$y = \frac{w}{Pk^2}\left[\frac{1 - \cos kL}{\sin kL}\sin kx + \cos kx - 1\right] + \frac{w}{2P}(x^2 - xL) \qquad (13\text{-}41)$$

Equation 13-41 for y may be compared with the previous equation 13-34 for y derived from the differential equation predicated on moment. If k^2EI is substituted for P, equation 13-41 is identical to equation 13-34. The student will thus note that the deflection curve may be found directly. The final equation for bending moment may be determined by taking the second derivative of y with respect to x starting with equation 13-41 and using the fundamental equation $EI\frac{d^2y}{dx^2} = -M$. This will confirm the moment equation 13-26 developed in the alternate manner. It is difficult to rationalize the merits of the two solutions. The student should be familiar with both fundamental approaches.

13–9. BEAM COLUMNS BY THE ENERGY METHOD

All of the ramifications of the energy approach to beam columns cannot be presented in this book, but an approximate procedure may be presented. In Fig. 17, let P and Q be simultaneously applied, and assume that the approximate deflection curve can be represented by $y = \Delta \sin \pi x/L$, where Δ is the undetermined maximum deflection. This assumption amounts to taking the first term of a Fourier series which would approximate the final configuration. In this assumed configuration, end B of the deflected beam has moved λ toward A, where

$$\lambda = \frac{1}{2}\int_0^L \left(\frac{dy}{dx}\right)^2 dx = \frac{\Delta^2}{2}\frac{\pi^2}{L^2}\int_0^L \cos^2\frac{\pi x}{L}\,dx$$

$$\lambda = \frac{\Delta^2\pi^2}{4L} \qquad (13\text{-}42)$$

The potential strain energy stored in the beam in the assumed configuration is

$$U = \int_0^L \frac{M^2\, dx}{2EI} \quad \text{or} \quad \frac{EI}{2} \int_0^L \left(\frac{d^2y}{dx^2}\right)^2 dx$$

We may substitute the second derivative of y with respect to x and obtain

$$U = \frac{EI\, \varDelta^2 \pi^4}{2L^4} \int_0^L \sin^2 \frac{\pi x}{L}\, dx$$

$$U = \frac{EI\, \varDelta^2 \pi^4}{4L^3} \tag{13-43}$$

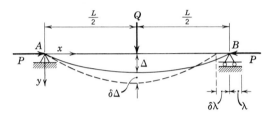

FIG. 13-17

If a small virtual displacement denoted as $\delta\varDelta$ is given to the center point of the beam, the external virtual work done by Q is $Q\, \delta\varDelta$.   Force P also does external work during this virtual displacement of $P\, \delta\lambda$ where, by differentiation of equation 13-42,

$$\delta\lambda = \frac{\pi^2\, \varDelta \delta\varDelta}{2L}$$

The virtual change in the potential strain energy due to the virtual displacement from equation 13-43 is

$$\delta U = \frac{\pi^4 EI\, \varDelta \delta\varDelta}{2L^3}$$

With both virtual external work and the change in potential strain energy known in terms of the configuration and the virtual displacement, we may equate the external work to the strain energy and obtain the equation

$$Q(\delta\varDelta) + P\left(\frac{\pi^2}{2L}\varDelta\delta\varDelta\right) = \frac{\pi^4 EI\, \varDelta\delta\varDelta}{2L^3}$$

solving

$$\varDelta = \frac{2QL^3}{\pi^4 EI(1 - PL^2/\pi^2 EI)}$$

and recognizing that the Euler critical load is $P_E = \pi^2 EI/L^2$, the center deflection is

$$\varDelta = \frac{2QL^3}{\pi^4 EI(1 - P/P_E)}$$

If the end load P equals zero, the approximate center deflection produced by Q alone would be

$$\varDelta_1 = \frac{2QL^3}{\pi^4 EI}$$

indicating that $\varDelta$ may be expressed as $\varDelta = \varDelta_1\left(\dfrac{1}{1 - P/P_E}\right)$ where the term in parentheses is the magnification factor, dependent upon the ratio of the end load to the critical load, by which the beam deflections, denoted as $\varDelta_1$, will be increased by the action of the axial load. The derived relation is only approximate but sufficiently close for most practical applications. The increase in deflection (δ) due to P may also be related to $\varDelta_1$.

The increase in deflection caused by P is

$$\delta = \varDelta - \varDelta_1$$

and by substitution

$$\delta = \frac{\varDelta_1}{1 - P/P_E} - \varDelta_1$$

or
$$\delta = \frac{\varDelta_1}{(P_E/P - 1)} \tag{13-44}$$

The approximate maximum bending moment at the center line of the span may be written as the sum of the moments produced by the transverse load and the axial load, namely,

$$M\text{ max} = \frac{QL}{4} + \frac{P\varDelta_1}{(1 - P/P_E)}$$

where $\varDelta_1$ may be calculated from $\varDelta_1 = QL^3/48EI$, the standard equation for beam deflection.

13–10. BEAM COLUMNS BY NUMERICAL APPROACH

The Newmark numerical approach provides a means for an approximate but satisfactory solution for beam columns with complex loadings. As an example, the problem of Fig. 18, a beam with two transverse loads Q and an end loading P, will be solved. The objective is to determine the maximum deflection at midspan so that maximum bending moment may be computed.   All calculations are given in Fig. 18.

The solution commences with a calculation of the simple beam deflections without the influence of P.   After determining Δ_1, the approximate

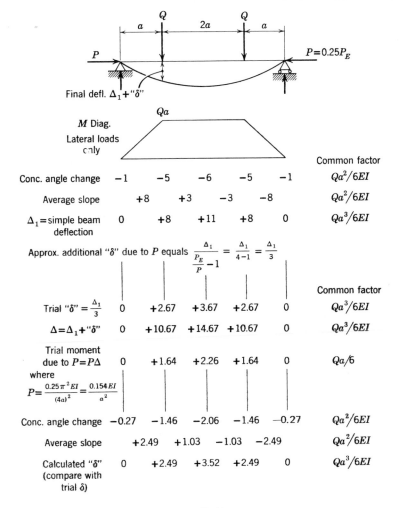

FIG. 13–18

value of additional deflection caused by P is calculated by equation 13-44 of the previous article.   This additional deflection combined with Δ_1 provides the moment arm for calculating the trial bending moments due to P. P has been taken as equal to one-quarter of P_E for this example.

With the trial bending moments the procedures of the Newmark method for deflections are again applied to obtain a calculated value for the extra deflection δ.   In calculating the concentrated angle changes, the bending moment curve (not shown) although curved is considered linear within each interval.   The calculated δ in this case compares favorably with the trial δ and no further refinement is necessary.   If the calculated δ was significantly different than the trial value, the final δ would be used to compute new moments to start an additional set of computations leading to an improved value of δ.   As many corrective cycles as necessary may be employed until close agreement between starting and final values of δ are secured.

Maximum unit fiber stresses at the critical section may now be calculated by $P/A + Mc/I$.   The maximum bending moment in the beam, without taking into account the end loads, was Qa.   The effect of the end loads is to augment this moment by approximately $2.26Qa/6$.   (See Fig. 18.)   This represents an approximate 38 per cent increase in bending moment.

13–11. MOMENT   DISTRIBUTION   MODIFIED   FOR   AXIAL COMPRESSION

Trends toward ultimate design and greater slenderness focus attention upon buckling of compression members which are component parts of the structure.   The axial compression in a slender member modifies the

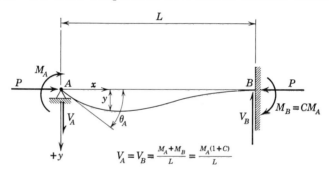

$$V_A = V_B = \frac{M_A + M_B}{L} = \frac{M_A(1+C)}{L}$$

FIG. 13–19

standard view of stiffness factors, carry-over factors, and fixed-ended moments as used in standard moment distribution procedures.

Figure 19 depicts a beam AB with B restrained and A free to rotate. An axial load P and a moment M_A are applied at A.   The moment at B

equals CM_A where C is the carry-over factor from A to B. When applied together, the deflection y is a result of both P and M_A. The load P acts to increase y and if P approached the critical column loading the deflections would become infinite. The differential equation $EI\dfrac{d^2y}{dx^2} = -M$ holds and may be expanded to a general expression. In terms of the quantities shown in Fig. 19, we obtain

$$EI\frac{d^2y}{dx^2} = -M_A + M_A\frac{(1+C)}{L}x - Py \qquad (13\text{-}45)$$

A solution of this differential equation follows: First, letting $k^2 = P/EI$, the equation reduces to

$$\frac{d^2y}{dx^2} + k^2y = -\frac{M_A}{EI} + M_A\frac{(1+C)}{EIL}x$$

which may be recognized as a differential equation having a general solution plus a particular solution leading to

$$y = A\cos kx + B\sin kx - \frac{M_A}{P} + M_A\frac{(1+C)}{PL}x \qquad (13\text{-}46)$$

The constants are evaluated as follows.
From the condition $y = 0$ when $x = 0$

$$A = \frac{M_A}{P}$$

and when $x = L$, $y = 0$

$$B = \frac{M_A}{P}\left[\frac{-\cos kL - C}{\sin kL}\right]$$

Equation 13-46 may now be restated, with $M_B/M_A = C$, as

$$y = \frac{M_A}{P}\cos kx - M_A\frac{\cos kL \cdot \sin kx}{P\sin kL} - \frac{M_B\sin kx}{P\sin kL} - \frac{M_A}{P} + \frac{M_Ax}{PL} + \frac{M_Bx}{PL}$$

which simplifies to

$$yP = M_A\left[\frac{\sin kL\cos kx - \cos kL\sin kx}{\sin kL} - \frac{(L-x)}{L}\right]$$

$$-M_B\left[\frac{\sin kx}{\sin kL} - \frac{x}{L}\right] \qquad (13\text{-}47)$$

Equation 13-47 may still be further simplified by employing the trigonometric substitution

$$\sin k(L - x) = \sin kL\cos kx - \cos kL\sin kx$$

Then equation 13-47 reduces to

$$y = \frac{M_A}{P} \left[\frac{\sin k(L-x)}{\sin kL} - \frac{(L-x)}{L} \right] - \frac{M_B}{P} \left[\frac{\sin kx}{\sin kL} - \frac{x}{L} \right] \quad (13\text{-}48)$$

We are now in a position to determine the carry-over factor by determining the ratio of M_B to M_A. One physical condition remains, namely that the slope at B equals zero. To ultilize this condition take the derivative of equation 13-48 as

$$\frac{dy}{dx} = +\frac{M_A}{P} \left[-\frac{k \cos k(L-x)}{\sin kL} + \frac{1}{L} \right] - \frac{M_B}{P} \left[\frac{k \cos kx}{\sin kL} - \frac{1}{L} \right] \quad (13\text{-}49)$$

with $\dfrac{dy}{dx}$ equal to zero when $x = L$, equation 13-49, in simplified form for

determining the carry-over factor, becomes

$$C = \frac{M_B}{M_A} = \frac{[-kL + \sin kL]}{[kL \cos kL - \sin kL]} \quad (13\text{-}50)$$

This complex expression may be checked by letting $P \to 0$ (or $k \to 0$) since without the end loads the carry-over factor will equal one-half. This requires finding the limit by L'Hospital's rule. Three differentiations are necessary before the finite limit of one-half is reached.

Before studying the effect of P on C let us consider the effect of P on stiffness. The rotation of the tangent at A, or θ_A, may be obtained from equation 13-49 by substituting $x = 0$, hence

$$\theta_A = \frac{M_A}{P} \left[-\frac{k \cos kL}{\sin kL} + \frac{1}{L} \right] - \frac{M_B}{P} \left[\frac{k}{\sin kL} - \frac{1}{L} \right] \quad (13\text{-}51)$$

using the relation of equation 13-50, M_B can be eliminated and θ_A stated in terms of M_A, or

$$\theta_A = \frac{M_A}{PL} \left[-\left(\frac{-kL + \sin kL}{kL \cos kL - \sin kL} \right) \left(\frac{kL}{\sin kL} - 1 \right) \right. $$
$$\left. + \left(\frac{-kL \cos kL}{\sin kL} + 1 \right) \right] \quad (13\text{-}52)$$

Equation 13-52 may be further simplified by substituting $k^2 = \dfrac{P}{EI}$,

reducing equation 13-52 to

$$M_A = \frac{\theta_A \cdot EI}{L} \left[\frac{(k^2 L^2 / 2) \sin 2kL - kL \sin^2 kL}{kL - 2 \sin kL - kL \cos^2 kL + \sin 2kL} \right] \quad (13\text{-}53)$$

For use in moment distribution *absolute stiffness* was defined as the moment at A required to rotate θ_A one radian (see Art. 4-2), or in general terms the *absolute stiffness* is

$$K_A = \frac{SEI}{L}$$

where S is a coefficient dependent upon end conditions and loading. The term in brackets in equation 13-53 represents the coefficient S for the given case of an end loaded strut.

The question of critical column loading can also be investigated. If P is equal to the critical load P_{cr} the deflections would become infinite and the angle θ_A would become infinite. If equation 13-53 is transformed to solve for θ_A it will be noted that $\theta_A \to \infty$ when $[(k^2L^2/2) \sin 2kL - kL \sin^2 kL]$ approaches zero, or by trigonometric substitution, we obtain

$$k^2L^2 \sin kL \cos kL - kL \sin^2 kL = 0$$

or
$$\tan kL = kL \tag{13-54}$$

Equation 13-54 is known as a transcendental equation whose solution for the roots kL is best obtained by a graphical solution or by trial and error. The smallest root is $kL = 4.493$. Restating that $k^2 = P/EI$, the critical load $P_{cr} = k^2EI = 20.16EI/L^2$. This is the same as found by equation 13-23.

For a member of length L that was pin-ended at A and B the Euler critical load was $P_E = \pi^2EI/L^2$. Thus, full end restraint at B increases the critical load by a factor of approximately 2. In the calculations to follow the ratio of the actual applied end load to P_E, or P/P_E, will be termed R.

To illustrate the effect of P assume that $P = 1.5P_E$ or $R = 1.5$. To calculate the carry-over factor C from equation 13-50 first find

$$k^2 = \frac{P}{EI} = \frac{1.5P_E}{EI} = \frac{1.5}{EI} \times \frac{\pi^2EI}{L^2} = \frac{1.5\pi^2}{L^2}$$

or
$$kL = 1.22\pi$$
$$= 3.84 \text{ radians}$$

or approximately 222°.
By slide rule
$$\cos kL = -0.743$$
$$\sin kL = -0.670$$

By equation 13-50
$$C = \frac{-kL + \sin kL}{kL \cos kL - \sin kL}$$
$$= \frac{-3.84 - 0.67}{3.84(-0.743) - (-0.67)} = 2.06$$

This may be an alarming result since C equals one-half in the usual case of no end load, but for highly loaded and slender struts the elementary view is shown to be insufficient.   If P is small relative to P_E then C is close to one-half.

The stiffness factor S for the same case is computed from the bracketed term of equation 13-53 where

$$kL = 1.22\pi$$
$$2kL = 2.44\pi$$
$$k^2L^2 = 1.5\pi^2$$

and substituting $2 \sin kL \cos kL$ for $\sin 2kL$

$$S = \frac{1.5\pi^2(-0.670)(-0.743) - 1.22\pi(-0.670)^2}{1.22\pi - 2(-0.670) - 1.22\pi(-0.743)^2 + 0.995}$$

$$S = 1.42$$

This stiffness factor may be compared with a factor of 4 in conventional moment distribution without end loads where the far end is fixed.

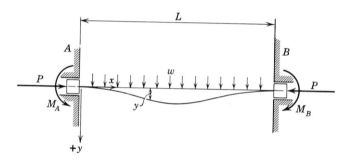

FIG. 13–20

As in conventional moment distribution the fixed-ended moments are required to establish the initial unbalanced joint moments.   Figure 20 is a sketch of a fixed-ended axially loaded beam.   The ends are restrained against rotation but free to move towards one another.

A uniform load of w is applied simultaneously with the axial load P in Fig. 20.   The differential equation of the elastic curve is

$$EI \frac{d^2y}{dx^2} = -M$$

which expands to

$$EI \frac{d^2y}{dx^2} = M_A - \frac{wLx}{2} + \frac{wx^2}{2} - Py \qquad (13\text{-}55)$$

Letting $k^2 = P/EI$, the solution of equation 13-55 is (see 13-40)

$$y = A \cos kx + B \sin kx + \frac{M_A}{P} + \frac{w}{2P}\left(x^2 - xL - \frac{2}{k^2}\right) \quad (13\text{-}56)$$

The constant A is determined from the condition that $y = 0$ when $x = 0$ as

$$A = \frac{1}{Pk^2}(-k^2 M_A + w)$$

To evaluate constant B first take the derivative of equation 13-56 as

$$\frac{dy}{dx} = -kA \sin kx + kB \cos kx + \frac{wx}{P} - \frac{wL}{2P} \quad (13\text{-}57)$$

Owing to symmetry $\dfrac{dy}{dx} = 0$ when $x = L/2$. This condition and the previously determined value of A requires

$$B = \frac{1}{Pk^2}(-k^2 M_A + w) \tan \frac{kL}{2}$$

To determine M_A, the fixed-ended moment, use the condition that $\dfrac{dy}{dx} = 0$ when $x = 0$.

Then from equation 13-57

$$M_A = \frac{-\dfrac{wL}{2} + \dfrac{w}{k}\tan\dfrac{kL}{2}}{k \tan \dfrac{kL}{2}} \quad (13\text{-}58)$$

The usual simplification is now made by letting $q = kL/2$ from which

$$k = \frac{2q}{L}$$

The equation 13-58 is restated as

$$M_A = \frac{wL^2}{12}\left(\frac{-q + \tan q}{\dfrac{q^2}{3}\tan q}\right) \quad (13\text{-}59)$$

It is to be noted that the coefficient of the above equation is $wL^2/12$, which is the fixed-ended moment for a beam uniformly loaded without axial load; thus, the term in parentheses may be defined as a magnification factor. As an illustration of the effect of P consider the example problem previously considered with A and B fixed and P equal to $1.5P_E$. Recalling that

$$q^2 = \frac{k^2 L^2}{4}$$

$$q^2 = \frac{1.5\pi^2}{L^2} \times \frac{L^2}{4} = 0.375\pi^2$$

$$q = 0.613\pi$$
$$= 1.926 \text{ radians}$$
$$= 110.36°$$
$$\tan q = -2.694$$

hence from equation 13-59

$$\frac{-q + \tan q}{\frac{q^2}{3} \tan q} = \frac{-1.926 - 2.694}{\frac{(1.926)^2}{3}(-2.694)} = 1.39$$

The magnification factor indicates that the fixed ended moments will be 39 per cent larger than the conventional value of $wL^2/12$.

An example of moment distribution involving the effect of axial load is the single braced column of Fig. 21. It is assumed that A is fixed and that B and C are immovable supports. It is further assumed that the axial load at C is small, and, in conjunction with the short length of span BC, no significant changes are produced in conventional moment distribution terms for span BC. However, span AB is subjected to an axial compression of $P = P_1 + P_2$. In order to use previously computed quantities assume that P equals $1.5P_E$ in this example problem. A consideration of *absolute stiffnesses* at joint B is as follows, considering C as a hinge and that P_1 is small relative to P_2.

$$K_{BC} = \frac{3EI}{L/2} = \frac{6EI}{L}$$

$$K_{BA} = \frac{1.42EI}{L} = \frac{1.42EI}{L}$$

$$\Sigma K = \frac{7.42EI}{L}$$

Leading to distribution factors at B,

$$\text{For } BC = \frac{6}{7.42} = 0.80$$

$$\text{For } AB = \frac{1.42}{7.42} = 0.20$$

The student should note that for conventional moment distribution, unmodified by the effect of axial compression, these factors would have been 0.6 and 0.4.

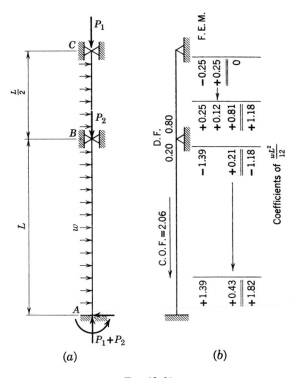

FIG. 13–21

In Fig. 21b the fixed-ended moments are established as

$$\text{For } BC = \frac{w \times (L/2)^2}{12} = \frac{wL^2}{48} = \frac{1}{4} \times \frac{wL^2}{12}$$

$$\text{For } AB = 1.39 \times \frac{wL^2}{12}$$

For distribution purposes only the coefficients of $wL^2/12$ have been used. With the previously computed carry-over factor of 2.06 from B to A distribution proceeds by conventional steps. The final moment at A is 58 per cent larger than the moment secured without modification of moment distribution procedures due to axial compression. The student should note that these modifications are chiefly of concern when members are slender and in the long column range. The entire subject of structural stability can not be treated in this textbook, but the foregoing should be sufficient to note the seriousness of the problem.

Modifications of moment distribution procedures for members under the influence of bending and axial tension are also often essential.

REFERENCES

1. Bleich, F., *Buckling Strength of Metal Structures*, Chapters 2, 6, and 7, New York: McGraw-Hill, 1952.
2. Newmark, N. M., "Numerical Procedures for Computing Deflections, Moments and Buckling Loads," *Trans. Am. Soc. Civil Engineers*, **108**, 1161 (1943).
3. Shanley, F. R., *Strength of Materials*, Chapter 24, New York: McGraw-Hill, 1957.
4. Timoshenko, S., *Strength of Materials*, Part II, Chapters 2 and 5, New York: Van Nostrand, 1956.
5. Timoshenko, S., *Theory of Elastic Stability*, New York: McGraw-Hill, 1936.
6. Wang, C. K., *Applied Elasticity*, Chapters 6, 9, 10, and 11, New York: McGraw-Hill, 1953.
7. Winter, G., P. T. Hsu, B. Koo, and M. H. Loh, *Buckling of Trusses and Rigid Frames*, Bulletin 36, Engineering Experiment Station, Cornell University, April 1948.

PROBLEMS

13–1. For purposes of showing the independence of the critical column load from Δ, assume the erroneous equilibrium configuration shown and determine the critical load. *Ans.* $P_{cr} = 12EI/L^2$.

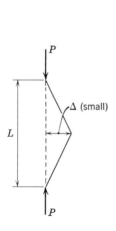

Prob. 13–1

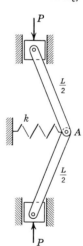

Prob. 13–2

13–2. A two-bar linkage is shown with guided ends free to move up and down, and point A is attached to a spring whose spring constant is k. The spring is unstretched when linkage is vertical. Determine the load P for a stable configuration.   *Ans. $P = kL/4$.*

13–3. A slender uniform mast supports two heavy antennas, each weighing W. By making an assumption for an approximate elastic curve determine the critical value of W. Neglect weight of the mast.   *Ans. $W_{cr} = 16EI/31a^2$.*

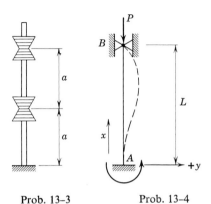

Prob. 13–3   Prob. 13–4

13–4. Determine the critical load for this slender column where the top is hinged and the base is fixed. EI is constant. Also determine the effective length.   *Ans. $P_{cr} = 20.16EI/L^2$ and $L_e \approx 0.7L$.*

13–5. Determine the critical load by the finite difference approximation method, using five intervals as shown. EI is constant. Compare your result with the exact Euler expression.

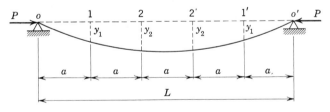

Prob. 13–5

13–6. Determine the approximate critical load P by the finite difference approximation approach. (See figure on next page.)

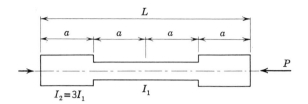

Prob. 13–6

13–7. Determine the approximate critical load by the finite difference approximation approach.   Note that deflections are not symmetrical.

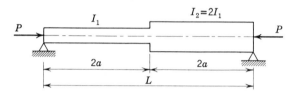

Prob. 13–7

13–8. Determine the approximate critical load by the Newmark numerical method.   Use six intervals and assume a polygonal configuration fitting a parabolic curve.   *EI* is constant.

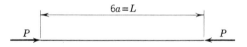

Prob. 13–8

13–9. Solve problem 13–6 by the Newmark numerical approach.

13–10. Solve problem 13–7 by the Newmark numerical approach.

13–11. Determine the approximate critical load for this slender column by using Newmark's numerical method.   Use six intervals and assume the initial deflected configuration as fitting a half-sine wave, or $y_a = \Delta \sin (\pi x/L)$, where Δ is the deflection at midlength.   State the result in terms of I at center line.

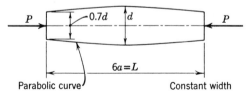

Prob. 13–11

13–12. Determine the approximate critical load for this slender column by using Newmark's numerical method. Use six intervals and assume initial configuration as fitting a parabolic curve. State results in terms of I at center line.

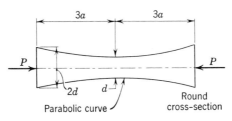

Prob. 13–12

13–13. Solve problem 13–6 for the approximate critical load using the energy method. Assuming the configuration as a parabolic curve, determine the critical load (*a*) by working with U from the expression $\dfrac{EI}{2}\int\left(\dfrac{d^2y}{dx^2}\right)^2 dx$, and (*b*) by working with U from the expression $\int\dfrac{M^2\,dx}{2EI}$. Compare your results with one another and with the result obtained in problem 13–6.

13–14. Determine the approximate load for this slender vertical mast by the energy method. Solve by assuming (*a*) that the deflection curve is a parabola, and (*b*) that the deflection curve is defined by a cosine curve.

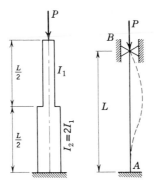

Prob. 13–14   Prob. 13–15

13–15. Determine the approximate critical load for this slender column by the energy method. Assume the deflected configuration to be similar to the elastic curve obtained by loading with a uniform transverse loading.

13–16. Determine the critical load P for this frame by the differential equation approach. Explore the value of critical load for $I = I_1$ and $L = L_1$, and for the case where I_1 is infinitely large.

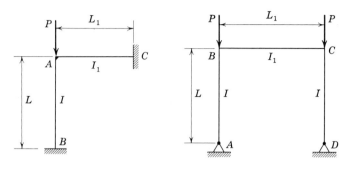

Prob. 13–16   Prob. 13–17

13–17. Determine the critical load P for this frame by the differential equation approach.

13–18. Determine an analytical expression by the differential equation approach for the maximum moment at center line. Let $k^2 = P/EI$.

$$Ans. \frac{QL}{4} \frac{\tan (kL/2)}{kL/2}.$$

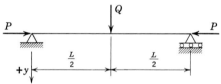

Prob. 13–18

13–19. Determine an analytical expression for the maximum moment under Q. Let $k^2 = P/EI$.

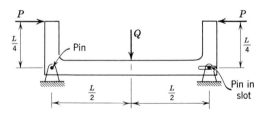

Prob. 13–19

13–20. Assuming that the deflection curve may be approximated by one term of a sine series, determine an expression for the deflection at the Q loads by the energy method.

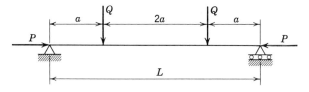

Prob. 13–20

13–21. Determine the approximate deflection at Q by Newmark's numerical method.   Assume that $P = P_E/3$.   Determine the deflection in terms of Q, a, and EI.   Use four intervals.   What is the percentage increase in maximum moment over the simple beam moment?

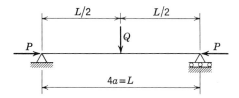

Prob. 13–21

13–22. For the beam column of problem 13–19, determine the deflection at the load Q by Newmark's numerical method. Use four intervals where $a = L/4$.   Assume that $P = P_E/4$, and that P also equals $Q/2$.

13–23. Determine the final deflection at the center line by Newmark's numerical method.   Determine the simple beam deflections for the uniform load by returning to standard equations to form the basis of the starting solution, Let $P = P_E/3$.   What is the percentage increase in maximum moment over the simple beam moment?

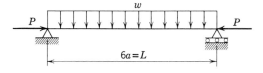

Prob. 13–23

13–24. Determine the moment at B by moment distribution theory considerations. A is free to rotate and move downward. The vertical column is a slender member having a uniform I. Assume $P = P_E$, where $P_E = \pi^2 EI/L^2$.

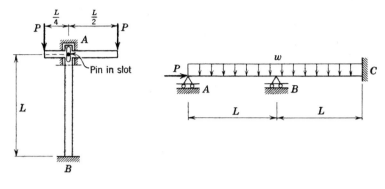

Prob. 13–24   Prob. 13–25

13–25. Determine the moment at B and C, using moment distribution procedures properly adjusted for the effect of axial load. Assume the beam to be of constant section and that $P = 1.2P_E$ where $P_E = \pi^2 EI/L^2$.

Index